Reviews of Physiology, Biochemistry and Pharmacology 138

Springer-Verlag Berlin Heidelberg GmbH

Reviews of

138 Physiology Biochemistry and Pharmacology

Special Issue on The Third Filament System
Edited by D. Pette and D. Fürst (Guest Editor)

Editors

M.P. Blaustein, Baltimore   R. Greger, Freiburg
H. Grunicke, Innsbruck   R. Jahn, Göttingen
W.J. Lederer, Baltimore   L.M. Mendell, Stony Brook
A.Miyajima, Tokyo   D. Pette, Konstanz   G. Schultz,
Berlin   M. Schweiger, Berlin

With 67 Figures and 4 Tables

Springer

ISSN 0303-4240
ISBN 978-3-662-31202-5   ISBN 978-3-540-49231-3 (eBook)
DOI 10.1007/978-3-540-49231-3

Library of Congress-Catalog-Card Number 74-3674

Originally published by Springer-Verlag Berlin Heidelberg New York in 1999
Softcover reprint of the hardcover 1st edition 1999

Production: PRO EDIT GmbH, D-69126 Heidelberg
SPIN: 10698774   27/3136-5 4 3 2 1 0 – Printed on acid-free paper

Preface

The ability to move is one of the fundamental properties of living organisms. It is based on the existence of two principally different molecular motors, the actomyosin and the microtubular systems. Within the actomyosin system, the regular sarcomeric organization of thick and thin filaments is one of the most fascinating achievements in the evolution of a tissue specialized in performing directed movements. As viewed from the highly ordered structures of thick and thin filaments in the myofibrils of cross-striated muscle, this arrangement can not be explained simply by the molecular properties of its major protein components. This difficulty became evident almost as early as the sliding filament mechanism had been proposed. Several observations pointed to the existence of an as yet unidentified ordering structure, for instance the maintained integrity of ghost fibers after the extraction of myosin (Huxley & Hanson, 1954), the elastic behaviour of those fibers, the strictly defined filament length, the integrity of the sarcomere after extension beyond thin/thick filament overlap (for review see Wang, 1985).

It was the pioneering work of Koscak Maruyama and collaborators which by the discovery of an only badly soluble proteinaceous residue called "connectin" provided for the first time a molecular explanation for these phenomena (Maruyama, 1976). This hitherto unknown gigantic protein, later on renamed as titin (Wang et al., 1979), was shown to be capable of fulfilling the demanded organizing role. Its molecular dimensions, sarcomeric arrangement and multiple functions deciphered since its discovery, justify its distinction as an independent structural element, the so-called third filament, and the notion of the sarcomere as a three-filament system (Fig. 1).

The large body of literature that has accumulated on titin and its associated proteins in the time since its discovery, made it appear timely to bring together the existing knowledge in a comprehensive form. This was the aim of the present volume.

Figure 1. Comprehensive sketch illustrating the so-called "three fila-ment model of the sarcomere". The upper box shows one sarcomere (with Z-discs, I band, A band, H zone, and M band indicated by arrows and brackets), as well as one representative thick filament and four thin filaments. An enlargement of thin filament substructure with an expla-nation of molecular symbols is shown in the left bottom box. The main features of thick filament structure (assembly of filaments based on myosin molecules, three-dimensional arrangement of myosin heads along the filament, cross section through the filament), and a titin molecule are given in the right bottom box. Note that exclusively titin provides continuity to the sarcomere; on the other hand, this molecule is required to be highly elastic in order not to impede the actin/myosin-based contractile machinery. For further details see the various contri-butions to this volume.

The history of the discovery of connectin/titin including comparative aspects of the third filament in diverse muscles of vertebrates and invertebrates is summarized in a chapter by K. Maruyama. Fundamental insights into molecular struc-ture/function relationships stem from the meticulous work of Siegfried Labeit's laboratory on the cloning and sequencing studies on titin cDNA and its gene. The results of this work and an outlook on the titin isoform family is presented in chapter 2. Robert Horowits highlights in chapter 3 the physi-ology of titin by its role in the mechanical properties, as elucidated by selective extractions and direct measurements of segmental elasticity. The complex pattern of titin interac-tions with other proteins of the myofibrillar apparatus along its length is dealt with by Mathias Gautel and colleagues in chapter 4. This chapter also includes available evidence for the regulation of these interactions by mechanisms under control of different signalling pathways. The role of the third filament as a ruler of the sarcomere during embryonic and in vitro myogenesis is discussed in chapter 5 by Alice Fulton. The interactions of specific titin regions with other protein ligands are described in detail in the following two chapters. Dieter Fürst and collaborators elaborate in chapter 6 interac-tions of the carboxy terminus of titin with M-band proteins,

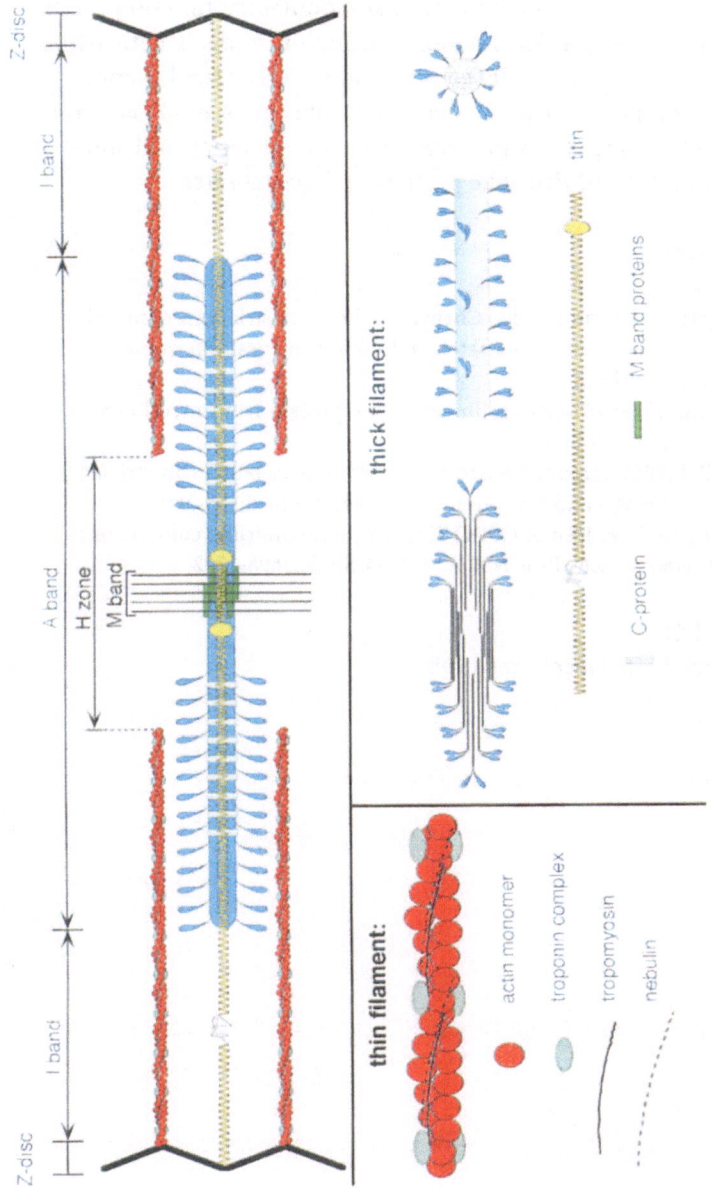

both with regard to the structure of the M-band and its assembly. The C-protein, its potential modulatory function on contractility and thick filament formation is dealt with by Pauline Bennett and colleagues in chapter 7. The final chapter of the volume by Guy Benian and colleagues emphasizes comparative aspects and focusses on the genetics and molecular biology of titin-like proteins in invertebrates.

References

Huxley H E, Hanson J (1954) Changes in the cross-striations of muscle during contraction and stretch and their structural interpretation. Nature 173:973-976

Maruyama K (1976) Connectin, an elastic protein from myofibrils. J Biochem (Tokyo) 80:405-407

Wang K (1985) Sarcomere-associated cytoskeletal lattices in striated muscle. Review and hypothesis. Cell Musc. Motil. 6:315-369

Wang K, McClure J,Tu A (1979) Titin: major myofibrillar components of striated muscle. Proc Natl Acad Sci USA 76:3698-3702

Dieter Fürst
and Dirk Pette December 1998

Contents

X

Indexed in Current Contents

Comparative Aspects of Muscle Elastic Proteins

K. Maruyama

Office of President, Chiba University, Chiba 263, Japan

Contents

1 Introduction

Contractility and elasticity are the major functions of striated muscle. Inter-actions of myosin and actin hydrolyzing ATP lead to muscle contraction and connectin/titin filaments are responsible for the elasticity of muscle.

In vertebrate striated muscle connectin/titin is the largest peptide so far known : 3000–4000 kDa and 1 μm long (for reviews, see Maruyama, 1986; 1994; 1997). Connectin links the myosin filament to the Z line and positions the filament at the center of a sarcomere. The N terminal 90 kDa portion of connectin binds to the Z line and the C terminal 120 kDa portion binds to myosin, myomesin, and M protein constituting the M line of the A band.

There are a number of isoforms of connectin. Especially, connectin iso-forms of cardiac, fast and slow striated muscles differ in size and elasticity.

In invertebrate muscle several connectin-related elastic proteins are characterized : 753 kDa twitchin (nematode, annelid and mollusc), 1200 kDa projectin (arthropod), and ~3000 kDa connectin-like protein (annelid and arthropod). In addition, ~2000 kDa connectin-like proteins are present in nematode, mollusc and prochordate.

2 Short History

When Reiji Natori (Jikei University, School of Medicine, Tokyo) first pre-pared "skinned fibers" of frog skeletal muscle in 1951, he observed reversible stretch and passive tension generation upon stretch. Natori assumed the presence of "internal elastic structure" in a sarcomere (Natori, 1954). Hugh Huxley and Jean Hanson also observed an elastic behavior of myosin-extracted myofibril and assumed the presence of s (stretch) filament linking two opposite actin filaments in a sarcomere (Huxley and Hanson, 1954).

It was Fritchof Sjöstrand (1962) that first saw "gap filaments" between separated thick and thin filaments in extremely stretched sarcomeres. Later, Graham Hoyle claimed that "thinner filaments" run in parallel with "thin (actin) filaments in a sarcomere and he named it T filament (McNail and Hoyle, 1967; Hoyle, 1983). Ferenc Guba also presented a similar view and called fibrillin (Guba et al. 1968).

The present writer started his work stimulated by Natori in 1975. Ghost skinned fiber treated with dilute alkaline solution was reversibly stretched and passive tension was generated upon stretch. The term connectin was given to a huge elastic protein (Maruyama et al. 1977). In 1979 Kuan Wang showed doublet band of very high molecular weight (~1000 kDa) in chicken skeletal muscle on an SDS gel electrophoresis and called it titin (Wang et al. 1979). In 1984 we estimated 2800 kDa for the molecular mass of connectin/titin (Maruyama et al. 1984), but nobody except for Wang and the present writer believed the presence of such a giant peptide!

Thanks to Richard Podolsky (NIH, USA) and Klaus Weber (Göttingen, Germany) for their entry to this field. Using the technique of ionized irradiation the former's group showed a protein with molecular weight of several million is responsible for passive tension generation (Horowits et al. 1986). The latter's group clearly demonstrated the localization of connectin molecules in a sarcomere under an electron microscope using more than ten kinds of monoclonal antibodies (Fürst et al. 1988). Since then the muscle elastic protein has become a current topic in cell biology.

It was very difficult to isolate the mother molecule of connectin from striated muscle. Its C terminal region bound to the M line of the A band is easily solubilized together with myosin in 0.6 M KCl, but the N terminal region firmly bound to the Z line is insoluble in most salt solutions. It was found that 0.1 M phosphate buffer, pH 7.0, partially set connectin free. α–Connectin was thus isolated from rabbit skeletal muscle in 1989 (Kimura and Maruyama).

Siegfried Labeit challenged to decipher connectin cDNA in 1989. Six years later he succeeded in the complete elucidation of human cardiac connectin cDNA, 82 kb coding 26,926 amino acids (2993 kDa) (Labeit and Kolmerer, 1995). To celebrate Labeit's monumental work, an international symposium on muscle elastic proteins was held at Chiba University, Chiba, Japan in 1995 (Maruyama and Kimura, 1996).

Quite recently, three groups have finally unraveled the mechanism of elasticity of a single connectin molecule: an entropy spring (more flexible PEVK region) and a series of unfolding and folding immunoglobulin domains (less flexible) (Tskhovrebova et al. 1997; Kellermayer et al. 1997; Rief et al. 1997).

3 Mystery of Doublet Band

As first described by Wang et al. (1979), there is doublet band of connectin at the top of polyacrylamide gel, when a piece of whole striated muscle is solublized in an SDS solution and electrophoresed. When freshly excised specimen is used, the upper of the doublet band is thicker than the lower. In aged muscle sample, the situation becomes opposite. Therefore, it is usually regarded that the lower band (β-connectin, titin 2) is due to the proteolytic product of the mother molecule, α-connectin (titin 1). In fact, β-connectin is prepared from chicken breast muscle after storage overnight at 4 °C (Kimura and Maruyama, 1983). The protease responsible for splitting α-connectin into β-connectin is inhibited by 1mM leupeptin and is thought to be a thiol

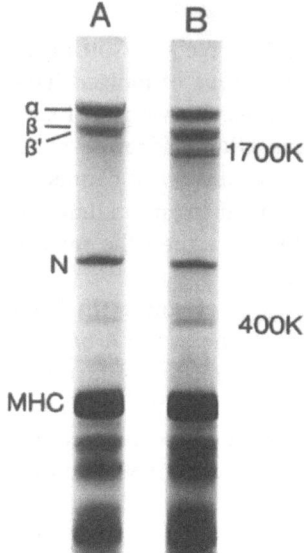

Fig. 1. SDS gel electrophoresis pattern of chicken breast muscle connectin. 1.8% polyacrylamide gel containing 6 M urea (Weber-Osborn system). A, control (myofibrils); B, treated for 2 min at pH 7.6 with 1/100 (w/w) clostripain in 0.3 M NaCl and 2.5 mM dithiothreitol. Note that β' band is not detected by Laemmli's system. α, β, β', α-, β-, β'- connectin; N, nebulin; MHC, myosin heavy chain. Hu D H, unpublished (1986) (cf. Hu et al. 1986; 1989)

protease, very probably, calpain (Kimura et al. 1993 ; cf. Sorimachi et al. 1996).

Question arises whether β-connectin is really a proteolytic product in vivo or not. Hu et al. (1986) observed that there are two β-connectin bands, β and β' in chicken breast muscle. The β' band was very sharp and this band remained as such on proteolysis of α-connectin, whereas the β band became thicker (see Fig. 1b, lanes 1 and 2 in Hu et al. 1989). As shown in Fig. 1, when washed myofibrils were briefly treated with clostripain, a thiol protease from *Clostrium*, the β band became thicker together with the appearance of 1700K and 400K bands. However, the β' band remained constant. Thus there is a possibility that at least β' exists in situ as such, not produced by proteolysis of α-connectin. If so, there could be a small isoform of connectin which just binds onto the myosin filament, corresponding to twitchin or projectin in invertebrate muscle (see section, 7.1 and 7.2).

It is to be pointed out that there is only a single band corresponding to α-connectin in chicken leg muscle (Hu et al. 1986; Hattori et al. 1995) and in lamprey skeletal muscle (Itoh et al. 1987).

It is probable that α-connectin might undergo proteolysis during preparation (Kimura and Maruyama, 1989). We have checked the isolated α-connectin by immunoblot tests using antibodies to the N and C terminal regions of connectin, respectively (Yajima et al. 1996; Soeno et al. 1997). The both antibodies reacted with isolated α-connectin indicating that the N and

C terminal regions of α-connectin were intact (Kimura and Maruyama, unpublished).

4 Connectin Isoforms

It is well known that a number of muscle structural proteins undergo changes of isoform expression during embryonic and postnatal development (cf. Obinata, 1993). We observed that an embryonic type of connectin was first expressed, then neonatal type and finally adult type appeared in chicken skeletal muscle during embryonic and postnatal development. The molecular size decreased in this order (Yoshidomi et al. 1985). However, Hattori et al. (1995) pointed out that there are only embryonic and adult types of connectin in chicken skeletal muscle distinguishable from electrophoretic mobility. We confirmed this conclusion (Kawamura et al. 1996). The size of embryonic isoform of connectin is similar to that of leg muscle connectin (~4000 kDa) compared with ~3000 kDa of breast muscle connectin.

Labeit and Kolmerer (1995) clearly demonstrated the presence of four types distinct isoforms of connectin in human striated muscle : two cardiac and two skeletal muscle isoforms. In the skeletal muscle isoforms, there are additional 53 imunoglobulin (motif II) domains as compared to cardiac isoforms. The elastic PEVK region differs in size: 163 amino acids (cardiac), 1400 (psoas, fast) and 2174 (soleus, slow).

5 Vertebrate Skeletal Muscle Connectin

Comparative studies using SDS gel electrophoresis revealed the ubiquitous presence of connectin bands in all the vertebrate skeletal muscles examined (Hu et al. 1986; Locker and Wild, 1986): Pisces: carp, goldfish; Amphibia; bull frog, newt; Reptilia: turtle, snake; Aves: chicken, thrush; Mammalia: rabbit, sheep, bovine. It is of interest to note that the band of snake or turtle skeletal muscle connectin showed much slower mobility than that of chicken breast muscle connectin. This was also the case with nebulin (Hu et al. 1986).

A comparative study was carried out with three species of fish using immunoblot and immunoelectron microscopy (Kawamura et al. 1994a). The connectin size of horse mackerel (Osteichthyes), electric ray (Chondrichthyes) and lamprey (Agnatha) was similar to that of chicken breast muscle. There was a single α-connectin band in lamprey muscle, not β-connectin (Itoh et al. 1987). Two kinds of antibodies to chicken skeletal connectin reacted with any fish connectin. As shown in Fig. 2, Pc1200 bound to the

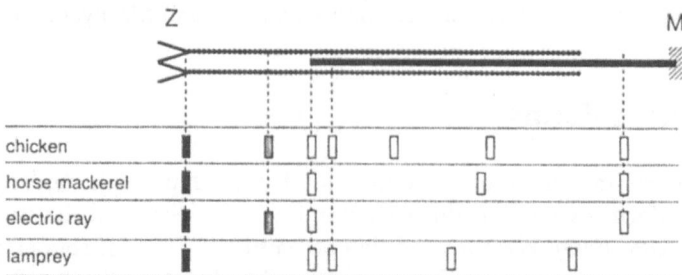

Fig. 2. A scheme of the positions of the epitopes to the antibodies to connectin in fish skeletal muscle sarcomeres. ⏢, 3B9; ▤, SM1; ▮, Pc1200. Z, Z line ; M, M line. Modified from Kawamura et al. (1994a). Courtesy of Dr. Hirohiko Yajima

edge of the Z lines in all the muscle sarcomeres. 3B9 bound to the common or different regions at the A band. SM1 bound to the same I band region of electric ray sarcomere as that of chicken breast sarcomere, but did not bind to horse mackerel and lamprey sarcomeres. Thus, some epitopes (e.g., to Pc1200) were preserved and the other ones (e.g., to SM 1) were present or missing. Epitopes to 3B9 were all present but their localization was different among species. These observations suggest that connectin cDNA sequences are largely varied except for functionally essential regions even in Pisces. This is understable for such a large peptide as connectin.

6 Smooth Muscle

Vertebrate smooth muscle cells contain a number of desmin intermediate filaments in addition to thin (actin) and thick (myosin) filaments. SDS gel electrophoresis patterns of smooth muscle show faint bands of approximately 1000 kDa above the thick band of ABP (filamin, 240 kDa). It is certain that ~3000 kDa connectin is not present in bovine aorta and chicken gizzard.

We have isolated the 1000 kDa protein from bovine aorta, but this did not crossreact with any antibodies to various regions of chicken skeletal muscle connectin (Kimura, S., unpublished).

7 Nonmuscle

Eilertsen and Keller (1992) reported that there are connectin-like 3000 kDa proteins (α-, and β-connectin) in brush borders of chicken intestinal epithelial cells. Antibodies to this protein (T protein) reacted with chicken breast muscle connectin. Partially purified protein showed fibrous appearance of approximately 1 μm long. Immunofluorescence revealed its localization in the brush border terminal web but not in the microvilli. This paper clearly shows the presence of connectin-like protein in nonmuscle cells. Eilertsem et al. (1994) further reported that cellular connectin colocalizes with myosin II filaments in stress fibers in the terminal web domain. Purified cellular connectin interacted with brush border myosin to form highly ordered myosin filaments in vitro.

Erythrocyte cell cortex fractions contain very faint bands of 500–800 kDa in SDS gel electrophoresis. This is also the case with liver cells. There is not any indication of ~3000 kDa connectin-like proteins, however (Maruyama, unpublished).

Connectin-like protein was present in sea-urchin egg cytomatrix and reactive with polyclonal antibodies to chicken skeletal muscle connectin. This protein is localized in the cleavage furrow but not in the mitotic apparatus (Pudles et al. 1990).

8 Invertebrate Muscle

Biodiversity is represented by a variety of connectin family in invertebrate muscle. At present at least five groups of long elastic proteins are known (cf. Ziegler, 1994; Maruyama et al. 1995; Kawamura et al. 1996). Recent cDNA sequence analyses have shown the common presence of repeating motifs I (fibronectin Type III) and/or II (immunoglobulin C2) in vertebrate connectin family.

Twitchin found in *C. elegans* bodywall muscle is the protein of connectin family first completely sequenced (Benian et al. 1989). This 753 kDa protein is distributed not only in nematode and annelid muscle but also in molluscan adductor muscle.

The 732 kDa protein coded by *unc*-89 gene is related to the M line formation in *C. elegans* bodywall muscle sarcomeres (Benian et al. 1996b).

Projectin, 1200 kDa protein, is found in arthropod muscle. In insect flight muscle projectin substitutes for connectin, but in leg muscle it is just bound onto the myosin filament as twitchin in nematode bodywall muscle.

Kettin, 400–700 kDa protein, is also arthropod muscle protein consisting of motif IIs with interdomains. Apart from its binding to the Z line, its function is unknown.

Invertebrate connectin, 2000–5000 kDa, found in annelid, mollusc, arthropod and prochordate muscles, are equivalent to vertebrate striated muscle connectin.

8.1 Twitchin

Twitchin, a prototype of connectin, 753 kDa, was discovered in *C. elegans* bodywall muscle, when *unc-22* gene was sequenced (Benian et al. 1989; 1993). Mutants defect in the *unc-22* gene results in malfunction of contractility. The sequence consists of a number of repeats, 31 motif Is (fibronectin Type III-like domains) and 30 motif IIs (immunoglobulin C2-like domains). Near the C terminus there is myosin light chain kinase (MLCK) domain . The sequence of twitchin is similar to that of connectin elucidated by Labeit and Kolmerer (1995). In fact, some antibodies to connectin (e.g. 3B9) cross-react with twitchin (Matsuno et al. 1989). Immunofluorescence observations located twitchin on the myosin filament in obliquely striated muscle.

Weber's school was able to isolate twitchin from the nematodes, *Ascaris lumbricoides* and *Caenorhabditis elegans* (Nave et al. 1991). Twitchin is a filamentous protein, 240–250 nm long and consists of mainly β-sheets. Twitchin is present both in bodywall muscle and pharynx. However, pharynx twitchin is an isoform different from bodywall one in the domain organization of the C terminal region (Benian et al. 1996a).

Twitchin was also purified from striated and smooth adductor muscles of scallop and from *Mytilus* catch (smooth) muscle (Vibert et al. 1993; 1996). The muscle suspension well washed with a saline solution was extracted with 0.15 M phosphate buffer, pH 7.0. After dialysis against low salt solution, myosin was sedimented away. The supernatant containing twitchin was subjected to DEAE-Sephacel column chromatography. Further purification was carried out with HPLC using Serva Si300 polyol column.

Molluscan twitchin is a rod-like molecule, 250 nm long and 3 nm wide with polarity. Immunoelectron microscopy showed that antibodies to MLCK region of *C. elegans* twitchin bound to scallop twitchin near one end of the molecule (Fig. 3) suggesting the presence of MLCK region near the C termi-

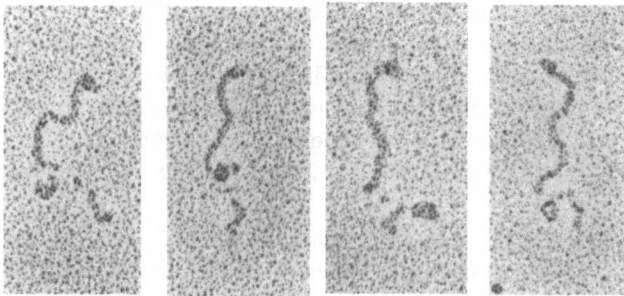

Fig. 3. Rotary shadowed scallop striated muscle twitchin treated with anti-MLCK kinase (*C. elegans* twitchin). Courtesy of Dr. Peter Vibert (Vibert et al. 1993). Permission obtained

nus of scallop twitchin. Twitchin is localized on the myosin filament of scallop striated muscle sarcomeres (Vibert et al. 1993).

Twitchin was also isolated from marine mollusc *Aplysia* (Heierhorst et al. 1994). *Aplysia* twitchin showed an autophosphorylation activity onto threonine. Bacteria-expressed kinase domain of *Aplysia* twitchin phosphorylates the 19 kDa myosin regulatory light chain of *Aplysia* myosin (Heierhorst et al. 1995). Furthermore, Probst et al. (1994) reported that twitchin is the major substrate for a cAMP dependent kinase and an increase in cAMP dependent phosphorylation of twitchin causes an increased rate of relaxation of *Aplysia* muscle. It was also found that Ca^{2+}/S100 calcium binding protein greatly activated the protein kinase activity of *Aplysia* twitchin (Heierhorst et al. 1996). Thus, twitchin may function as a signal transmitter in addition to its myosin filament organizing action.

8.2 UNC 89

In *C. elegans* bodywall muscle of *unc* 89 gene mutants, the myosin filaments are disorganized without the M line. Benian and associates cloned cDNA and completely sequenced (Benian et al. 1996b). The product of *unc* 89 gene, UNC 89 (731,897 dalton), was located at the M line. It is regarded that the tandem rows of 7 and 46 motif IIs bind to the myosin filament at the M line. Between the motif II rows there is KSP domain containing 44 KSP residues.

8.3 Projectin

Since Auber and Couteaux (1963) first described in fly flight muscle, it is well established that a connecting filament links the myosin filament to the Z line in insect fibrillar flight muscle (cf. Pringle, 1977). In 1981 Saide showed that 360 kDa protein was the connecting filament protein and this filament could be elongated by stretch based on immunofluorescence observations. Saide (1981) named it projectin. The molecular mass of projectin was reestimated to be larger than 600 kDa (Saide et al. 1990). Projectin has been revived since the concept of muscle elastic protein was established in late 80's.

Nave and Weber (1990) isolated a proteolytic product of projectin from locust flight muscle. Washed myofibrils were extracted with 0.6 M KCl and after removal of myosin the protein was purified by DEAE-cellulose and TSK 6000 PW chromatography. MW of the protein was estimated to be 600,000 and had a rod-like shape, 260 nm long. Immunofluorescence and immnoelectron microscopy studies clearly indicated that the protein in question links the myosin filament to the Z line in locust and honeybee flight muscle.

The antibodies to this protein crossreacted with twitchin of *C. elegans* body-wall muscle, and therefore, did not distinguish projectin from twitchin. Although Nave and Weber (1990) called this protein "mini-titin", it is to be called projectin in order to avoid confusion.

Intact projectin was isolated from crayfish claw muscle (Hu et al. 1990). Projectin was solubilized from washed myofibrils by 0.1 M phosphate buffer, pH 6.6. After removal of myosin projectin was purified by DEAE-Sephadex A50. The proteolytic product of projection was separated by this column chromatography with a great loss of projectin. The molecular mass of projectin was estimated to be 1200 kDa. The value is much larger than previously reported one, 600 kDa (Nave and Weber, 1990). Polyclonal antibodies to crayfish projectin bound to honeybee thoracic muscle sarcomere from the Z line to the edge of the M line.

However, in giant sarcomeres (sarcomere length (SL), 10 μm at rest) of crayfish claw muscle, projectin bound to the A band alone. What links the Z line to the A band in giant sarcomeres? The question was solved by Manabe et al. (1993). Connectin-like 3000 kDa protein links the Z line to the myosin filament in crayfish giant sarcomeres. This is also the case with insect leg muscle (SL, 7-8 μm, at rest), as shown by Ohtani et al. (1996).

It was revealed that *Drosophila* projectin has repeated sequences, II-I-I, very similar to those of twitchin (Ayme-Southgate et al. 1991). This tandem alignment of motifs I and II is present in the myosin binding region of vertebrate skeletal muscle connectin (Labeit and Kolmerer, 1995). This partial sequence was independently shown in *Drosophila* cDNA (Fyrberg et al. 1992). In addition, it was observed that a lethal mutant has stop codon insertions suggesting that projectin is essential for life. Projectin has a MLCK region (Ayme-Southgate et al. 1995).

8.4 Kettin

In 1990 Lakey et al. reported that a high molecular weight protein called p700 is localized in the Z line of waterbug *(Lethocerus)* flight muscle and the corresponding protein in leg muscle is called p500. Later, it was shown that partial cDNA sequence of *Drosophila* p500 consists of repeats of 95 amino acids corresponding to connectin motif II separated by linker sequences of 35 amino acids (Lakey et al. 1993). Kettin thus termed is regarded as a scafford of insect muscle Z lines (Bullard and Leonard, 1996).

Kettin, 540 kDa, was isolated from crayfish claw muscle (Maki et al. 1995). It is a filamentous protein, 300–360 nm long. Antibodies to crayfish kettin crossreacted with *Lethocerus* p500. It is of interest to note that Pc1200 to the Z line binding region of chicken connectin crossreacted with both

Table 1. Reactivity of arthropod elastic proteins to antibodies to chicken skeletal muscle connectin

| | | Antibodies | |
Protein	Pc1200	3B9	SM1
Connectin	−	−	+
Projectin	−	+	−
Kettin	+	−	−

Compiled from Hu et al. (1990), Manabe et al. (1993) and Maki et al. (1995). For antibodies, see Fig. 2.

waterbug and crayfish kettin. However, Pc1200 did not react with arthropod projectin and connectin (Table 1). Immunoelectron microscopy suggested that crayfish kettin was extended up to 500 nm in the I band starting from the Z line in stretched crayfish muscle sarcomeres. It may bind to the actin filament in the sarcomere.

8.5 Connectin

An SDS gel electrophoresis survey of invertebrate muscle suggested presence of connectin-like high molecular weight proteins in various but not all the muscles tested (Locker and Wild, 1986 ; Hu et al. 1986). However, in order to identify them, at least immunoblot tests must have been done.

At present there are two groups of connectin-like proteins in invertebrate muscle: ~2000 kDa and ~3000 kDa. ~2000 kDa protein was detected in body wall muscle of *C. elegans* (Maruyama et al. 1995) and of ascidian (Nakauchi and Maruyama 1992; Vibert et al. 1996) and scallop adductor muscle (Maruyama et al. 1995) using a monoclonal antibody 3B9 to chicken breast muscle connectin (see Fig. 2). The content is very low and therefore, a large amount of muscle dissolved in an SDS solution has to be applied to gel electrophoresis. Any attempt to isolate it has not yet been done .

3000 kDa protein was found in polychaete bodywall muscle (Kawamura et al. 1994b) and *Amphioxus* striated muscle (Kimura et al. 1997). This connectin-like protein is widely distributed in arthropod muscle: insect flight and leg muscle (Lakey et al. 1990; Nave and Weber, 1990; Hu et al. 1990; Ohtani et al. 1996), crayfish claw muscle (Hu et al. 1990; Manabe et al. 1993) and barnacle muscle (Maki et al. 1995).

In arthropod muscle, there are giant sarcomeres in some type of muscles such as crayfish claw opener (SL at rest, 10 µm), insect leg muscle (SL at rest,

(a)

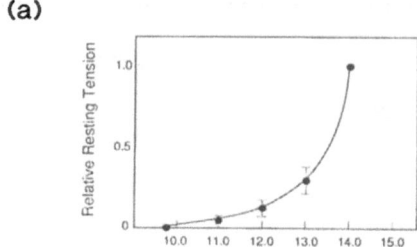

(b)

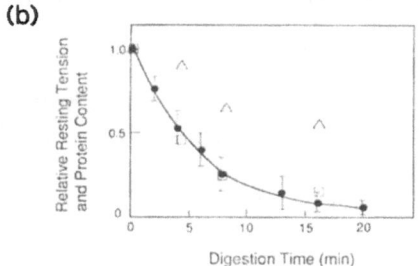

(c)

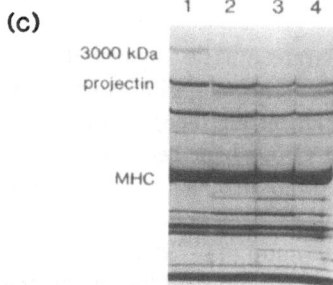

Fig. 4. Resting tension generation of skinned fibers of crayfish claw opener and the effect of trypsin treatment. **(a)**, tension generation developed at varied sarcomere length. ●, tension; bar, SD. **(b)**, effects of trypsin treatment on the resting tension generation at a sarcomere length of 14 μm. Skinned fibers were treated with trypsin, 0.5 μg ml^{-1}, in a relaxing solution (90 mM KCl, 5.2 mM MgCl$_2$, 4.3 mM ATP, 4.0 mM EGTA and 10 mM MOPS (pH 7.0)) at 23 °C. ●, tension; bar, SD; □, relative amount of crayfish connectin obtained by densitometry of the band in (C); Δ, relative amount of projectin.(C); effects of trypsin treatment on the proteins of skinned fibers. 1, control; 2, treated for 4 min; 3, 8 min; 4, 16 min. SDS gel electrophoresis was carried out using 2–8% polyacrylamide gels and the relative amounts of crayfish connectin and projectin were estimated by densitometry. From Manabe et al. (1993). Permission obtained

7–8 µm) and barnacle adductor (SL at rest, 10 µm). In these muscles 3000 kDa or even larger protein is always present. Insect protein reacted with monoclonal antibody SM1 but not with 3B9 (Table 1). In giant sizes of sarcomeres the 3000 kDa connectin links the Z line to the myosin filament and is responsible for passive tention generation upon stretch (Fig. 4). It appears that projectin binds to connectin on the myosin filament. The 3000 kDa protein was partially purified from crayfish claw muscle and was shown to be very long filament (Manabe et al. 1993).

It is of interest to note that 2000 kDa protein is present in ascidian body-wall muscle, while 3000 kDa protein is found in *Amphioxus* striated muscle (Kimura et al. 1997). On the other hand, 3B9 reactive with fish connectin does not react with *Amphioxus* connectin, but Pc1200 reacts with both connectins.

Sequence data are not available for invertebrate connectin except for a preliminary report describing the presence of motif II and PEVK region in crayfish claw muscle connectin (Fukuzawa et al. 1997).

9 Perspectives

Connectin is the largest single peptide ever known (3000–4000 kDa). Its main function is to keep the myosin filament at the center of a sarcomere as elastic spring attached to the Z line. How has such a giant exon as 80–90 kb evolved?

Comparative biology has revealed that at least there are three groups of smaller elastic proteins in invertebrate muscle (Table 2). The smallest is 753 kDa twitchin and projectin is 1200 kDa. Small connectin is as large as

Table 2. Distribution of muscle elastic proteins in invertebrate muscle

Animal	Twitchin	Projectin	Connectin	
			2000 kDa	3000 kDa
nematode	*C. elegans*	–	*C. elegans*	–
annelid	polychaete	–	–	polychaete
mollusc	scallop	–	scallop	–
arthropod	–	insect crayfish barnacle	–	insect crayfish barnacle
prochordate	–	–	acidian	*Amphioxus*

2000 kDa. First of all, it is hoped to elucidate cDNA sequences of projectin and 2000 kDa connectin (the complete sequences of twitchin and vertebrate connectin are known). Then we may speculate the evolution of connectin very probably as following:

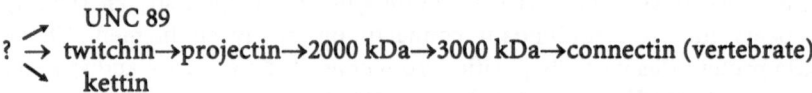

Now a large variety of monoclonal and polyclonal antibodies to vertebrate connectin is available. Antibodies to twitchin, projectin and invertebrate connectin are also available. By applying immunoblot tests, it may be possible to detect proteins smaller than twitchin and other than kettin or UNC 89 in lower invertebrates. By cDNA sequencing, we may find a candidate for the putative common precursor (?) of connectin.

Acknowledgements

The writer is most grateful to Dr. Sumiko Kimura for her ever lasting collaboration, and his former students at the Department of Biology, Faculty of Science, Chiba University for their productive work described in this review.

References

Auber J, Couteaux R (1963) Ultrastructure de la strie dans des muscle de Dipteres. J Microsc Paris 2:309–324

Ayme-Southgate A, Vigoreaux J, Benian G M, Pardue M L (1991) *Drosophila* has a twitchin/titin-related gene that appears to encode projectin. Proc Natl Acad Sci USA 88:7973–7977

Ayme-Southgate A, Southgate R, Saide J D, Benian G M, Pardue M L (1995) Both synchronous and asynchronous muscle isoforms of projectin (the *Drosophila* bent locus product) contain functional kinase domains. J Cell Biol 128:393–403

Benian G M, Kiff J E, Neckelmann N, Moerman D G, Waterston R H (1989) Sequence of an unusually large protein implicated in regulation of myosin activity in *C. elegans*. Nature 342:45–50

Benian G M, L'Hernault S W, Morris M E (1993) Additional sequence complexity in the muscle gene, unc-22, and its encoded protein, twitchin, of *C. elegans*. Genetics 134:1097–1104

Benian G M, Tang X, Tinly T L (1996a) Twitchin and related giant Ig super family members of *C. elegtans* and other invertebrates. Adv Biophys 33:175–198

Benian G M, Tinley T L, Tang X, Borodovsky M (1996b) The *Caenorhabditis elegans* gene unc-89, required for muscle M-line assembly, encodes a giant modular protein composed of Ig and signal transduction domains. J Cell Biol 132:835–848

Bullard B, Leonard K (1996) Modular proteins of insect muscle. Adv Biophys 33:211–222

Eilertsen K J, Keller T C S (1992) Identification and characterization of two huge protein components of the brush border cytoskeleton: Evidence for a cellular isoform of titin. J Cell Biol 119:549–557

Eilertsen K J, Kazmierski S T, Keller T C S (1994) Cellular titin localization in stress fibers and interaction with myosin II filaments in vitro. J Cell Biol 126:1201–1210

Fukuzawa A, Kohei S, Yajima H, Kimura S, Maruyama K (1997) Partial sequence of crayfish claw muscle connectin and its localization. Zool Sci 14:54

Fürst D O, Osborn M, Nave R, Weber K (1988) The organization of titin filaments in the half-sarcomere revealed by monoclonal antibodies in immunoelectron microscopy: A map of ten nonrepetitive epitopes starting at the Z line extends close to the M line. J Cell Biol 106:1563–1572

Fyrberg C C, Labeit S, Bullard B, Leonard K, Fyrberg E (1992) *Drosophila* projectin: relatedness to titin and twitchin and correlation with lethal (4) 102 CDa and bent-dominant mutants. Proc Roy Soc Lond B249:33–40

Guba F, Harsasnyi V, Vajda E (1968) Ultrastructure of myofibrils after selective protein extraction. Acta Biochim Biophys Acad Sci Hung 3:435–442

Hattori A, Ishii T, Tatsumi R, Takahashi K (1995) Changes in the molecular types of connectin and nebulin during development of chicken skeletal muscle. Biochim Biophys Acta 1244:179–184

Heierhorst J, Probst W C, Vilim F S, Buku A, Weiss K R (1994) Autophosphorylation of molluscan twitchin and interaction of its kinase domain with calcium/calmodulin. J Biol Chem 269:21086–21093

Heierhorst J, Probst W C, Kohanski R A, Buku A, Weiss K R (1995) Phosphorylation of myosin regulatory light chains by the molluscan twitchin kinase. Eur J Biochem 233:426–431

Heierhorst J, Kobe B, Feil S C, Parker M W, Benian G M, Weiss K R Kemp B E (1996) Ca^{2+}/S100 regulation of giant protein kinases. Nature 380:636–639

Hoyle G (1983) Muscle and their neural control. John Wiley and Sons, New York, pp 90–100

Horowits R, Kempner E S, Bisher M E, Podolsky R J (1986) A physiological role for titin and nebulin in skeletal muscle. Nature 323:160–164

Hu D H, Kimura S, Maruyama K (1986) Sodium dodecyl sulfate gel electrophoresis studies of connectin-like high molecular weight proteins of various types of vertebrate and invertebrate muscles. J Biochem 99:1485–1492

Hu D H, Kimura S, Kawashima S, Maruyama K (1989) Calcium-activated neutral protease quickly converts α-connectin to β-connectin in chicken breast muscle myofibrils. Zool Sci 6:797–800

Hu D H, Matsuno A, Terakado K, Matsuura T, Kimura S, Maruyama K (1990) Projectin is an invertebrate connectin (titin): Isolation from crayfish claw muscle and localization in crayfish claw muscle and insect flight muscle. J Muscle Res Cell Motil 11:497–511

Huxley H E, Hanson J (1954) Changes in the cross striations of muscle during contraction and stretch and their structural interpretations. Nature 173:973–976

Itoh Y, Hu D H, Ohashi K, Kimura S, Maruyama K (1987) Lamprey connectin. Zool Sci 4:379–380

Kawamura Y, Ohtani Y, Maruyama K (1994a) Biodiversity of the localization of the epitopes to connectin antibodies in the sarcomeres of lamprey, electric ray, and horse mackerel skeletal muscles. Tissue & Cell 26:677–685

Kawamura Y, Suzuki J, Kimura S, Maruyama K (1994b) Characterization of con-nectin-like proteins of obliquely striated muscle of a polychaete (Annelida). J Muscle Res Cell Motil 15:623–632

Kawamura Y, Ohtsuka H, Murata H, Maki S, Ohtani Y, Manabe T, Kimura S, Maruyama K (1996) Comparative aspects of muscle elastic protein. Adv Biophys 33:175–181

Kellermayer M S, Smith S B, Granzier H L, Bustamante C (1997) Folding-unfolding transtitions in single titin molecules characterized with laser tweezers. Science 276:1112–1116

Kimura S, Maruyama K (1983) Preparation of native connectin from chicken breast muscle. J Biochem 94:2083–2085

Kimura S, Maruyama K (1989) Isolation of α-connectin, an elastic protein, from rabbit skeletal muscle. J Biochem 106:952–954

Kimura S, Maki S, Maruyama K (1993) The role of a thiol protease in the proteolysis of connectin in rabbit skeletal muscle myofibrils. Biomed Res 14 Suppl 2:89–92

Kimura S, Kawamura Y, Kubokawa K, Watanabe A, Maruyama K (1997) Connectin-like protein in *Amphioxus* striated muscle. Zool Sci 14:78

Labeit S, Kolmerer B (1995) Titins: Giant proteins in charge of muscle ultrastructure and elasticity. Science 270:293–296

Lakey A, Ferguson C, Labeit S, Reedy M, Larkins A, Butcher G, Leonard K, Bullard B (1990) Identification and localization of high molecular weight proteins in insect flight and leg muscle. EMBO J 9:3459–3467

Lakey A, Labeit S, Gautel M, Ferguson C, Barlow D P, Leonard K, Bullard B (1993) Kettin, a large modular protein in the Z-disc of insect muscles. EMBO J 12:2863–2871

Locker R H, Wild D J C (1986) A comparative study of high molecular weight pro-teins in various types of muscle across the animal kingdom. J Biochem 99:1473–1484

Maki S, Kimura S, Maruyama K (1994) Localization of connectin-like proteins in the giant sarcomeres of barnacle muscle. Zool Sci 11:821–824

Maki S, Ohtani Y, Kimura S, Maruyama K (1995) Isolation and characterization of a kettin-like protein from crayfish claw muscle. J Muscle Res Cell Motil 16:579–585

Manabe T, Kawamura Y, Higuchi H, Kimura S, Maruyama K (1993) Connectin, giant elastic protein, in giant sarcomeres of crayfish claw muscle. J Muscle Res Cell Motil 14:654–665

Maruyama K, Matsubara S, Natori Y, Nonomura Y, Kimura S, Ohashi K, Murakami F, Handa S, Eguchi G (1977) Connectin, an elastic protein of muscle: characteri-zation and function. J Biochem 82:317–337

Maruyama K, Kimura S, Yoshidomi H, Sawada H, Kikuchi M (1984) Molecular size and shape of β-connectin, an elastic protein of striated muscle. J Biochem 95:1423–1433

Maruyama K (1986) Connectin, an elastic filamentous protein of striated muscle. Int Rev Cytol 104:81–114

Maruyama K (1994) Connectin, an elastic protein of striated muscle. Biophys Chem 50:73–85

Maruyama K, Ohtani Y, Maki S, Kawamura Y. Benian G M, Kagawa H Kimura S (1995) Connectin-related phenomena and biodiversity of the connectin family. In: Maruyama K, Nonomura Y, Kohama K (eds) Calcium as cell signal. Igakushoin, Tokyo, pp 73–79

Maruyama K, Kimura S (eds) (1996) Muscle elastic proteins. Adv Biophys 33:1–241

Maruyama K (1997) Connectin/Titin, giant elastic protein of muscle. FASEB J 11:341–345

Matsuno A, Takano-Ohmuro H, Itoh Y, Matsuura T, Shibata M, Nakae H, Kaminuma T, Maruyama K (1989) Anti-connectin monoclonal antibodies that react with the unc-22 gene product bind dense bodies of Caenorhabditis (nematode) bodywall muscle cells. Tissue & Cell 21:495–505

McNail P A, Hoyle G (1967) Evidence for superthin filaments. Am Zoologist 7:483–498

Nakauchi Y, Maruyama K (1992) Immunoblot detection of vertebrate-type of connectin (titin) in ascidian bodywall muscle and tadpole. Zool Sci 9:219–221

Natori R (1954) The property and contraction process of isolated myofibrils. Jikeikai Med J 1:18–23

Nave R, Weber K (1990) A myofibrillar protein of insect muscle related to vertebrate titin connects Z-band and A-band:purification and molecular characterization of invertebrate mini-titin. J Cell Sci 95:535–544

Nave R, Fürst D O, Vinkemeier U, Weber K (1991) Purification and physical properties of nematode mini-titins and their relation to twitchin. J Cell Sci 98:491–496

Obinata T (1993) Contractile proteins and myofibrillogenesis. Int Rev Cytol 143:153–189

Ohtani Y, Maki S, Kimura S, Maruyama K (1996) Localization of connectin-like proteins in leg and flight muscles of insects. Tissue & Cell 28:1–8

Pringle J W S (1977) The mechanical characterization of insect fibrillar muscle. In: Tregear R T (eds) The mechanical characterization of insect fibrillar muscle. Elsevier, Amsterdam, pp 177–196

Probst W C, Cropper E C, Heierhorst J, Hooper S L, Jaffe H, Vilim F, Beushausen S, Kupfermann I, Weiss K R (1994) cAMP-dependent phosphorylation of Aplysia twitchin may mediate modulation of muscle contractions by neuropeptide cotransmitters. Proc Natl Acad Sci USA 91:8487–8491

Pudles J, Moudjou M, Hisanaga S, Maruyama K, Sakai H (1990) Isolation of a giant protein from sea-urchin egg cytomatrix. Exp Cell Res 189:253–256

Rief M, Gautel M, Oesterhelt F, Fernandez J M, Gaul H E (1997) Reversible unfolding of individual titin immunoglobulin domains by AFM. Science 276:1109–1112

Saide J D (1981) Identification of a connecting filament protein in insect fibrillar flight muscle. J Mol Biol 153:661–679

Saide J D, Chin-Bow S, Hogan-Sheldon J, Busquets-Turner L (1990) Z-band proteins in the flight muscle and leg muscle of the honeybee. J Muscle Res Cell Motil 11:125–136

Sjöstrand F S (1962) The connections between A-and I-band filaments in striated frog muscle. J Ultrastruct Res 7:225–246

Soeno Y, Yajima H, Kawamura Y, Kimura S, Maruyama K (1998) Organization of connectin/titin filaments in sarcomeres of differentiating chicken skeletal muscle cells. Mol Cell Biol in press

Sorimachi H, Kinbara K, Kimura S, Takahashi M, Ishiura S, Sasagawa N, Sorimachi N, Shimada H, Tagawa K, Maruyama K, Suzuki K (1995) Muscle-specific calpain, p94, responsible for limb girdle muscular dystrophy type 2A, associates with connectin through IS2, a p94-specific sequence. J Biol Chem 270:31158–31162

Tskhovrebova L, Trinick J, Sleep J A, Simmons R M (1997) Elasticity and unfolding of single molecules of the giant protein titin. Nature 387:308–312

Vibert P, Edelstein S M, Castellani L, Elliott B W (1993) Mini-titins in striated and smooth molluscan muscles – structure, location and immunological crossreactivity. J Muscle Res Cell Motil 14:598–607

Vibert P, York M L, Castellani L, Edelstein S M, Elliot B, Nyitray L. (1996) Structure and distribution of minititins. Adv Biophys 33:199–210

Wang K, McClure J, Tu A (1979) Titin: Major myofibrillar components of striated muscle. Proc Natl Acad Sci USA 76:3698–3702

Yajima H, Ohtsuka H, Kawamura Y, Kume H, Murayama T, Abe H, Kimura S, Maruyama K (1996) A 11.5-kb 5'-terminal cDNA sequence of chicken breast muscle connectin/titin reveals its Z line binding region. Biochem Biophys Res Comm 223:160–164

Yoshidomi H, Ohashi K, Maruyama K (1985) Changes in the molecular size of connectin, an elastic protein, in chicken skeletal muscle during embryonic and neonatal development. Biomed Res 6:207–212

Ziegler, C (1994) Titin-related proteins in invertebrate muscles. Comp Biochem Physiol 109A:823–833

The Titin cDNA Sequence and Partial Genomic Sequences: Insights Into the Molecular Genetics, Cell Biology and Physiology of the Titin Filament System

B. Kolmerer[1*], C.C. Witt[1*], A. Freiburg[2*], S. Millevoi[1], G. Stier[1],
H. Sorimachi[3], K. Pelin[4], L. Carrier[5], K. Schwartz[5], D. Labeit[2],
C.C. Gregorio[6], W.A. Linke[7], S. Labeit[1,2]

[1] EMBL Heidelberg, Meyerhofstraße 1, D-69012 Heidelberg, Germany
[2] Universitätsklinikum Mannheim, Theodor-Kutzer-Ufer, D-68167 Mannheim,
Germany
[3] Department of Applied Biological Chemistry, Graduate School of Agricultural
and Life Sciences, The University of Tokyo, 1-1-1 Yayoi, Bunkyo-ku,
Tokyo 113–8657, Japan
[4] Department of Medical Genetics, University of Helsinki and the Folkhälsan
Institute of Genetics, Mannerheimintie 97, FIN-00280 Helsinki, Finland
[5] Unite de Recherches 153 de l'INSERM, Institut de Myologie,
Rue du Mur des Fermiers Genereaux, Groupe Hospitalier Pitié-Salpetrière,
47 Boulevard de l'Hopital, 75651 Paris cedex 13, France
[6] Department of Cell Biology and Anatomy, University of Arizona,
School of Medicine, 1501 N. Campbell Avenue, Tucson, AZ 85724–5044, USA
[7] Physiologie II, Universität Heidelberg, Im Neuenheimer Feld 326,
D-69120 Heidelberg, Germany
* The first three authors equally contributed to the work

Contents

1 Introduction

Titin, or connectin (Maruyama et al. 1977; Wang et al. 1979), is the third most abundant protein of vertebrate striated muscle, and therefore one of the most abundant proteins of our body. Assuming that titin comprises approximately 10% of the protein mass of a myofibril, a human adult body will contain several hundred grams of titin. Considering this abundancy, it is surprising that the protein was discovered so late. Even after its identification, the molecular properties of this giant protein remained characterized very little for the first few years. The molecular weight had been a matter of debate: values ranging from 1.2 to 4 megadalton were proposed. During the 1980s, the laboratories of K. Maruyama, K. Wang, K. Weber and J. Trinick developed methods to extract titin proteins from muscle. These preparations were used to raise a number of polyclonal and monoclonal anti-titin-specific antibodies. Immunoelectron microscopy studies then demonstrated that titin is a filamentous protein which connects the Z-disc to the A-band and therefore spans the I-band (Maruyama et al. 1984; Maruyama et al. 1985; Wang et al. 1984). Subsequently, it was demonstrated by immunoelectron microscopy, using a set of ten monoclonal antibodies, that titin in fact spans complete half-sarcomeres (Fürst et al. 1988).

Within the muscle field, the potential elastic properties of titin as well as its predicted role as a molecular ruler of the sarcomere have attracted most interest. Immunohistochemical studies had shown that epitopes in the I-band change their position relative to the A-band during myofibril extension (Itoh et al. 1988; Fürst et al. 1988). Moreover, degradation of titin by radiation, proteolytic attack, or its removal by extraction was shown to result in decreased passive tension of muscle fibers (Horowits et al. 1986; Salviati et al. 1990; Funatsu et al. 1990; Granzier and Irving, 1995). Thus, the titin filament, or more precisely, the I-band portion of titin, was assumed to behave as an elastic spring. However, outside the titin field, some researchers felt it was not proven beyond doubt that titin indeed consisted of single giant chains, instead of a number of not-yet-characterized subunits. As a further argument against the concept of a third, elastic filament, it was pointed out that the invertebrate myofibrils also develop passive tension, but do not have a titin filament (for a current view on this issue, see the contribution of K. Maruyama in this volume). Similarly, if titin is predicted to act as a molecular ruler, why would the thick filaments of invertebrate muscles without titin be as precisely organized as those of vertebrate muscles (Whiting et al. 1989)? Despite these potential problems, the pioneering studies from the laboratories of K. Maruyama, K. Wang, K. Weber, and J. Trinick clearly established by the late 1980s that striated muscle contains a third filament system that is presumably composed of elastic titin molecules. It seemed to

Lay-out of the vertebrate sarcomere

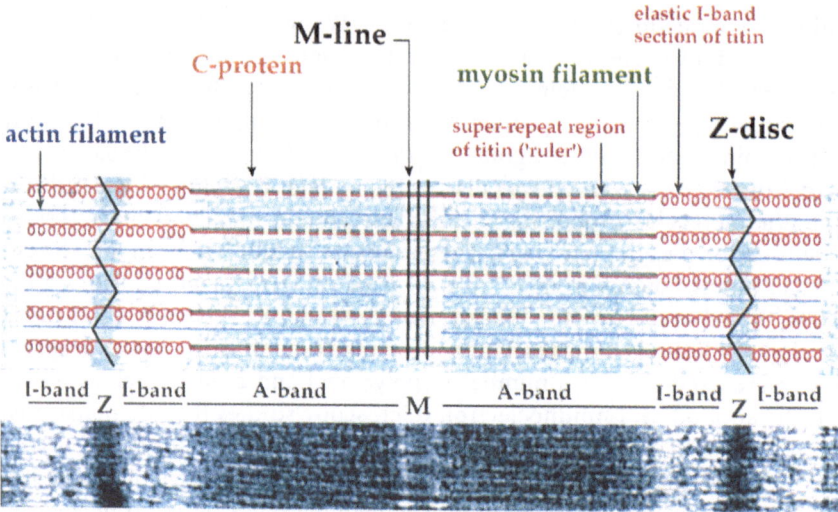

Fig. 1. Overview on the titin-filament lay-out in the vertebrate sarcomere. Immuno-histochemical studies with titin-specific antibodies had shown between 1984 and 1989 that single titin molecules span from Z-discs to M-lines (Maruyama et al. 1984; Wang et al. 1984; Fürst et al. 1988; Whiting et al. 1989). Upon myofibril stretch, titin epitopes in the A-band remained stationary relative to the M-line, whereas epitopes in the I-band translated away from the center of the sarcomere, which suggested that the I-band portion of titin is extensible (Itoh et al. 1988; Fürst et al. 1988)

be clear that the elastic filaments consist of an extensible I-band section and an A-band portion which is stiff at physiological degrees of stretch (see Fig. 1). Based upon the general doubts about the existence of megadalton-sized proteins, the concept of giant titin molecules forming this elastic filament system remained unpopular until more recently.

Our team started to work on titin in 1989, as a spin-off from the stay of John Trinick as an EMBO fellow at EMBL in 1988 (for reviews, see Trinick 1994 and 1996; and his contribution to this volume). Our initial attempts focussed on the cloning of partial titin cDNAs, then as a joint effort with Mathias Gautel (see also his contribution to this volume). We hypothesized that if titin was indeed a megadalton-sized protein, then Northern blots using titin probes should demonstrate the presence of a giant mRNA species in muscle RNAs. Next, we expected that the molecular cloning of titin cDNAs would allow a further molecular characterization of its filament system.

2 The Analysis of the cDNA Sequence of Titin

2.1 General Considerations on the Representation of Titin cDNAs

The cloning of the first titin cDNAs was complicated by the fact that it was not known how the cDNA cloning of a potentially giant mRNA should be approached. In the absence of reliable data, we expected that if a three-megadalton-sized titin chain indeed existed, a full-length titin cDNA would be 70 kb or larger. Based on this assumption, we were concerned that conventional poly-A$^+$-selected libraries may not contain full titin cDNAs, only the extreme 3' end region. Since none of the antibodies available for immunoscreenings (BD6, CH11, CE12, AB5; Whiting et al. 1989) appeared to be directed against the end regions of the titin filament, i.e. against the Z-disc or the M-line regions, we decided to construct unconventional cDNA libraries (S.L.). These were initially prepared from RNAs, which were not poly-A$^+$-selected, but rather, subjected to size-selection procedures. The first strand synthesis was then performed with random primers only and not with poly-dT-primers. The first two partial titin cDNAs were in fact isolated from a "self-made" rabbit psoas muscle cDNA library. However, the subsequent screening of commercially available muscle cDNA libraries both with partial titin cDNA probes and with titin antibodies showed, to our surprise, that titin cDNAs are well represented in conventional cDNA libraries. The cloning and the sequence determination of the full-length titin cDNA has now made it possible to determine the representation of partial titin cDNAs in commercial cDNA libraries in more detail. Notably, more statistical insights into such representation came from the identification of EST entries derived from titin and their assignment to the titin sequence (Fig. 2, from Kolmerer et al. 1996a). It appears that there is an approximate five-fold lower representation of titin cDNAs from the extreme 5' end, compared with the frequency of EST entries from the 3'-end titin region. However, since the titin message is abundantly expressed in heart and skeletal muscles, this means that partial titin cDNAs, which map 50–100 kb 5' from the poly-A-tail, are still reasonably well represented in conventional poly-A$^+$-selected cDNA libraries. Therefore, all partial titin cDNAs (and also nebulin cDNAs) that have been cloned in our laboratory since 1991 have been isolated from commercially available cDNA libraries.

At present, it is unclear why titin cDNAs are well represented in poly-A$^+$-selected cDNA libraries. A possible reason might be that the ~100 kb titin mRNAs stick to the cellulose resigns, which are commonly used for the poly-A$^+$-selection of RNA species, in a non-specific manner. Alternatively, the

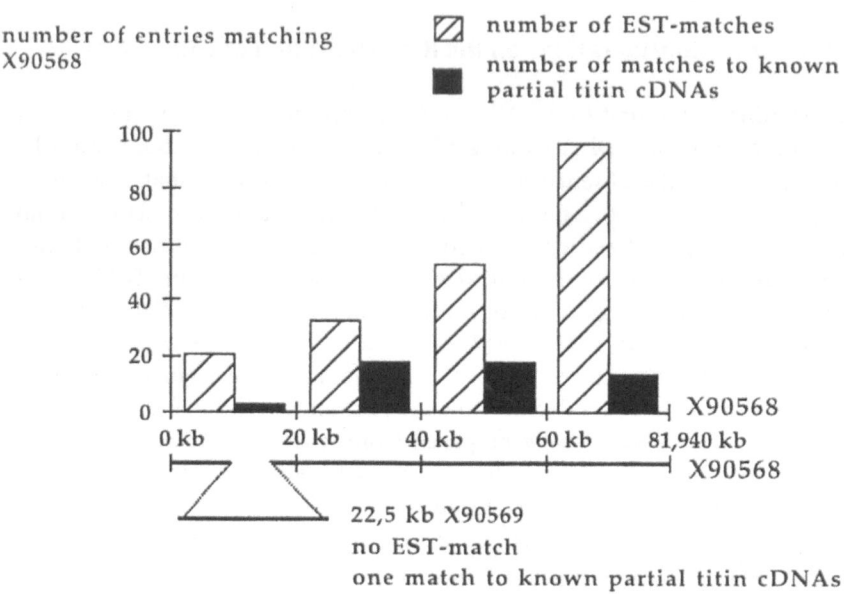

Fig. 2. The frequency of titin cDNAs in the ESTs present in the data libraries was determined by searching the human titin sequences (accessions X90568 and X90569) versus the EMBL and Genbank data sets (releases 43 and 90, respectively, from September 1995). A total of about 200 titin ESTs were identified, i.e., on average one EST every 500 bp of the titin sequence. From the 3' end towards the extreme 5' end an about fivefold drop in representation occurs. Interestingly, no single EST match to the skeletal-specific partial titin sequence was identified (EMBL data library accession X90569, EMBL data library). This probably reflects the preferential sequence analysis of clones from cardiac libraries in search of genes implicated in cardiovascular diseases (modified from Kolmerer et al. 1996a)

~100 kb titin RNAs may contain a significant amount of incompletely spliced pre-mRNAs. Then, AT-rich intron sequences which are, for example, present in the determined partial genomic A-band titin sequences (Kolmerer et al. 1996b) will be retained. These sequences may also attach to the oligo-dT resigns. The assumption that poly-A-tails do not appear to be required for retaining titin mRNAs on oligo-dT-columns is also supported by the fact that the sequenced 3'-end titin mRNAs either contain no poly-A-tail, or only a very short one (S.L., unpublished data). Similar observations have also been made for the nebulin 3' cDNAs sequenced in our laboratory (S.L., unpublished data). These observations raise the possibility that muscle

cells either do not efficiently polyadenylate the giant titin and nebulin transcripts or alternatively, that these transcripts' poly-A-tails are under dynamic control and are truncated by exonucleases. Clearly, a more thorough
investigation of the poly-A-tail lengths of the titin and nebulin mRNAs, and
of the effect of myocyte growth and differentiation is likely to be rewarding.

2.2 Cloning of Partial Titin cDNAs

By 1989, several laboratories had produced a variety of titin-specific monoclonal antibodies that had been epitope-mapped by immunoelectron microscopy (for an overview, see Maruyama, 1994). As discussed above, it
subsequently became clear that partial titin cDNAs are well-represented in
conventional, randomly primed cDNA libraries. Therefore, our laboratory
launched a systematic effort to isolate partial cDNAs originating from different regions of the titin molecule. As summarized in Fig. 3, expression
screenings with the titin monoclonal antibodies BD6, CE12, CH11 and AB5
(Whiting et al. 1989), and with the T12 antibody (Fürst et al. 1988), identified
partial titin cDNAs that reacted specifically with the respective antibodies. In
addition, sera from myasthenia gravis patients, who also have a tumor of the
thymus, had been shown to contain autoantibodies to titin directed mostly
against the A/I-junction region of the molecule (Aarli et al. 1990). By performing expression screenings, these patients' sera were then used to identify partial titin cDNAs, which presumably code for the A/I-juncton region
of titin (Gautel et al. 1993a). In summary, by using expression screening
methods, our laboratory cloned six partial titin cDNAs derived from different regions of the titin molecule (Fig. 3).

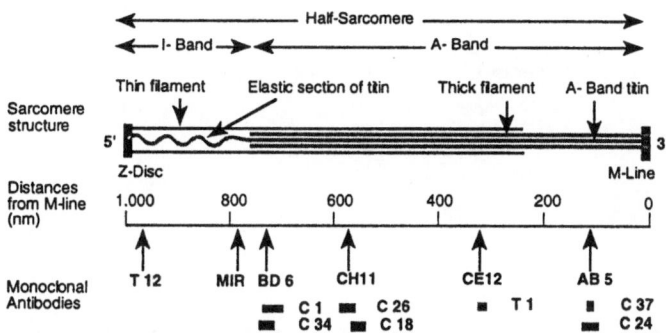

Fig. 3. Overview on the cloning of the full-length titin cDNA. Different titin-specific
monoclonal antibodies were used to screen skeletal and cardiac muscle cDNA expression libraries. A set of corresponding partial cDNAs was identified and sequenced. Their partial sequences were used to initiate cDNA extension walks
(modified from Labeit et al. 1992)

2.3 Is Titin Really a Three-Megadalton Protein?

As mentioned above, there had been some reservations whether titin was indeed a three-megadalton protein, rather than a composition of a number of smaller subunits (for a discussion, see e.g. Fulton and Isaacs, 1991). The cloning of partial titin cDNAs allowed us to investigate the properties of these cDNAs when used as specific probes on Northern blots. This approach demonstrated that the different partial titin cDNAs hybridize to a giant mRNA species (Fig. 4). Also, gene mapping studies assigned the different partial titin cDNAs to a single locus on the long arm of the human chromosome 2 (for details, see section 3.1). Finally, the DNA extension methods showed that all partial cDNAs are part of a single, giant, full-length mRNA. Therefore, we demonstrated that indeed titin consists of a single three to four megadalton chain. It will be challenging to understand how the synthesis, assembly, and turnover of this huge protein is controlled in living cells.

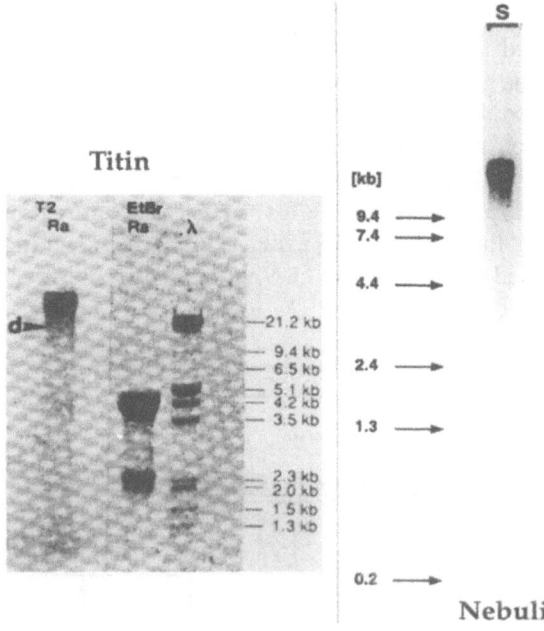

Fig. 4. Northern blot analysis of muscle RNAs with titin (left) and nebulin (right) probes. In both cases, giant mRNA species are detected that migrate on conventional 1% agarose gels with limited mobility (left part from Labeit et al. 1990; right part from B.K., unpublished results)

2.4 Molecular Cloning of the Full-Length Cardiac Titin cDNA

When it became clear that the six isolated partial titin cDNAs were part of an abundantly expressed, full-length, giant titin mRNA, we decided to extend all six partial cDNAs in both the 5' and 3' directions (see Fig. 3). While performing this work between 1990 and 1994, we introduced a number of methodological improvements. First, with the introduction of the polymerase chain reaction (PCR) method we could overcome the need for conventional cDNA insert preparations by plaque-purification of identified titin clones, and their restriction-digest analysis (Saiki et al. 1985). Thus, it became possible to characterize non-plaque-purified primary positive clones from a library screen using the PCR technique. For the amplification of extending inserts by anchored PCR techniques, a combination of a specific titin primer with a vector-derived primer was used. However, it should be noted that our attempts to completely overcome the need for library screens failed when we used the anchored PCR technique in liquid on total cDNA as a template. Rather, we found it necessary to identify single clones by conventional filter hybridizations and to employ the PCR technique only for the identification and amplification of extending cDNA fragments. This approach combined the security of conventional cDNA cloning methods (by using a stringent hybridization for the clone-selection step) with the speed of PCR-based methods (when using the PCR method during the subsequent clone-analysis steps). A second significant improvement resulted from the introduction of cycle-sequencing methods, which allowed for the direct analysis of PCR fragments without the need to subclone them for sequence analysis.

Application of the initial conventional technique and the later PCR-based cDNA extension method allowed us to isolate a set of 54 partial titin cDNAs from a human cardiac cDNA library (Stratagene, catalogue number 936208). At the 5' end, the presence of an ATG codon, which is preceeded in frame by a stop codon, suggested that titin's N-terminus had been reached (Labeit and Kolmerer, 1995). Consistent with this idea, Maruyama and co-workers have isolated a chicken connectin/titin cDNA, which is homologous to the 5' end of the human cardiac titin message (Yajima et al. 1996). As for the opposite end of the titin sequence, the presence of a non-coding region and of a poly-A-tail indicated that the cDNA walks had reached the 3' end of the titin mRNA (Gautel et al. 1993b). However, it should be noted that seven out of the eight analyzed 3'-end titin clones did not contain a poly-A-tail, whereas the remaining clone had a short poly-A-tail only 15 bp long (S.L., unpublished data). This raises the possibility that the polyadenylation of the titin mRNA, or the maintainance of the poly-A-tail length is under tight control.

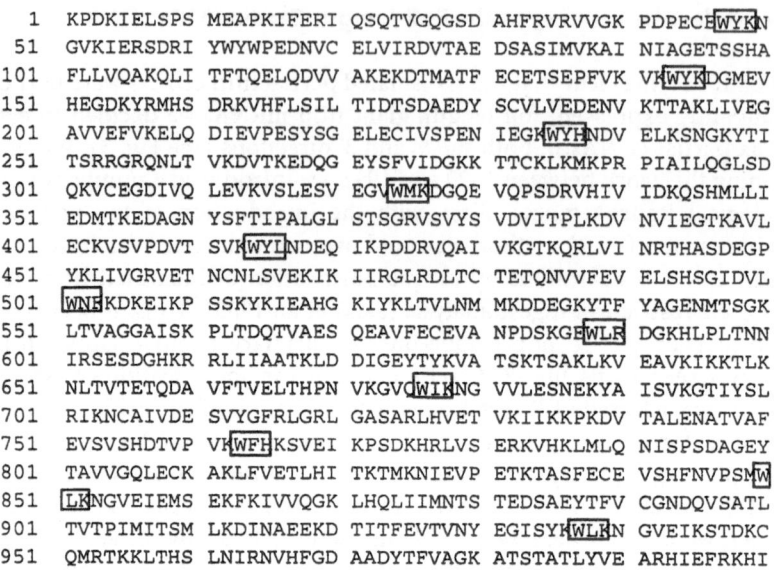

```
    1   KPDKIELSPS  MEAPKIFERI  QSQTVGQGSD  AHFRVRVVGK  PDPECHWYKN
   51   GVKIERSDRI  YWYWPEDNVC  ELVIRDVTAE  DSASIMVKAI  NIAGETSSHA
  101   FLLVQAKQLI  TFTQELQDVV  AKEKDTMATF  ECETSEPFVK  VKWYKDGMEV
  151   HEGDKYRMHS  DRKVHFLSIL  TIDTSDAEDY  SCVLVEDENV  KTTAKLIVEG
  201   AVVEFVKELQ  DIEVPESYSG  ELECIVSPEN  IEGKWYHNDV  ELKSNGKYTI
  251   TSRRGRQNLT  VKDVTKEDQG  EYSFVIDGKK  TTCKLKMKPR  PIAILQGLSD
  301   QKVCEGDIVQ  LEVKVSLESV  EGVWMKDGQE  VQPSDRVHIV  IDKQSHMLLI
  351   EDMTKEDAGN  YSFTIPALGL  STSGRVSVYS  VDVITPLKDV  NVIEGTKAVL
  401   ECKVSVPDVT  SVKWYLNDEQ  IKPDDRVQAI  VKGTKQRLVI  NRTHASDEGP
  451   YKLIVGRVET  NCNLSVEKIK  IIRGLRDLTC  TETQNVVFEV  ELSHSGIDVL
  501   WNEKDKEIKP  SSKYKIEAHG  KIYKLTVLNM  MKDDEGKYTF  YAGENMTSGK
  551   LTVAGGAISK  PLTDQTVAES  QEAVFECEVA  NPDSKGEWLR  DGKHLPLTNN
  601   IRSESDGHKR  RLIIAATKLD  DIGEYTYKVA  TSKTSAKLKV  EAVKIKKTLK
  651   NLTVTETQDA  VFTVELTHPN  VKGVQWIKNG  VVLESNEKYA  ISVKGTIYSL
  701   RIKNCAIVDE  SVYGFRLGRL  GASARLHVET  VKIIKKPKDV  TALENATVAF
  751   EVSVSHDTVP  VKWFHKSVEI  KPSDKHRLVS  ERKVHKLMLQ  NISPSDAGEY
  801   TAVVGQLECK  AKLFVETLHI  TKTMKNIEVP  ETKTASFECE  VSHFNVPSMW
  851   LKNGVEIEMS  EKFKIVVQGK  LHQLIIMNTS  TEDSAEYTFV  CGNDQVSATL
  901   TVTPIMITSM  LKDINAEEKD  TITFEVTVNY  EGISYKWLKN  GVEIKSTDKC
  951   QMRTKKLTHS  LNIRNVHFGD  AADYTFVAGK  ATSTATLYVE  ARHIEFRKHI
```

Fig. 5. Modular structure of the titin peptide. A section from the I-band corresponding to about 3% of the full-length titin sequence is shown. A characteristic WYK motif is repeated every 90 residues. Upon extraction and alignment of the 90-residue repeats, a data library search reveals their homology to the Ig-domain superfamily. Structural studies of expressed I-band Ig repeats have confirmed this prediction (Politou et al. 1995; Improta et al. 1996)

From the sequence of the 54 partial cardiac titin cDNAs, we assembled an 82 kb cDNA, which appears to be a full-length version of the human cardiac titin cDNA (EMBL data library accession X90568). The determined 82 kb sequence contains a single open reading frame of 81 kb length, predicting a 27,000-residue protein. Analysis of the predicted peptide reveals characteristic motifs that occur repeatedly at regular intervals (Fig. 5). Within the titin sequence, two different classes of repeats can be identified, which code for domains belonging to either the fibronectin-type III (FN3) or the immunoglobulin (Ig) domain superfamilies (also referred to as class I and class II repeats, respectively, and first identified as components of intracellular proteins in the twitchin protein; Benian et al. 1989). Similarly, a search for the two motifs in the skeletal titin sequence reveals a total of up to 298 copies of these two distinct repeat families. Therefore, the up to 33,000-residue sequence information of the titin can well be described by giving a schematic overview of the domain architecture (Fig. 6). The Ig and FN3 repeats together make up 90% of titin's mass. Non-repetitive sequences comprise

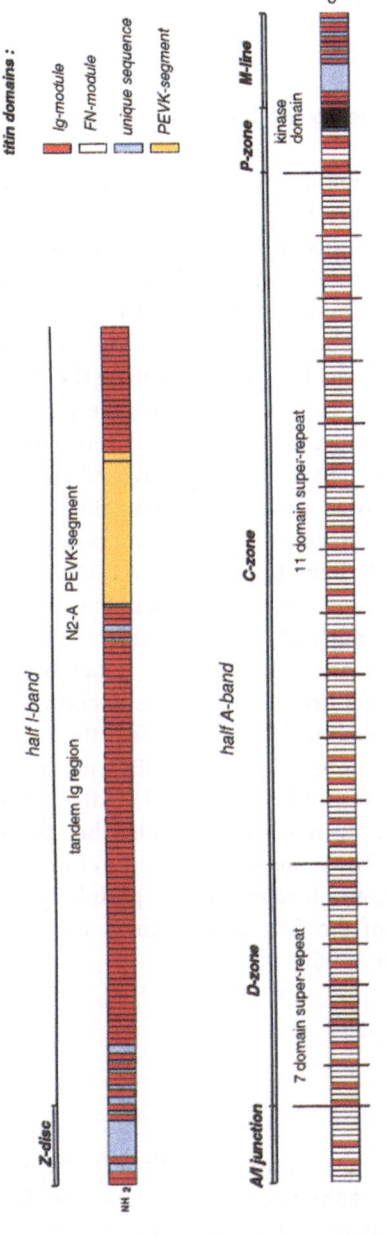

Fig. 6. Domain architecture of the skeletal titin. Different kinds of motif arrangements are present in different regions of the titin filament. Specific for the extensible I-band region of the titin are tandemly repeated Ig motifs, specific for the central region of the thick filaments are ordered "super repeat" patterns of the FN3 and Ig repeats. Unique sequences comprise about 10% of the titin molecule. These sequences occur preferentially in those regions, in which the titin filament binds a number of ligands, thereby forming a complex protein meshwork (Z-disc, M-line, and perhaps the N2-region). The Z-disc, I-band, A-band and M-line portions of the titin filament have been determined by immunoelectron microscopy with epitope-mapped titin-specific antibodies (Yajima et al. 1996; Bennett and Gautel, 1996; Obermann et al. 1996)

about 10% of the titin peptide. The non-repetitive sequences also encode a serine/threonine kinase domain (Labeit et al. 1992) and three potential phosphorylation domains (Gautel et al. 1993b; Sebestyen et al. 1995; Labeit and Kolmerer, 1995; Gautel et al. 1996).

2.5 Lay-Out of the Titin Sequence in the Sarcomere

Immunoelectron microscopy labeling studies with a variety of titin antibodies demonstrated that titin makes up part or all of a filament system that extends from the Z-disc to a region close at the M-line (Maruyama et al. 1984; Maruyama et al. 1985; Wang et al. 1984; Fürst et al. 1988; Whiting et al. 1989). In these studies, the antibody recognizing an epitope nearest to the M-line (T33) was shown to map about 80 nm away from the center of the M-line (Fürst et al. 1988). Over the last two years following the determination of titin's coding sequence, a more refined model for the lay-out of the titin filament has become available (Fig. 6). An initial overview on the lay-out of the titin sequence in the half-sarcomere was obtained by us by epitope-localization studies with a limited number of monoclonal titin antibodies (Labeit and Kolmerer, 1995). Recently, novel antibodies to selected expressed titin domain fragments have been generated. Ultrastructural localization of the selected titin epitopes on muscle fibers has further refined the model lay-out of the titin filament within the sarcomere (Yajima et al. 1996; Obermann et al. 1996; Bennett and Gautel, 1996). It has become apparent that ~100 kDa of the N-terminal domain of titin is anchored in the Z-disc, whereas another ~100 kDa stretch of I-band titin near the Z-disc is inextensible under physiological conditions (Fürst et al. 1988), possibly reflecting its association with actin filaments (Linke et al. 1997). After spanning the remainder of the (half) I-band, titin filaments enter the A-band, with approximately 2000 kDa of the titin sequence involved in binding to the thick filaments. Titin's kinase domain locates at the periphery of the M-line structure; a 200 kDa section of the C-terminal titin is involved in forming this structure (Obermann et al. 1996).

3 The Titin Gene

3.1 Gene Mapping Studies

Following the cloning of partial mammalian titin cDNAs, studies on the structure and the localization of the titin gene became possible. Hybridization of partial A-band titin cDNAs to Southern blots of different vertebrate

DNAs suggested a high degree of interspecies conservation. In the human genome, the gene coding for titin was assigned to a single locus on the long arm of Chromosome 2 (Labeit et al. 1990). Subsequent breeding studies and in-situ hybridization experiments mapped the titin gene with higher resolution in the genomes of man and mouse, respectively (Müller-Seitz et al. 1993; Rossi et al. 1994). The genetic and FISH mapping studies demonstrated that in both man and mouse, the genes for titin and nebulin are in physical proximity. The mapping studies also indicated that the two genes are members of a group of genes with syntenie which has remained conserved during

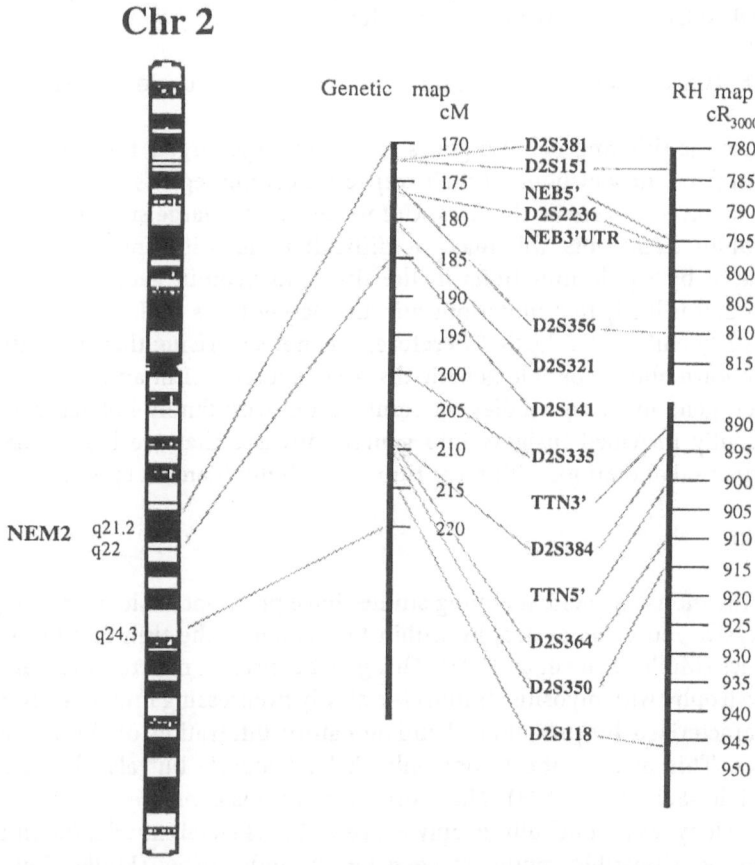

Fig. 7. Localization of the human titin and nebulin genes and the flanking genetic markers on the long arm of chromosome 2 (Pelin et al. 1997). The titin and nebulin genes are physically linked and part of a group of syntenic genes (Rossi et al. 1994). The map position of nebulin, but not titin, overlaps with a locus causing the nemaline rod disease (Wallgren et al. 1995; Pelin et al. 1997)

evolution since the divergence of man and rodents. More recently, the availability of full-length titin and nebulin cDNA sequences has allowed the development of primer pairs specific to the 5' and 3' ends of both genes. This has made it possible to map the position of these genes with high resolution by the radiation hybrid method (Cox, 1995). By using this approach, the titin gene has been assigned to 2q24, the nebulin gene to 2q22 (Pelin et al. 1997). The physical distance between both genes appears to be in the range of a few megabases (see Fig. 7). The orientation of the transcriptional units could be assigned for the titin gene from 5' to 3' as from the telomer to the centromer, whereas that of the nebulin gene is opposite, i.e. the nebulin gene is transcribed from the centromer to the telomer.

3.2 The Titin and Nebulin Filament Systems in Muscle Disease

Since the titin and nebulin genes represent large targets for mutations – at least from the viewpoint of their respective coding regions – both genes may be involved in genetic diseases. Unfortunately, the large size of the titin and nebulin transcripts will make it difficult to identify any genetic disease caused by subtle mutations in the titin and nebulin genes. On the other hand, it is likely that mutations affecting larger parts of the genes will result in early embryonic death. Therefore, it is not surprising that presently little is known about possible genetic diseases involving titin and nebulin. However, gene mapping studies, in combination with family linkage data, have recently provided insights into genetic diseases that are likely caused by structural alterations within the titin and nebulin filament systems.

Titin

In the mouse, genetic mapping studies have positioned a locus causing progressive muscular dystrophy within the region of the titin and the nebulin genes (Müller-Seitz et al. 1993). This genetic disease, referred to as muscular 'dystrophy with myositis' (mdm) is a slowly progressing muscular dystrophy characterized by pronounced inflammatory infiltration of the muscle tissues. This disease affects not only skeletal muscle but also heart muscle (Müller-Seitz et al. 1993). Therefore, since titin is expressed in both striated muscle types and nebulin is only expressed in skeletal muscle, the titin gene may be a possible candidate gene for the mdm locus (Müller-Seitz et al. 1993). If mutations in the titin gene of the mdm mouse are indeed identified, it will be interesting to see how these mutations are linked to the progressive muscular dystrophy and the prominent myositis which is present in the affected mdm muscles during the later stages of the disease.

Nebulin

Family studies in humans have identified a recessive locus within the 2q region of the human genome, which causes a nemaline rod myopathy (Wallgren et al. 1995). Previously, a mutation in the alpha-tropomyosin gene TPM3 had been identified as the genetic cause for nemaline myopathy. The identified point mutation is inherited in an autosomal dominant fashion (Laing et al. 1995). Therefore, more than one gene is involved in the nemaline myopathy disease. The two genetically different forms causing a nemaline myopathy have been termed NEM1 (autosomal dominant form, caused by TPM3) and NEM2 (autosomal recessive form, linkage to 2q), respectively. Recently, the location of the titin and nebulin genes have been mapped with high resolution within the 2q region of the human genome (Pelin et al. 1997) by using the radiation hybrid method (Cox, 1995). Also, by molecular genetic studies on NEM2-affected families, higher resolution linkage data have been obtained (Pelin et al. 1997). These studies have shown that the nebulin gene, but not the titin gene, is within the region of the NEM2 locus. Therefore, the nebulin gene is a good candidate gene for the NEM2-form of the nemaline myopathy. Interestingly, the nemaline rod-like inclusion bodies found in the patients' muscles are regular assemblies of Z-disc proteins, but the assemblies are distinct from those found in normal Z-bands (Schultheiss et al. 1992b; Schroeter et al. 1996). Therefore, it will be interesting to see how mutations in alpha-tropomyosin and perhaps in nebulin may disturb the precise arrangement of the structural Z-disc proteins.

The Titin Ligand Cardiac C-Protein

C-proteins belong to a family of homologous isoforms, which are encoded by a gene family (see contribution of D.A. Fischman to this volume). The members of this family, C-, H- and X-proteins bind to the central regions of the thick filament at 43 nanometer intervals (Craig and Offer, 1976; Craig, 1977; Sjöström and Squire, 1977). The structural precision of the repetitive binding of the C-proteins to the thick filament appears to be caused by the presence of regularly spaced C-protein binding sites in the A-band titin (Labeit et al. 1992). Expressed domains from the A-band portion of titin contain up to eleven sites that interact with the members of the C-protein family in in vitro binding assays (Freiburg and Gautel, 1996). Heart muscle expresses a member of the C-protein family, which is somewhat structurally different from that found in skeletal muscles, and encoded by a different gene (Gautel et al. 1995). During the last two years, molecular genetic studies

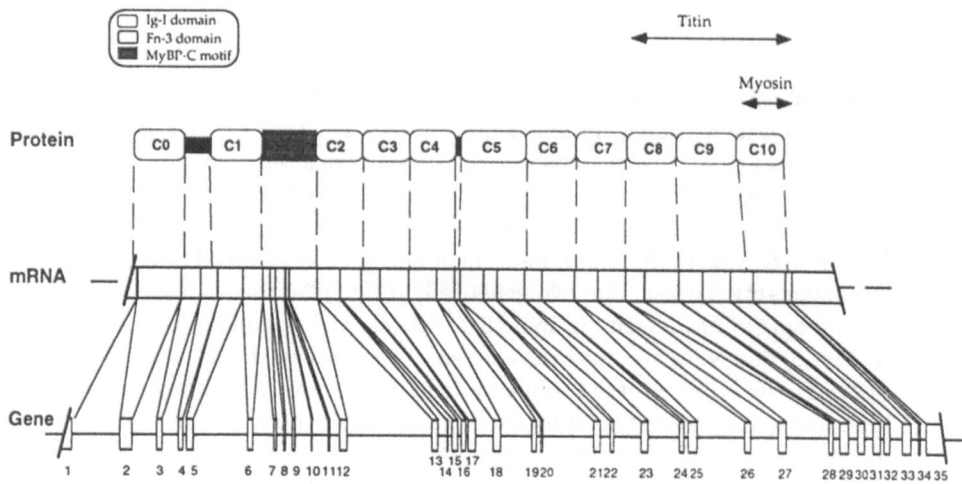

Fig. 8. Modular architecture of the cardiac C-protein in correlation with the exon-intron structure of its gene (Carrier et al. 1997). At its N-terminus, the cardiac C-protein contains an Ig repeat – designated C0 – and three phosphorylation sites (indicated as a red box), which are specific for the cardiac C-protein isoform, whereas the binding sites for titin and myosin at the C-terminal portion are common to cardiac and skeletal muscle C-proteins (Carrier et al. 1997 and references therein). The cDNA and the genomic sequence of the human cardiac C-protein are available from the EMBL data library (accession numbers X84075 and Y10129 from S. L. und L.C., respectively)

on the cardiac C-protein have implicated an involvement of this titin ligand in cases of autosomally dominant familial hypertrophic cardiomyopathy (FHC). Four genetically distinct subgroups have been identified so far. The three sarcomeric proteins cardiac beta-myosin heavy chain, cardiac troponin T and alpha-tropomyosin were all shown to be involved in hypertrophic cardiomyopathy when a mutation is present in one allele of the respective genes. This has led to the concept that FHC is a disease of the sarcomere (Thierfelder et al. 1994; for reviews, Schwartz et al. 1995; Watkins et al. 1995; Vikstrom and Leinwand, 1996). Genetic family linkage studies also identified an FHC-causing locus on the short arm of the chromosome 11 (Carrier et al. 1993). Since the gene encoding the cardiac C-protein was shown to localize to the short arm of chromosome 11 (Gautel et al. 1995), the cardiac C-protein gene has emerged as a candidate for the sought-after fourth gene which is involved in the genesis of FHC. Several groups have now confirmed that mutations are indeed present in the cardiac C-protein gene in the 11p-linked FHC families (Bonne et al. 1995; Watkins et al. 1995; Carrier et al. 1997; Rottbauer et al. 1997). Notably, the complete genomic sequence coding for the full-length 4.5 kb cardiac C-protein mRNA has been reported (Carrier et al. 1997). This accomplishment will greatly facilitate future searches for mutations in the cardiac C-protein gene (Fig. 8). Interestingly, the ten mutations observed in the four family studies reported so far (Bonne et al. 1995; Watkins et al. 1995; Carrier et al. 1997; Rottbauer et al. 1997) all predict that the mRNA structure transcribed from the mutated allele is altered significantly. Nine of the ten mutations directly alter the mRNA size, because they represent splice donor, splice acceptor, insertion or deletion mutations. Only one mutation observed so far is a point mutation, but this point mutation locates at the last base of an exon and is therefore likely to give rise to a mis-splice product (Carrier et al. 1997). It will be interesting to see if indeed, a change in mRNA structure is relevant to the mechanism of the pathology in the 11p-linked FHC families. If so, this would represent a novel pathomechanism. In this context, it is also noteworthy that a recent study failed to observe a truncated translation product, as would be expected from a mutated cardiac C-protein allele (Rottbauer et al. 1997). Therefore the expression of the truncated mutant cardiac C-protein may not be required for the genesis of the FHC-phenotype.

3.3 Partial Gene Structure of the Titin

The full-length human titin cDNA also provides the information needed to generate a set of probes to clone fragments of the titin gene, and for the detailed analysis of the exon/intron organization of the gene. As a first step towards this goal, a pool of cardiac and skeletal human titin cDNAs have been prepared and hybridized to a commercially available human genomic lambda-fix II library (Stratagene catalog number 946206). As a result, a set of 22 lambda phages were isolated spanning the coding region of the titin gene (B.K., unpublished data). The assignment of the individual titin cDNAs to the 22 lambda phages indicates that the coding region of the titin gene is contained in about 300 kb of genomic DNA. Overall, the titin gene appears to have a very different organization in different regions. Partial genomic data (unpublished) indicate that the A-band-titin-encoding region of the gene is extremely compact and contains about 80% of coding DNA. The compactness of the titin gene in its A-Band region is illustrated by the fact, that a single BAC-clone was found to span the entire A-band section (C.C.W. unpublished data). In contrast, the PEVK-encoding region of the titin gene contains very small exons, and perhaps less than 10% of the genomic DNA represent coding information (B.K., unpublished data).

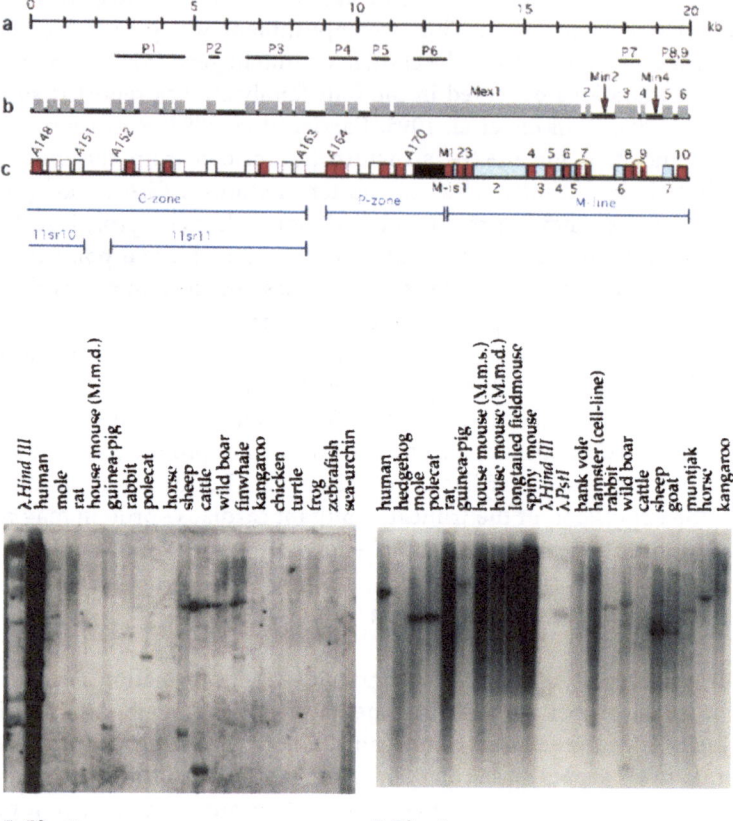

Min2   **Min4**

Fig. 9. *Top:* Exon-intron organization of the titin gene at its 3′ end (from Kolmerer et al. 1996b). The sequenced 20 kb partial genomic titin sequence contains 23 exons. For 21 exons, the junctions with intron sequences locate at domain junctions; only the two exons Mex2 and Mex4 correspond to the N-terminal and C-terminal halfs, repectively, of an Ig repeat. The correlation of exon-intron organization with the modular organization of the titin protein raises the possibility that functional titin isoforms may be generated by shuffling of domains as a consequence of alternative splicing events. The M-line region of the titin filament is coded for by six exons, referred to as Mex1 to Mex 6, respectively. The five introns present in the M-line region have been termed Min1 to Min5 (Kolmerer et al. 1996b). *Bottom:* Not only the coding exon sequences, but also the sequences of the two introns Min2 and Min4 are highly conserved between species, as detected in interspecies cross-hybridization experiments

As a more long-term effort, we have begun to systematically determine the nucleotide sequences from the human titin gene, with the ultimate goal to determine its complete sequence (B.K., S.L.). For the 3'-end region of the titin gene, a 20 kb contig has been completed, and was shown to contain the 13.5 kb of titin cDNA from the 3' end of the cardiac titin mRNA. Analysis of the exon/intron organization within this partial sequence has shown that the exon/intron junctions in general locate at domain junctions (Kolmerer et al. 1996b). The M-line region of the titin filament is coded for by six exons, referred to as Mex1 to Mex6. These six exons are separated by five introns, termed Min1 to Min5 (counting from 5' to 3'; Fig. 9). Interestingly, the sequence of two introns, Min2 and Min4, has been found to be significantly conserved in different mammalian species (Fig. 9). For Min4, this may relate to the fact that the downstream exon, Mex5, is alternatively expressed (see section 4). The functional constraints that lead to a conservation of the Min2 intronic sequence are not known at present. One conclusion from this partial sequencing project on the titin gene is that the determination of the genomic titin sequence appears to be worthwhile also from a functional perspective.

4 The Titin Isoform Family

4.1 Identification of Differentially Expressed Titin Isoforms

Before the cloning of titin and the detailed characterization of differential titin expression, gel electrophoretic mobility studies showed that titins from different striated muscle tissues have different apparent mobilities on low percentage SDS gels (Wang et al. 1991; Horowits, 1992). This led to the suggestion that different types of striated muscle tissues express different titin isoforms. It was also suggested that the isoform expression may be related to the well-known differences in passive mechanical properties between relaxed skeletal and heart muscles (Wang et al. 1991; Horowits, 1992). Later, the knowledge of titin's primary structure allowed investigation of the types of titin sequence expressed in various muscle tissues.

Different strategies are possible to search for differentially expressed isoforms of titin. Since the gene-mapping studies have identified a single locus for titin on chromosome 2, titin-isoform diversity is likely to be a result of differential splicing of the gene. Therefore, one sound approach to search for possible isoforms of titin will be the complete sequencing of its gene. As described above, this approach has already led to the identification of two conserved introns in the M-line titin, and of one exon that appears to be alternatively expressed. Another approach to search for alternatively expressed titin sequences is to hybridize titin cDNA probes to slot-blotted RNAs from different muscle types. Although this approach is qualitative it provides a rapid overview of the segments of the titin gene which are expressed (Labeit and Kolmerer, 1995). In the future, an extremely powerful extension of this method might be the hybridization of muscle RNAs isolated from different tissue sources to oligonucleotide-coated microchips that display titin sequences from different exons and their junctions. In more general terms, a characterization of the differentially expressed isoforms of titin may extend our understanding on how the titin structure is related to the sarcomeric structure, and will further test and possibly extend the "molecular ruler" hypothesis (Whiting et al. 1989).

Previously, electron microscopic studies have shown that all vertebrate thick filaments characterized so far possess eleven 43 nanometer structural repeats, made up by a regular C-protein/myosin arrangement (Craig and Offer, 1976; Craig, 1977; Sjöström and Squire, 1977). This arrangement appears to correlate with the conserved domain architecture of the central A-band titin. All titins characterized so far contain eleven copies of an eleven-domain super-repeat sequence, and no alternative splicing events have been detected in this titin region. In contrast, the Z-disc, I-band, and the M-line titin portions have a variable structure, depending on the type of muscle tissue that express the respective isoform. Therefore, an emerging concept is that structural adaptations of specific sarcomeric regions are incorporated tissue-specifically through the differential expression of the titin gene (Labeit and Kolmerer, 1995).

4.2 Isoforms of the Z-Disc Titin

Vertebrate Z-discs share features common to all Z-discs of striated muscle tissues and species, such as the tetragonal packaging of the overlapping thin filaments from opposite sarcomeres, and their cross-linking by alpha-actinin. However, some tissue and species variability of the Z-disc structure is also apparent (for reviews, Squire, 1981; Vigoreaux, 1994; Squire, 1997). When longitudinal sections of different vertebrate muscles are analyzed at the ultrastructural level, Z-discs are visualized to differ in width. In cross-sections, the overlapping actin filaments and the alpha-actinin cross-linking Z-molecules are observed to be packed together in different configurations. The mammalian cardiac Z-disc has the largest width; it appears as a dense ~ 100 nm-wide structure. The soleus red skeletal muscle Z-disc has a width of about 80 nm, whereas white skeletal muscles with predominantly fast fiber types (e.g., psoas muscle) have a quite narrow about 40 nm-wide Z-disc (Yamaguchi et al. 1985). These differences in the Z-disc structure have been

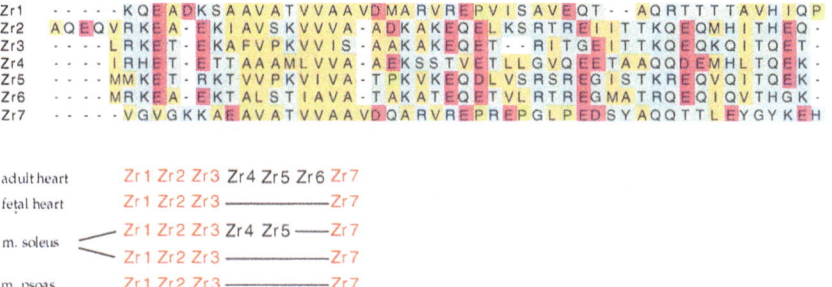

```
Zr1    · · · · · K Q E A D K S A A V A T V V A A V D M A R V R E P V I S A V E Q T · · A Q R T T T T A V H I Q P
Zr2    A Q E Q V R K E A · E K I A V S K V V V A · A D K A K E Q E L K S R T R E I I T T K Q E Q M H I T H E Q ·
Zr3    · · · · L R K E T · E K A F V P K V V I S · A A K A K E Q E T · · R I T G E I T T K Q E Q K Q I T Q E T ·
Zr4    · · · · I R H E T · E T T A A A M L V V A · A E K S S T V E T L L G V Q E E T A A Q Q D E M H L T Q E K ·
Zr5    · · · · M M K E T · R K T V V P K V I V A · T P K V K E Q D L V S R S R E G I S T K R E Q V Q I T Q E K ·
Zr6    · · · · M R K E A · E K T A L S T I A V A · T A K A T E Q E T V L R T R E G M A T R Q E Q I Q V T H G K ·
Zr7    · · · · · V G V G K K A E A V A T V V A A V D Q A R V R E P R E P G L P E D S Y A Q Q T T L E Y G Y K E H
```

adult heart	Zr1 Zr2 Zr3 Zr4 Zr5 Zr6 Zr7
fetal heart	Zr1 Zr2 Zr3 ————————Zr7
m. soleus	⟨ Zr1 Zr2 Zr3 Zr4 Zr5 ————Zr7
	Zr1 Zr2 Zr3 ————————Zr7
m. psoas	Zr1 Zr2 Zr3 ————————Zr7

Fig. 10. The Z-repeat family and the differential expression of Z-repeats. *Top:* Shown are the seven copies of Z-repeats expressed in the rabbit cardiac titin. Comparison of both the differentially and the constitutively expressed Z-repeats with their human homologs reveals a high degree of interspecies conservation (data not shown). *Bottom:* Summary of the Z-repeats expressed in different types of muscles. The repeats Zr1, Zr2, Zr3 and Zr7 were found to be expressed in all striated muscles, whereas inclusion of Zr4, Zr5 and Zr6 was tissue-type dependent and also reflected the developmental stage of the respective muscle

recently found to be correlated with the expression of titin isoforms in the Z-disc region. Gautel and co-workers reported that three clones amplified from a human cardiac cDNA library contained five, six, or seven copies of a 45-residue repeat, referred to as Z-repeat. The authors speculated that this may account for different Z-disc widths in different muscle tissues (Gautel et al. 1996). A more detailed analysis of the Z-disc titin transcripts expressed in different muscles revealed that rabbit cardiac titin encompasses seven copies of Z-repeats (Sorimachi et al. 1997); only a few species with five to six copies of Z-repeats were detectable in this muscle. In the soleus skeletal muscle, two species were abundantly expressed, which contained either four or six copies of the Z-repeats (Fig. 10). In the psoas skeletal muscle, a species with four copies of Z-repeats was detected. Similarly, the expression of the Z-repeats in different copy numbers was observed in chicken skeletal muscles (Ohtsuka et al. 1997b). At present it is difficult to determine whether the differential expression of the Z-repeats correlates with, and possibly accounts for, the variable Z-disc widths observed in different tissues, because outside the Z-disc region, additional alternative splice events occur (Fig. 11). For a discussion of a possible functional significance of the differential expression of the Z-repeats, see section 5.1.

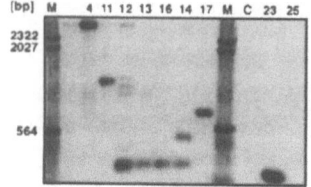
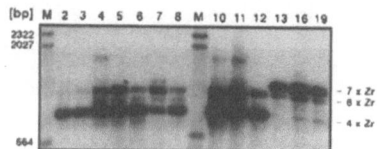

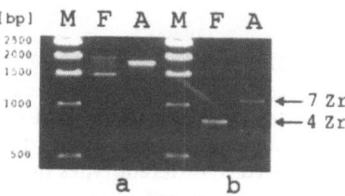

Fig. 11. RT-PCR analysis of the differentially expressed Z-disc titin segment (taken from Sorimachi et al. 1997). Primer pairs from the Z-disc titin region were used to amplify cDNA from a collection of striated muscle tissues. Amplified fragments were separated on agarose gels, and detected with specific probes on Southern blots. Tissues: 2, M. psoas; 3, M. longissimus dorsi; 4, M. soleus; 5, M. gastrocnemius; 6, M. plantaris longus; 7, diaphragm; 8, M. extensor digitorum longus; 10; M. rectus femoris; 11, tongue; 12, M. pectoralis major; 13, cardiac left ventricle; 14, proximal oesophagus; 16, cardiac right ventricle; 17, distal oesophagus; 19, left cardiac atrium; 23, uterus; 25, bladder, c negative controls, M, λ Hind III size marker. *Top left:* Differently sized variants of the Z-disc titin domain are detected in 4, 11, 12, 14, 17 and 23. In the smallest species amplified from uterus cDNA (23), a 900-residue segment is excluded. A full-length version of this domain is expressed in the heart (13, 16), but this fragment is too large to be amplified under the PCR conditions used. *Top right:* Different copy numbers of Z-repeats are expressed in the striated muscles, and the three major length variants observed have seven, six, and four repeats, respectively, as indicated on the right. Skeletal muscle tissues containing predominantly fast fibers (psoas, longissimus dorsi, extensor digitorum longus) express the four-Z-repeat variant. Cardiac tissues (left ventricle, right ventricle, left atrium) express mainly the seven-Z-repeat version. Other muscle tissues mostly co-express mixtures of a four-repeat and of a six-repeat isoform. *Bottom:* Developmental regulation of the cardiac Z-repeat copy number. Total cDNAs obtained from human fetal (F) and human adult (A) heart cDNA libraries were amplified with a primer pair flanking the Z-repeat region (a) and with a primer pair within the Z-repeat region (b). In adult muscle, a single species comprising seven copies of Z-repeats is detected. In cDNA from embryonic heart, mixtures of isoforms are detected. Within the Z-repeat region, a four-copy version is the most abundant species (taken from Sorimachi et al. 1997)

4.3 Isoforms of the I-Band Titin

The I-band region of titin has been the inital focus of our isoform studies, because in the sarcomeres of resting cardiac and skeletal muscles, I-bands are of quite different length (Fig. 12). Therefore, we aimed to compare the sequence of cardiac I-band titin with that of skeletal I-band titin. For an initial comparison, a human cardiac muscle library (Stratagene, catalogue number 936208), and a human skeletal muscle cDNA library (Clontech HL1024) were purchased, plated out, and total lambda phage DNA was prepared for both libraries. Although the DNA from these two phage pools are not template sources and not comparable to a pure cDNA or a first strand synthesis, these two-phage DNA pools, termed hh (for human heart) and hsk (for human skeletal muscle) were found to be sufficient for a qualitative screening (B.K.). Comparison of the amplification patterns obtained with a set of I-band primer pairs by the PCR technique on the hh and hsk templates indicated that the central region of the human cardiac I-band titin could not be amplified from the hsk template DNA (B.K.). Therefore, we subsequently

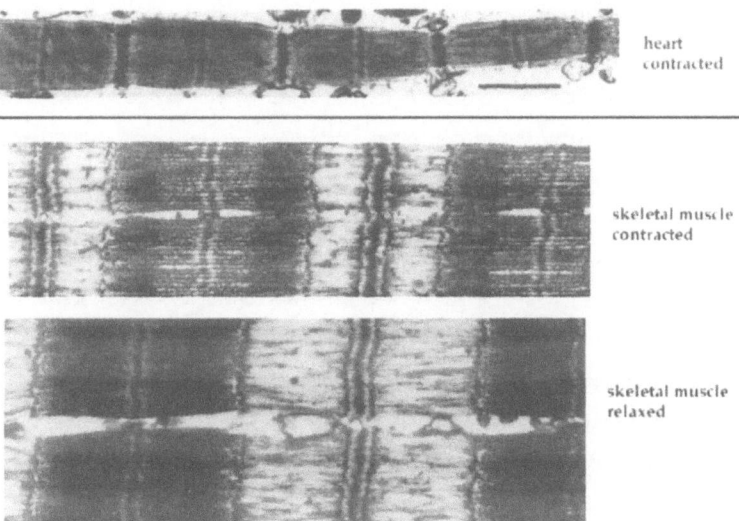

heart
contracted

skeletal muscle
contracted

skeletal muscle
relaxed

Fig. 12. Comparison of I-band lengths in sarcomeres of mammalian cardiac and skeletal muscle tissue. At the top, a cardiac myofibril is shown in the contracted state (scale bar: 1 μm; taken from Linke et al. 1993). The bottom part shows mammalian skeletal myofibrils in the contracted and relaxed state (taken from Trombitas & Pollack 1993). The thick filaments have about the same length in both types of muscle, whereas the I-band region is much longer in the skeletal muscle

determined the cDNA sequence of the central I-band region from both cardiac and skeletal titin cDNA clones by conventional library screenings and cDNA extension methods (see sections 2.2 to 2.4). As predicted, this showed that the I-band titin structure is substantially different in both tissue types (Fig. 12). Two distinct sequence families, tandemly repeated Ig domains and a region of unusual protein composition are expressed in different length

Fig. 13. The sequence of the PEVK domain from the human soleus muscle. In the central region of the I-band, the titin filament is composed of an element with unusual amino acid composition; the residues proline (P), glutamate (E), valine (V), and lysine (K) constitute about 70% of its mass. In the different striated muscle types investigated so far, the length of this domain varies by a factor of up to 13

variants (Labeit and Kolmerer, 1995). As for the tandem-Ig modules, structural studies have indicated that the fold of expressed I-band-Ig domains differs somewhat from that of a previously studied M-line-titin Ig repeat (Pfuhl et al. 1995; Politou et al. 1995; Improta et al. 1996). The second region of the I-band titin, which is also expressed in different lengths, is very unusual in its amino acid composition (see Fig. 13): about 70% of this sequence consist of the four residues proline (P), glutamate (E), valine (V), and lysine (K). Therefore, this I-band titin segment was termed the PEVK region (Labeit and Kolmerer, 1995). The relevance of the differential expression of these two distinct motif families for muscle elasticity are discussed in section 5.2.

4.4 Isoforms of the M-Line Titin

Determination of the partial genomic sequence coding for the M-line titin portion, in combination with RT-PCR studies, showed that a single exon from the M-line titin is differentially expressed in different types of striated muscle (Kolmerer et al. 1996b). This exon, referred to as Mex5, has no significant homology to other protein sequences in the data bases. The interspecies conservation of this exon and of the upstream intron Min4 is high, suggesting an important function that has been conserved during evolution. Comparison of the titins expressed in the different types of striated muscle from human, mouse and rabbit has shown that the Mex5 exon is either included or skipped during the splicing of the titin pre-mRNA. Therefore, the alternative splicing of the Mex5 exon leads to two distinct isoforms of the M-line titin, which are referred to as Mex5(+) and Mex5(−) titins. In the human heart muscle, only Mex5(+) titins are expressed, whereas in the rabbit and mouse heart, also small amounts of Mex5(−) titins are present. It remains to be seen, which factors account for these subtle differences. In the skeletal muscles, very different ratios of Mex5(+) to Mex5(−) titins are expressed. For example, in the soleus muscle, approximately equimolar amounts of both forms are detected, whereas in the rectus femoris, essentially Mex5(−) titins are only expressed (Fig. 14). A possible functional significance of the alternative splicing of Mex5 is discussed in section 5.3.

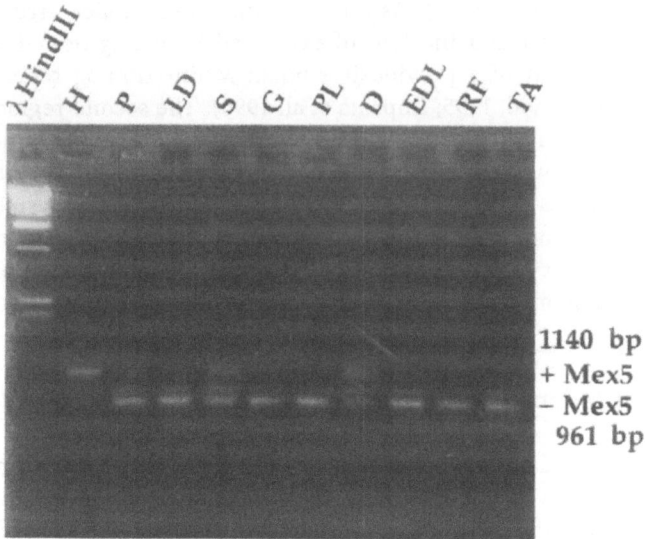

Fig. 14. Differential expression of the M-line titin in two distinct isoforms in different rabbit muscles. Primer pairs that flank the Mex5 exon (see Fig. 9) detect its tissue-specific alternative splicing. The heart muscle includes this exon and therefore expresses a Mex5(+) titin, whereas skeletal muscles skip this exon to different extents (taken from Kolmerer et al. 1996b). Tissues: H, heart; P, M. psoas; LD, M. longissimus dorsi; S, M. soleus; G, M. gastrocnemius; PL, M. plantaris longus; D, diaphragm; EDL, M. extensor digitorum longus; RF, M. rectus femoris; TA, M. tibialis anterior

5 Functional Consequences of the Differential Titin Expression

5.1 Insights Into the Function of the Differentially Expressed Z-Disc Titin Segment

RT-PCR studies have indicated that a 900-residue segment of the Z-disc titin region is expressed in different length variants in different muscle tissues (Sorimachi et al. 1997). This prompted us to search for possible ligands that may interact with this titin region. Therefore, yeast two-hybrid screens were performed (Fields and Song, 1989) using as a "bait" the 900 residues from the differentially expressed titin region. These screens identified alpha-actinin cDNAs derived from the ACTN2 gene (Beggs et al. 1992). Detailed mapping demonstrated that the C-terminal domain of the alpha-actinin-2 isoform is responsible for the interaction with titin (Sorimachi et al. 1997).

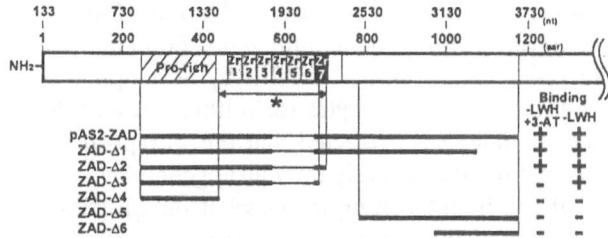

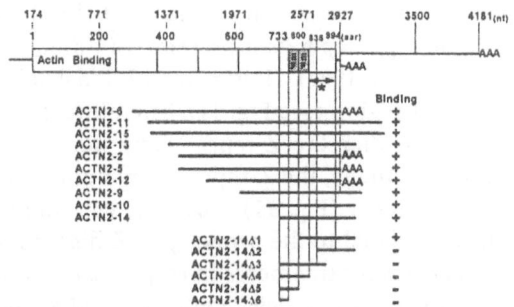

Fig. 15. Identification and fine-mapping of the alpha-actinin binding sequences in the Z-disc titin. *Top:* Using as a bait the differentially expressed Z-disc titin sequences, yeast two-hybrid screens have identified the alpha-actinin isoform encoded by the ACTN2 gene as a ligand of the titin filament. Detailed mapping by a series of deletions identifies a 90-residue domain at the C-terminus of the ACTN2 as the titin-binding domain. *Bottom:* In the Z-disc titin, the region of alpha-actinin binding assigns to the titin Z-repeat family. The most C-terminal Z-repeat, referred to as Zr7, has alpha-actinin binding properties. Possible further alpha-actinin binding sites locate at the N-terminal Z-repeats Zr1, Zr2, Zr3 (taken from Sorimachi et al. 1997)

Fine-mapping of the alpha-actinin-2 binding site on titin assigned the interaction to the region of titin's Z-repeats (Gautel et al. 1996), and suggested that this family of 45-residue motifs may have alpha-actinin binding properties (Fig. 15). These data are in excellent agreement with the data of the Chiba group that independently searched for the ligands of the N-terminal region of the chicken connectin/titin, using the recently reported N-terminal titin sequences of this species (Yajima et al. 1996). The C-terminus of alpha-actinin was shown to bind to a C-terminal Z-repeat from the chicken titin/connectin (Ohtsuka et al. 1997a, b). Previously, it has been shown that during early muscle cell differentiation in 9-somite stage chick embryonic hearts (Tokuyasu and Maher, 1987), Z-disc titin epitopes become organised into cross-striational patterns at the time of first myofibril formation, and to co-localise with alpha-actinin spots. Similarly, a cross-striated organisation of Z-disc titin epitopes at an early developmental stage was also found in cultured cardiac myocytes (Schultheiss et al. 1992a). Based on all the findings to date, it is predicted that the Z-repeats of titin co-assemble with alpha-actinin during early myofibrilogenesis.

It will be interesting to see what the functional consequences of the expression of the Z-repeats in different copy numbers are. Gautel and co-workers observed a variability in the number of Z-repeats expressed in cardiac cDNAs and speculated that this may be important for the tissue-specific width of the Z-disc (Gautel et al. 1996). Experimental determination of the Z-repeat copies expressed in cardiac and skeletal muscles has demonstrated the muscle-type-dependent presence of different copy numbers of Z-repeats (Sorimachi et al. 1997). However, additional splice events also occur outside the Z-repeat region (see Fig. 11). Therefore, further studies will be required to decide whether the Z-repeat copy number is indeed related to Z-disc width. Interestingly, the Z-repeat copy number is also regulated during cardiac development (Fig. 11). Possibly, variations in the copy number of Z-repeats are related to the property of Z-discs to dynamically transmit forces between adjacent sarcomeres in response to mechanical stress (Goldstein et al. 1986; 1989; for a review, Vigoreaux, 1994; Squire, 1997); myofibrils subjected to higher forces may have a larger copy number of alpha-actinin-binding Z-repeats (maximally seven), whereas other myofibrils will have less (four to six) Z-repeats.

5.2 Control of Myofibrillar Elasticity by the Differential Expression of I-Band Titin Isoforms

Characterization of the different I-band titins expressed in cardiac and skeletal muscles has led to the identification of two distinct motif families that are expressed in different length variants. First, the number of Ig repeats contained within the more N-terminal tandem-Ig-segments varies between 15 and 68 copies. Second, the length of the PEVK titin varies between 163 and almost 2,200 residues. These findings led us to speculate that both elements are important for the elasticity of the titin filament system (Labeit and Kolmerer, 1995). To test this assumption, two different experimental strategies were applied. Using the knowledge of the sites at which differential expression occurs, Gautel and co-workers generated monoclonal antibodies that flank the regions of differential splicing. Our group decided to generate polyclonal antibodies against the N2-A and the cardiac-specific N2-B sequences. These two segments separate the N-terminal tandem-Ig region and the PEVK segment of the I-band titin as linker-like elements. The N2-A and N2-B antibodies, as well as previously described antibodies from the end regions of the elastic I-band-titin section, were then used for immunofluorescence microscopic studies on stretched single myofibrils, to monitor the extension behavior of the I-band segments on either side of the N2-titin sequence (W.L.).

By comparing the translational movement of the N2-A epitope with that of the T12 epitope, which is located about 100 nm away from the center of the Z-disc (Fürst et al. 1988), we could draw conclusions about the elastic properties of the N-terminal tandem-Ig segment (Fig. 16). It was found that initial myofibril stretch from the slack length resulted in a threefold to fourfold elongation of the T12/N2-A segment. During this phase of Ig-domain-segment elongation, skeletal myofibrils did not develop significant passive forces. In addition, since structural studies have indicated that expressed tandem-Ig modules adopt a stable fold (Politou et al. 1995; Improta et al. 1996), it was considered likely that the extensibility of the N-terminal tandem-Ig region is due to straightening of the Ig-domain chain, rather than to unfolding of individual repeats (Linke et al. 1996). Once the Ig-domain chain is straightened out, I-band-titin lengthening with additional sarcomere stretch is mostly restricted to the region C-terminal to the N2-A segment. In this context, it is pointed out that unfolding of tandem-Ig domains was recently monitored by atomic force microscopy on recombinant titin fragments (Rief et al. 1997). This is certainly a remarkable technological achievement, but the high forces necessary to induce unfolding (250–300 picoNewton) suggest that Ig-domain unfolding may not occur under conditions relevant to the physiological situation.

The I-band titin region C-terminal to the N2-A and N2-B sequences contain the PEVK region. In analogy to the approach described in the previous paragraph, we obtained information about PEVK-segment extensibility by comparing the translational movement of the N2-A epitope with that of titin epitopes at the A/I junction (MIR, BD6). The results of those studies are consistent with the idea that the PEVK domains represent highly elastic elements of the skeletal titin molecules (Fig. 17). The part of the titin filament containing the PEVK sequences can extend substantially at physiological degrees of myofibril stretch, and such extension is correlated with the generation of passive tension (Linke et al. 1996). In the maximally stretched state, each residue of the PEVK domain may span about 0.3 nanometer. Then, with the overall amount of PEVK-segment extension calculated (approximately 600 nm in soleus muscle), we would predict that at the maximum sarcomere lengths a skeletal muscle may experience in vivo, the

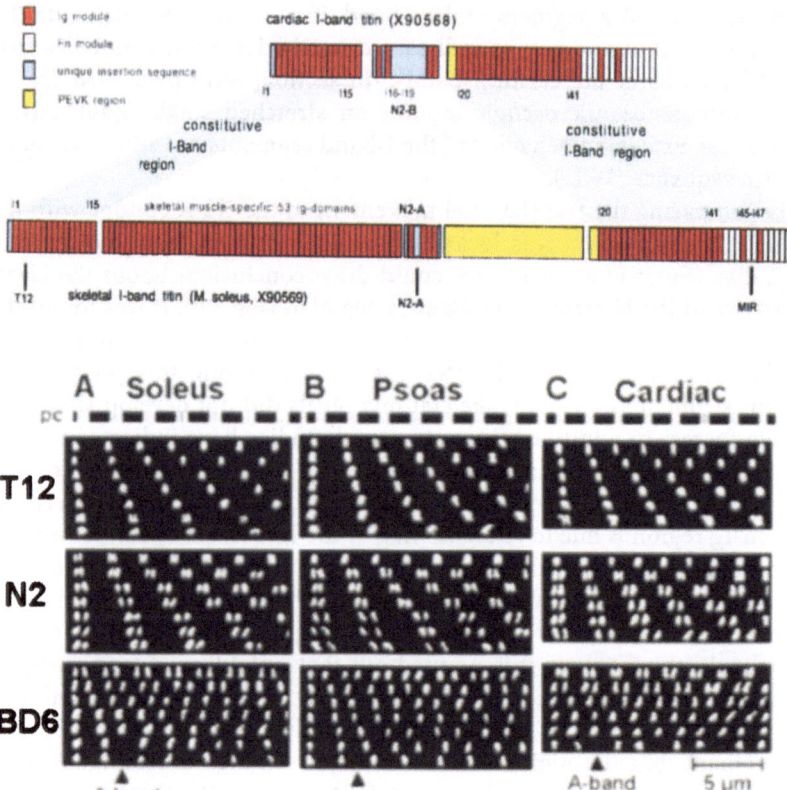

Fig. 16. Mechanical studies on the properties of the tandem-Ig and PEVK regions contained in the I-band titin. *Top:* Overview on the domain architecture of the soleus and cardiac titin, and on antibodies used. The epitopes for the T12 and the BD6 antibodies (Fürst et al. 1988; Whiting et al. 1989) locate at the N-terminus of the tandem-Ig region and near the end of the thick filament, respectively. The mapping position of the BD6 antibody on the titin filament is given in Fig. 3. For the N2-A and N2-B region, polyclonal antibodies were generated by us. *Bottom:* Comparison of the mobilities of the T12, BD6 and N2-A antibodies during stretch reveals non-uniform extensibilities of the different parts of the titin filament within its I-band section (taken from Linke et al. 1996)

PEVK domain unfolds to a linearly extended conformation (Linke et al. 1996).

Similar models to explain the extensibility of I-band titin have also been proposed independently by Gautel and Goulding (1996) and, more recently, by Tskhovrebova and Trinick (1997). It will be interesting to see, which tertiary fold allows the PEVK domain to undergo readily reversible transitions from a compact folded conformation to a linear extended state. Structural studies will be necessary to elucidate how PEVK unfolding can lead to a passive tension response.

5.3 Possible Functions of the Alternative Splicing of the M-Line Exon Mex5

The M-line region of titin is encoded by six exons, referred to as Mex1 to Mex6 (Kolmerer et al. 1996b). Five of the six exons of the M-line titin are expressed in all striated muscles, whereas a single exon is alternatively spliced. This exon, referred to as Mex5, can be either skipped or included in the titin pre-mRNA during the splicing (Kolmerer et al. 1996b). First insights into a possible functional relevance of the differential expression of the Mex5 exon came unexpectedly from the characterization of muscle-specific calpain proteases. Family studies of patients from the "Ile de la Reunion" showed that mutations in the calpain protease-encoding gene CAPN3 were associated with a loss of muscle function, and thus, represent the genetic

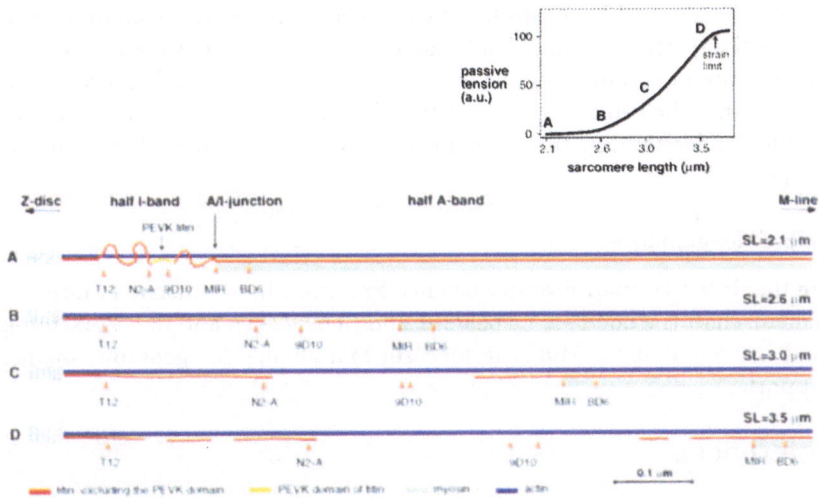

Fig. 17. Model for the elastic properties of the skeletal (psoas) I-band titin during stretch. A: During small amounts of stretch (from slack sarcomere length to approximately 2.5 micrometer), an about threefold to fourfold extension is observed in the tandem-Ig segment. This presumably corresponds to the straigthening but not to the unfolding of the Ig domains. During this process, no significant passive tension is generated (see resting tension curve on inset top right). B to D: At larger amounts of stretch, most of the extension occurs in the PEVK region, and a progressively increasing passive tension response is developed. The amount of extension detected in the PEVK region suggests that this domain completely unfolds (taken from Linke et al. 1996)

cause for familial limb-girdle muscular dystrophy, type LGMD type 2A (Richard et al. 1995). This finding prompted studies to search for factors that may regulate the activity of the 94 kDa protease encoded by the CAPN3 gene. Surprisingly, yeast two-hybrid studies using the p94 sequences as a bait identified two specific binding sites in the titin filament (Sorimachi et al. 1995). Within the I-band titin, the N2-A segment provides one binding site for the p94 protease. A second binding site is present near the C-terminus of titin, corresponding to the region of the Mex5 domain (Sorimachi et al. 1995). Recently, it was shown that the minimal binding site for the p94 protease in the M-line region of titin consists of Mex5 and its flanking Ig-modules M9 and M10 (Kinbara et al. 1997). This indicates that skipping of the Mex5 exon abolishes the binding of the p94 protease. At present, the exact functional consequences of the differential expression of Mex5 for muscle function are unclear. The half-life of the free soluble p94 protease is apparently very short, whereas that of p94 bound to the titin filament could be much longer (Sorimachi et al. 1995). Therefore it is possible that the M-line-bound p94 corresponds to a more stable form of p94. Then, skipping of the Mex5 exon could perhaps be a mechanism to increase the amount of free (and active?) versus bound (and inactive?) p94 protease. Clearly, a detailed study of the interactions of p94 with titin may reward us with more detailed insights into the cell biology of the titin filament system, which may be relevant for understanding the mechanisms involved in the pathology of muscular dystrophies.

Acknowledgements

We thank the Human Frontier Science Program, the Deutsche Forschungsgemeinschaft (La 668/3–3; La 668/5–1 and Li 690/2–1), and the "Forschungsfond der Fakultät für klinische Medizin Mannheim" for generous financial support.

References

Aarli JA, Stefansson K, Marton LSG & Wollmann RL (1990) Patients with myasthenia gravis and thymoma have in their sera IgG autoantibodies against titin. Clin Exp Immunol 82:284–288

Beggs AH, Byers TJ, Knoll JHM, Boyce FM, Bruns GAP & Kunkel LM (1992) Cloning and characterization of two human skeletal muscle alpha-actinin genes located on chromosomes 1 and 11. J Biol Chem 267:9281–9288

Bennett PM & Gautel M (1996) Titin domain patterns correlate with the axial disposition of myosin at the end of the thick filament . J Mol Biol 255:604–616

Bonne G, Carrier L, Bercovici J, Cruaud C, Richard P, Hainque B, Gautel M, Labeit S, James M, Weissenbach J, Vosberg HP, Fiszman M, Komajda M & Schwartz K (1995) Cardiac myosin binding protein-C gene splice acceptor site mutation is associated with familial hypertrophic cardiomyopathy. Nature Genetics 11:438–440

Carrier L, Bonne G, Bährend E, Yu B, Richard P, Niel F, Hainque B, Cruaud C, Gary F, Labeit S, Bouhour J-B, Dubourg O, Desnos M, Hagège AA, Trent RJ, Komajda M, Fiszman M & Schwartz K (1997) Organization and sequence of human cardiac myosin binding protein C gene (MYBPC3) and identification of mutations predicted to produce truncated proteins in familial hypertrophic cardiomyopathy. Circ Res 80:427–434

Cox DR (1995) Mapping with radiation hybrids. Genome Digest 4:14–15

Craig R & Offer G (1976) The location of C-protein in rabbit skeletal muscle. Proc R Soc Ser B 192:325–332

Craig R (1977) Structure of A-segments from frog and rabbit skeletal muscle. J Mol Biol 109:69–81

Fields S & Song O (1989) A novel genetic system to detect protein-protein interactions. Nature 340:245–246

Freiburg A & Gautel M (1996) A molecular map of the interactions between titin and myosin-binding protein C. Implications for sarcomeric assembly in familial hypertrophic cardiomyopathy. Eur J Biochem 235:317–323

Fürst DO, Osborn M, Nave R & Weber K (1988) The organization of titin filaments in the half-sarcomere revealed by monoclonal antibodies in immunoelectron microscopy: a map of ten nonrepetitive epitopes starting at the Z line extends close to the M line. J Cell Biol 106:1563–1572

Fulton AB & Isaacs WB (1991) Titin, a huge, elastic sarcomeric protein with a probable role in morphogenesis. Bioessays 13:157–61

Funatsu T, Higuchi H & Ishiwata S (1990) Elastic filaments in skeletal muscle revealed by selective removal of thin filaments with plasma gelsolin. J Cell Biol 110:53–62

Gautel M & Goulding D (1996) A molecular map of titin/connectin elasticity reveals two different mechanisms acting in series. FEBS Lett 385:11–4

Gautel M, Leonard K & Labeit S (1993a) Phosphorylation of KSP-motifs in the C-terminal region of titin in differentiating myoblasts. EMBO J 12:3827–3834

Gautel M, Lakey A, Barlow DP, Holmes Z, Scales S, Leonard K, Labeit S, Mygland A, Gilhus N E & Aarli J (1993b) Titin antibodies in myasthenia gravis: Identification of a major auto-immunogenic region of titin. Neurology 43:1581–1585

Gautel M, Zuffardi O, Freiburg A & Labeit S (1995) A cooperative phosphorylation switch in human cardiac myosin-binding protein C specific for the cardiac isoform: A modulator of cardiac contraction? EMBO J 14:1952–1960

Gautel M, Goulding D, Bullard B, Weber K & Fürst DO (1996) The central Z-disk region of titin is assembled from a novel repeat in variable copy numbers. J Cell Sci 109:2747–2754

Goldstein MA, Michael LH, Schroeter JP & Sass RL (1986) The Z-band lattice in skeletal muscle before, during and after tetanic contraction. J Muscle Res Cell Motil 7:527–536

Goldstein MA, Michael LH, Schroeter JP & Sass RL (1989) Two structural states of Z-bands in cardiac muscle. Am J Physiol (Heart Circ Physiol) 256:H552–H559

Granzier HL & Irving TC (1995) Passive tension in cardiac muscle: contribution of collagen, titin, microtubules and intermediate filaments. Biophys J 68:1027–1044

Horowits R, Kempner ES, Bisher ME & Podolski RJ (1986) A physiological role for titin and nebulin in skeletal muscle. Nature 323:160–164

Horowits R (1992) Passive force generation and titin isoforms in mammalian skeletal muscle. Biophys J 61:392–398

Itoh Y, Suzuki T, Kimura S, Ohashi K, Higuchi, H, Sawada H, Shimizu TM & Maruyama K (1988) Extensible and less-extensible domains of connectin filaments in stretched vertebrate skeletal muscle as detected by immunofluorescence and immunoelectron microscopy using monoclonal antibodies. J Biochem 104:504–508

Improta S, Politou AS & Pastore A (1996) Immunoglobulin-like modules from titin I-band: extensible components of muscle elasticity. Structure 4:323–337

Kinbara K, Sorimachi H, Ishiura S & Suzuki K (1997) Muscle-specific calpain, p94, interacts with the extreme C-terminal region of connectin, a unique region flanked by two immunoglobulin C2 motifs. Arch Biochem Biophys 342:99–107

Kolmerer B, Olivieri N, Herrmann BG & Labeit S (1996a) A systematic search of the data bases for sequences homologous to titin/connectin. Advances in Biophysics 33:3–11

Kolmerer B, Olivieri N, Herrmann BG & Labeit S (1996b) Genomic organization of the M-line titin and its tissue-specific expression in two distinct isoforms. J Mol Biol 256:556–563

Labeit S, Barlow DP, Gautel M, Gibson T, Holt J, Hsieh CL, Francke U, Leonard K, Wardale J, Whiting A & Trinick J (1990) A regular pattern of two types of 100-residue motif in the sequence of titin. Nature 345:273–276

Labeit S, Gautel M, Lakey A & Trinick J (1992) Towards a molecular understanding of titin. EMBO J 11:1711–1716

Labeit S & Kolmerer B (1995) Titins, giant proteins in charge of muscle ultrastructure and elasticity. Science 270:293–296

Labeit S, Kolmerer B & Linke WA (1997) The giant protein titin: Emerging roles in physiology and pathophysiology. Circ Res 80:290–294

Laing NG, Wilton SD, Akkari PA, Boundy K, Kneebone C, Blumbergs P, White S, Watkins H, Love DR & Haan E (1995) A mutation in the a-tropomyosin gene TPM3 associated with autosomal dominant nemaline myopathy. Nature Genet 9:75–79

Linke WA, Bartoo, ML & Pollack GH (1993) Spontaneous sarcomeric oscillations at intermediate activation levels in single isolated cardiac myofibrils. Circ Res 73:724–734

Linke WA, Ivemeyer M, Olivieri N, Kolmerer B, Rüegg JC & Labeit S (1996) Towards a molecular understanding of the elasticity of titin. J Mol Biol 261:62–71

Linke WA, Ivemeyer M, Labeit S, Hinssen H, Rüegg JC & Gautel M (1997) Actin-titin interaction in cardiac myofibrils: probing a physiological role. Biophys J 73:905–919

Maruyama K, Matsubara S, Natori R, Nonomura Y, Kimura S, Ohashi K, Murakami F, Handa S & Eguchi G (1977) Connectin, an elastic protein of muscle: characterization and function. J Biochem (Tokyo) 82:317–337

Maruyama K, Sawada H, Kimura S, Ohashi K, Higuchi H & Umazume Y (1984) Connectin filaments in stretched skinned fibers of frog skeletal muscle. J Cell Biol 99:1391–1397

Maruyama K, Yoshioka T, Higuchi H, Ohashi K, Kimura S & Natori R (1985) Connectin filaments link thick filaments and Z lines in frog skeletal muscle as revealed by immunoelectron miscroscopy. J Cell Biol 101:2167–2172

Müller-Seitz M, Kaupmann K, Labeit S & Jockusch H (1993) Chromosomal localization of the mouse titin gene and its relation to "muscular dystrophy with myositis" and nebulin genes on chromosome 2. Genomics 18:559–561

Obermann WMJ, Gautel, M, Steiner, F, vanderVeen, PFM, Weber, K & Fürst DO (1996) The structure of the sarcomeric M band: localization of defined domains of myomesin, M protein, and the 250 kD carboxy terminal region of titin by immunoelectron microscopy. J Cell Biol 134:1441–1453

Ohtsuka H, Yajima H, Maruyama K & Kimura S (1997a) Binding of the N-terminal 63 kDa portion of connectin/titin to alpha-actinin as revealed by the yeast two-hybrid system. FEBS Lett 401: 65–67

Ohtsuka H, Yajima H, Maruyama K & Kimura S (1997b) The N-terminal Z repeat 5 of connectin/titin binds to the C-terminal region of alpha-actinin. Biochem Biophys Res Commun. 235: 1–3

Pelin K, Ridanpää M, Donner K, Wilton, S, Krishnarajah J, Laing N, Kolmerer B, Millevoi S, Labeit S, de la Chapelle A & Wallgren-Pettersson C (1997) Refined localization of the genes for nebulin and titin on chromosome 2q allows the assignment of nebulin as a candidate gene for autosomal recessive nemaline myopathy. Eur J Hum Genet 5:229–234

Pfuhl M & Pastore A (1995) Tertiary structure of an immunoglobulin-like domain from the giant muscle protein titin: a new member of the I set. Structure 3:391–401

Politou A, Thomas DJ & Pastore A (1995) The folding and stability of titin immunoglobulin-like modules, with implications for the mechanism of elasticity. Biophys J 69:2601–2610

Richard I, Broux O, Allamand V, Fougerousse F, Chiannilkulchai N, Bourg N, Brenguier L, Devaud C, Pasturaud P, Roudaut C, Hillaire D, Passos-Bueno MR, Zatz M, Tischfield JA, Fardeau M, Jackson CE, Cohen D & Beckmann JS (1995) Mutations in the proteolytic enzyme calpain 3 cause limb-girdle muscular dystrophy 2A. Cell 81: 27–40

Rossi E, Faiella A, Zeviani M, Labeit S, Floridia G, Brunelli S, Cammarata M, Boncinelli E & Zuffardi O (1994) Order of six loci at 2q24–q31 and orientation of the HOXD locus. Genomics 24:34–40

Rottbauer W, Gautel M, Zehelein J, Labeit S, Franz WM, Fischer C, Vollrath B, Mall G, Dietz R, Kübler W & Katus H (1997) A novel splice donor site mutation in the cardiac myosin binding protein-C gene in familial hypertrophic cardiomyopathy: Characterization of cardiac transcript and protein. J Clin Invest 100:475–482

Saiki RK, Scharf SJ, Faloona F, Mullis GT & Erlich HA (1985) Enzymatic amplification of beta-globin genomic sequences and restriction site analysis for diagnosis of sickle cell anemia. Science 230:1350–1354

Salviati G, Betto R, Ceoldo S & Pierobon-Bormioli S (1990) Morphological and functional characterization of the endosarcomeric elastic filament. Am J Physiol 259:C144–C149

Schroeter JP, Bretaudiere JP, Sass RL & Goldstein MA (1996) Three-dimensional structure of the Z band in a normal mammalian skeletal muscle. J Cell Biol 133:571–583

Schultheiss T, Lin, Z, Lu M-H, Murray J, Fischman DA, Weber K, Masaki M, Imamura M & Holtzer H (1992a) Differential distribution of subsets of myofibrillar proteins in cardiac nonstriated and striated myofibrils. J Cell Biol 110:1159–1172

Schultheiss T, Choi J, Lin ZX, DiLullo C, Fischman DA & Holtzer H (1992b) A sarcomeric alpha-actinin truncated at the carboxyterminal end induces the breakdown of stress fibers in PtK2 cells and the formation of nemaline-like bodies and breakdown of myofibrils in myotubes. Proc Natl Acad Sci USA 89:9282–9286.

Schwartz K, Carrier L, Guicheney P & Komajda M (1995) The molecular basis of cardiomyopathies. Circulation 91:532–540

Sebestyan MG, Wolff JA & Greaser ML (1995) Characterization of a 5.4 kb cDNA fragment from the Z-line region of rabbit cardiac titin reveals phosphorylation sites for proline-directed kinases. J Cell Sci 108:3029–3037

Sjöström M & Squire JM (1977) Fine structure of the A-band in cryosections. J Mol Biol 109:49–68

Sorimachi H, Kinbara K, Kimura S, Takahashi M, Ishiura S, Sasagawa N, Sorimachi N, Shimada H, Tagawa K, Maruyama K & Suzuki K (1995) Muscle-specific calpain, p94, responsible for limb girdle muscular dystrophy type 2A, associates with connectin through IS2, a p94-specific sequence. J Biol Chem 270:31158–31162

Sorimachi H, Freiburg A, Kolmerer B, Ishiura S, Stier G, Gregorio CC, Labeit D, Linke WA, Suzuki K & Labeit S (1997) Tissue-specific expression and alpha-actinin binding properties of the Z disc titin. Implications for the nature of vertebrate Z discs. J Mol Biol 270:688–695

Squire JM (1981) The structural basis of muscular contraction. New York, Plenum Press

Squire JM (1997) Architecture and function in the muscle sarcomere. Curr Opinion Struct Biol 7: 247–257

Thierfelder L, Watkins H, McRae C, Lamas R, McKenna W, Vosberg HP, Seidman JG & Seidman CE (1994) Alpha-tropomyosin and cardiac troponin T mutations cause familial hypertrophic cardiomyopathy: a disease of the sarcomere. Cell 77:701–712

Tokuyasu KT & Maher PA (1987) Immunocytochemical studies of cardiac myofibril-logenesis in early chick embryos. II. Generation of alpha-actinin dots within titin spots at the time of the first myofibril formation. J Cell Biol 105:2795–2801

Trinick J (1994) Titin and nebulin: protein rulers in muscle? Trends in Biochem Sciences 19: 405–409

Trinick J (1996) Cytoskeleton – titin as a scaffold and a spring. Curr Biol 6:258–260

Trombitas K & Pollack GH (1993) Elastic properties of the titin filament in the Z-line region of vertebrate striated muscle. J Muscle Res Cell Motil 14:416–422

Tskhovrebova L & Trinick J (1997) Direct visualization of extensibility in isolated titin molecules. J Mol Biol 265:100–106

Vigoreaux JO (1994) The muscle Z band: lessons in stress management. J Muscle Res Cell Motil 15:237–255

Vikstrom KL & Leinwand LA (1996) Contractile protein mutations and heart disease. Curr Opin Cell Biol 8:97–105

Wallgren-Pettersson C, Avela K, Marchand S, Kolehmainen J, Tahvanainen E, Juul Hansen F, Muntoni F, Dubowitz V, de Visser M, Van Langen IM, Laing NG, Faure S & de la Chapelle A (1995) A gene for autosomal recessive nemaline myopathy assigned to chromosome 2q by linkage analysis. Neuromusc Disord 5:441–443

Wang K, McClure J & Tu A (1979) Titin: Major myofibrillar components of striated muscle. Proc Natl Acad Sci USA 76:3698–3702

Wang K, Ramirez-Mitchell R & Palter D (1984) Titin is an extraordinarily long, flexible, and slender myofibrillar protein. Proc Natl Acad Sci USA 81:3685–3689

Wang K, McCarter R, Wright J, Beverly J & Ramirez-Mitchell R (1991) Regulation of skeletal muscle stiffness and elasticity by titin isoforms: A test of the segmental extension model of resting tension. Proc Natl Acad Sci USA 88:7101–7105

Watkins H, Conner D, Thierfelder L, Jarcho JA, MacRae C, McKenna WJ, Maron BJ, Seidman JG & Seidman CE (1995) Mutations in the cardiac myosin binding protein-C gene on chromosome 11 cause familial hypertrophic cardiomyopathy. Nature Genet 11: 434–437

Watkins H, Seidman JG & Seidman CE (1995) Familial hypertrophic cardiomyopathy: a genetic model of cardiac hypertrophy. Hum Mol Genet 4:1721–1727

Whiting A, Wardale J & Trinick J (1989) Does titin regulate the length of muscle thick filaments? J Mol Biol 205:163–169

Yajima H, Ohtsuka H, Kawamura Y, Kume H, Murayama T, Abe H, Kimura S & Maruyama K (1996) A 11.5 kb 5'-terminal cDNA sequence of chicken breast muscle connectin/titin reveals its Z line binding region. Biochem Biophys Res Commun 223:160–164

Yamaguchi M, Izumimoto M, Robson RM & Stromer MH (1985) Fine Structure of wide and narrow vertebrate muscle Z-lines. A proposed model and computer simulation of Z-line architecture. J Mol Biol 184:621–644

The Physiological Role of Titin in Striated Muscle

R. Horowits

National Institute of Arthritis and Musculoskeletal and Skin Diseases
National Institutes of Health, Bethesda, MD 20892

Contents

I Introduction

The elastic molecule of striated muscles known as titin [97] or connectin [61] is now recognized to be one of the largest polypeptides known. Connectin was initially described as the elastic residue remaining in myofibrils extracted with potassium iodide or highly alkaline solutions [63], while titin was more precisely defined as two extremely high molecular weight proteins (titin-1 and titin-2) present in striated muscles [97] (Fig. 1). When titin was recognized as the major component of connectin preparations [61], the two terms became synonymous. It was later established that the smaller titin polypeptide (titin-2, also called ß-connectin) was in many cases a proteolytic fragment of the larger, intact molecule (titin-1, also called α-connectin) [58, 94].

This article will deal with the role of titin in the physiology of vertebrate striated muscles. Following a brief review of basic aspects of titin morphology and its organization in the sarcomere, I will discuss the evidence relating to titin's role in determining the mechanical properties of relaxed and active muscle fibers, as well its role in maintaining the regular order of the myofilaments in the sarcomere. Finally, I will discuss the implications for muscle function *in vivo*. Aspects of titin function related to biochemical activity,

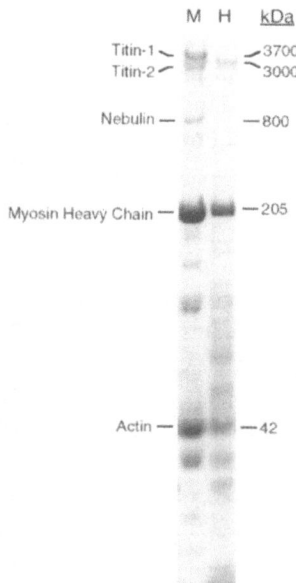

Fig. 1. Proteins from mouse skeletal muscle (M) and heart (H) separated on an SDS polyacrylamide gel. Note that the intact skeletal muscle titin (titin-1) is significantly larger than the cardiac titin. The molecular weights are estimates based on the cDNA sequence of human titin from cardiac and soleus muscles [46, 47]

myofibrillogenesis, and regulation of myofilament length are reviewed elsewhere in this volume.

II Titin Morphology and Organization in the Sarcomere

A Rotary Shadowed Images of Isolated Titin

Early studies by several groups showed that rotary shadowed titin molecules appeared to be long, flexible string-like structures [62, 85, 98]. These studies established that the titin molecule is a filamentous structure having a diameter of approximately 4 nm. The titin filaments also appeared to have a beaded structure, with an axial periodicity on the order of 5 nm [85, 98]. Random bending of the isolated filaments complicated the analyis of these images, but also revealed that the titin molecule is highly flexible along its entire length. The measured distribution of filament lengths was very broad, with individual filaments apparently ranging from 0.2 μm to more than 1 μm in length [62, 85, 98].

Nave et al. [68] eliminated the problems inherent in analyzing images of randomly bent, string-like titin molecules by developing a technique to straighten the filaments using centripetal force. These authors measured a much narrower distribution of filament lengths, with a mean length of approximately 0.9 μm. They also found that the isolated titin filaments had a globular head at one end.

B Antibody Studies

In the original paper describing titin, Wang et al. [97] described a polyclonal anti-titin antibody that stained myofibrils at the A-bands and A-I junctions, as well as at the Z-lines and M-lines. Subsequent studies using polyclonal antibodies localized titin exclusively to the A-band [9, 73], to both the A-band and I-band [61, 64, 72], and to the A-band, I-band, and Z-lines [65].

Titin was soon found to be associated with the so-called gap filaments [9, 87, 94], the ultra-thin filaments found in the gap between the thick and thin filaments in highly stretched sarcomeres [3, 33, 55, 57, 82]. In addition, flexible, string-like structures morphologically similar to isolated titin were found located at the ends and alongside isolated native thick filaments [85, 86]. Binding of anti-titin antibodies confirmed that these structures were titin filaments [9, 25].

The picture became much clearer once monoclonal antibodies were generated that bound at various points along the titin molecule. By observing the location and mobility of site-specific anti-titin antibodies in sarcomeres

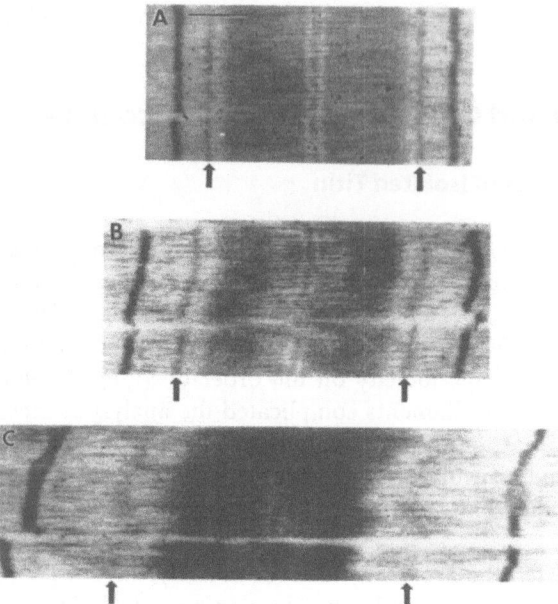

Fig. 2. Electron micrographs of relaxed sarcomeres from rabbit psoas muscle stretched to varying lengths and stained with monoclonal anti-titin antibody SM1. In sarcomeres stretched to lengths of 2.3 or 2.8 μm, antibody density is sharply defined (arrows in A and B, respectively). Staining is more diffuse in the 4.0 μm long sarcomere (C). Note that the distances from the Z-line to the antibody and from the antibody to the thick filaments increase as the sarcomere is stretched. Calibration bar: 0.5 μm. Previously published in [29]

at various lengths, conclusions could be drawn regarding the organization of titin within the sarcomere, as well as the extensibility of distinct regions of the titin molecule. Figure 2 shows a typical result from such experiments: A sarcomere stained with a monoclonal anti-titin antibody contains two extra transverse electron dense stripes that are organized symetrically around the center of the sarcomere. In this example, the antibody stains an epitope in the I-band. When the sarcomere is passively stretched, this epitope moves away from both the edges of the A-band as well as from the Z-line (Fig. 2). Plots of antibody position versus sarcomere length show that these spacings increase linearly with stretch (Fig. 3). The behavior is consistent with that of an elastic element anchored near the Z-line and the end of the thick filaments.

Figure 4 schematically shows a compilation of the results from several independent groups that have localized titin epitopes utilizing monoclonal antibodies [7, 8, 11, 18, 29, 36, 38, 40, 87, 89, 90, 91, 95, 96, 101, 102, 103, 105].

Fig. 3. Position of antibody SM1 expressed as the distance to the Z-line (open circles) or to the edge of the A-band (closed circles) as a function of sarcomere length. Each point represents the mean value from a single relaxed fiber from rabbit psoas muscle. The standard error of the mean is less than two times the diameter of each point. The linear behavior is consistent with that of an elastic element anchored near the Z-line and the end of the thick filaments. The curves are replotted from previously published data [29]

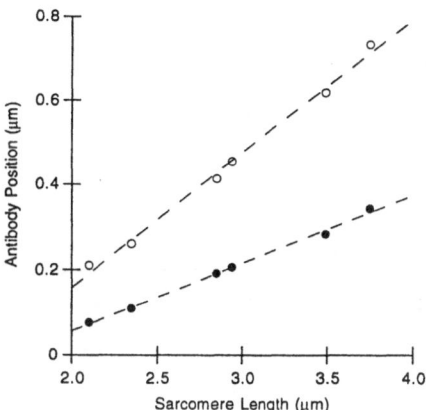

The results show that single titin molecules stretch from the Z-line to the center of the sarcomere. Furthermore, the results are remarkably consistent regarding the elasticity of titin epitopes. All epitopes in the A-band that were assayed for movement remained fixed relative to the thick filaments upon moderate stretching or shortening. Likewise, those epitopes within 100 nm of the Z-line or within 50 nm of the edge of the A-band remained a fixed distance relative to the Z-line or the A-band, respectively, under these conditions (Fig. 4, asterisks). In contrast, titin epitopes located in the I-band between the short inelastic regions near the Z-line and the ends of the thick filaments increased their distance from both the Z-line and the A-band as the sarcomere length increased (Fig. 4, pluses). The results provide strong evidence that titin molecules are anchored at the Z-lines and at the ends of the thick filaments, and may be anchored at other points along the A-band. The region between the A-bands and the Z-lines serves as an elastic link between these structures.

Confirmation that the ends of the titin molecule are located at the Z-line and the M-line comes from a variety of sources. First, monoclonal antibodies that bind near the Z-line (T20 and T21 in Fig. 4) detect only the intact titin polypeptide, titin-1, on immunoblots, but do not detect the large proteolytic fragment, titin-2 [8]. Second, a monoclonal antibody that binds very close to the M-line in intact sarcomeres (T33 in Fig. 4) binds to the head-rod junction in isolated titin molecules [68]. Finally, correlation of titin epitope locations within the sarcomere with their locations in the titin cDNA sequence established that the amino terminus of titin is anchored at the Z-line, while the carboxy-terminus is near the M-line [45, 46].

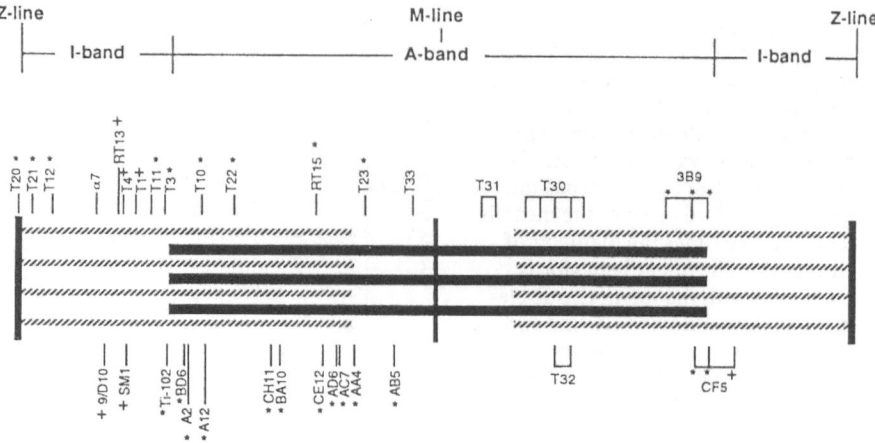

Fig. 4. Schematic diagram showing the positions in the sarcomere stained by mono-clonal anti-titin antibodies from a variety of sources. Asterisks (*) indicate epitopes that maintain their position relative to the A-band or the Z-line upon moderate stretch or shortening of the sarcomere. Pluses (+) indicate epitopes that increase their spacing from both the Z-line and the A-band as sarcomere length is increased. Antibodies that produce only one electron dense stripe per half sarcomere are shown on the left, while antibodies that stain repetitive epitopes are shown on the right. Relative positions of antibodies used in different studies should be considered estimates, since the species and specific muscles used may vary. Antibodies T20, T21, T12, T4, T1, T11, T3, T10, T22, T23, T33, T31, T30, and T32 were described by Furst et al. [7, 8]. Antibodies BD6, CH11, BA10, CE12, AD6, AC7, AA4, AB5, and CF5 were described by Whiting et al. [105]. Antibodies RT13 and RT15 were described by Wang et al. [95, 96, 101]. Antibodies SM1 and 3B9 were described by Itoh et al. [36]. Antibodies A2 and A12 were described by Wang et al. [103]. Antibodies 9/D10, Ti-102, and α7 were described by Wang and Greaser [102], Jin [38], and Kawamura et al. [40], respectively. The diagram incorporates information from subsequent studies using SM1 [29], RT13 [91],T11 [40, 87, 90], T12 [11, 18, 89, 90], 9/D10 and Ti-102 [11, 18, 89], and 3B9 [40]

III Titin's Role in the Mechanical Properties of Relaxed Muscle Fibers

A Selective Extraction or Fragmentation of Myofibrillar Proteins

1 Selective Fragmentation of Titin

One strategy for studying the physiological function of titin has involved devising methods to selectively cleave titin in single muscle fibers. Early studies showed that myofibrillar titin was extremely susceptible to proteolytic degradation [60, 94]. Using this general property, Yoshioka et al. [108] found that when single skinned frog muscle fibers were treated with very small concentrations of trypsin, titin-1 (α-connectin) was rapidly cleaved to yield titin-2 (ß-connectin). Further proteolysis resulted in cleavage of titin to yield smaller fragments. These effects occurred before any significant degradation of myosin or smaller myofibrillar proteins could be observed by gel electrophoresis. The passive tension produced by these muscle fibers upon stretch also decreased during this period, supporting the hypothesis that stretch of the titin filaments gave rise to the resting tension. These results were later extended to show that the decrease in passive tension produced by single skinned rabbit muscle fibers quantitatively matched the extent of titin degradation by trypsin treatment [5]. Similar results have been obtained using single cardiac myocytes from rats [12]. Trypsin treatment has also been used as a means to study the role of titin in restoring sarcomeres

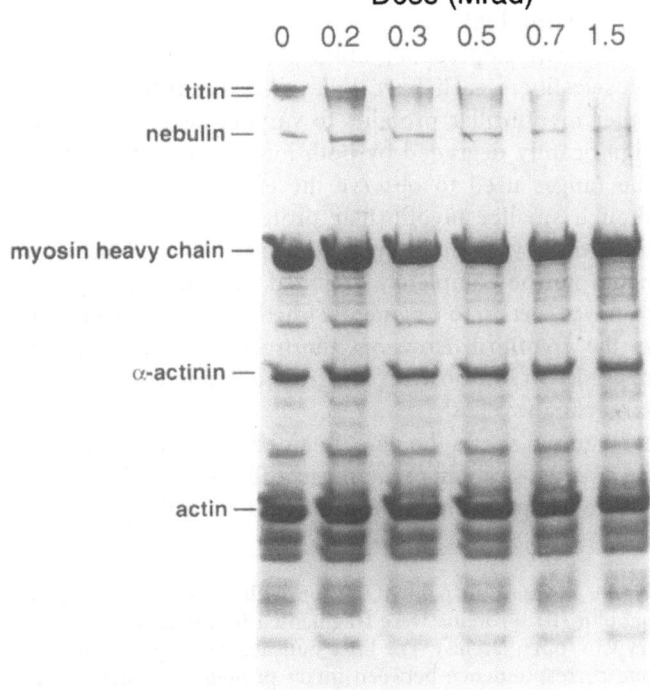

Fig. 5. SDS polyacrylamide gel electrophoresis ot proteins from skinned rabbit psoas fibers that had been frozen and then exposed to the indicated dose of high energy electrons. Note that titin and nebulin are preferentially degraded at these doses, as predicted by target analysis. Previously published in [28]

to their original length following shortening. Helmes et al. found that selective degradation of titin by trypsin abolished the ability of isolated cardiac myocytes to relengthen to their original slack length following shortening [18]. The results suggest that titin functions as a bidirectional elastic element that resists shortening as well as lengthening.

In an alternate approach, Horowits et al. [28] utilized low doses of ionizing radiation to fragment the largest proteins present in single fibers from rabbit psoas muscle. This approach was useful for studying titin function because the radiation sensitivity of macromolecules is directly related to their mass, provided that irradiation occurs in the absence of liquid water [44]. Therefore, irradiation of dry or frozen samples can be used to study the function of the largest components involved in specific cellular processes. As shown in Fig. 5, this resulted in the selective fragmentation of titin and nebulin, while leaving the other myofibrillar proteins relatively intact. Passive tension was reduced over the same range of radiation doses that fragmented titin [28].

The two approaches utilized to selectively cleave titin polypeptides in muscle fibers each have their own disadvantages. Both techniques affect other myofibrillar proteins to varying degrees. Nebulin, in particular, was significantly degraded by both ionizing radiation and trypsin treatment in the ranges used to observe the effects of titin fragmentation [5, 28]. Although smaller myofibrillar proteins appeared intact on denaturing gels after the trypsin treatment, cleavage of small pieces from these proteins could not be completely excluded. In contrast, exposure to ionizing radiation was certain to fragment a fraction of each species of polypeptide present in the myofibril. However, the fraction of any particular protein that was degraded at low doses of ionizing radiation could be accurately predicted by target analysis [28].

Figure 6 shows the data obtained by the two approaches to destroying titin in single rabbit psoas muscle fibers. The data were extracted from the previously published works and plotted as the percentage of passive tension

———▶

Fig. 6. Decrease in passive tension with increasing fragmentation of titin (A) and nebulin (B). Proteins were fragmented by ionizing radiation (closed circles) or by trypsin (open circles). The lines indicate the response predicted if there is a one to one correspondence between intact protein and passive tension. Passive tensions were measured in skinned rabbit psoas fibers at sarcomere lengths of 3.3 μm for trypsin digestion and at 3.5 μm for radiation inactivation. Note that passive tension declines in parallel with titin fragmentation (A), but faster than nebulin fragmentation (B). (However, passive tension measured at shorter sarcomere lengths is somewhat less sensitive to radiation dose [28] and trypsin treatment [108]). The data for trypsin treated fibers is replotted from Funatsu et al.'s Fig. 9 [5]; the data for irradiated fibers was published in different form by Horowits et al. [28]

A. Passive Tension vs. Remaining Titin

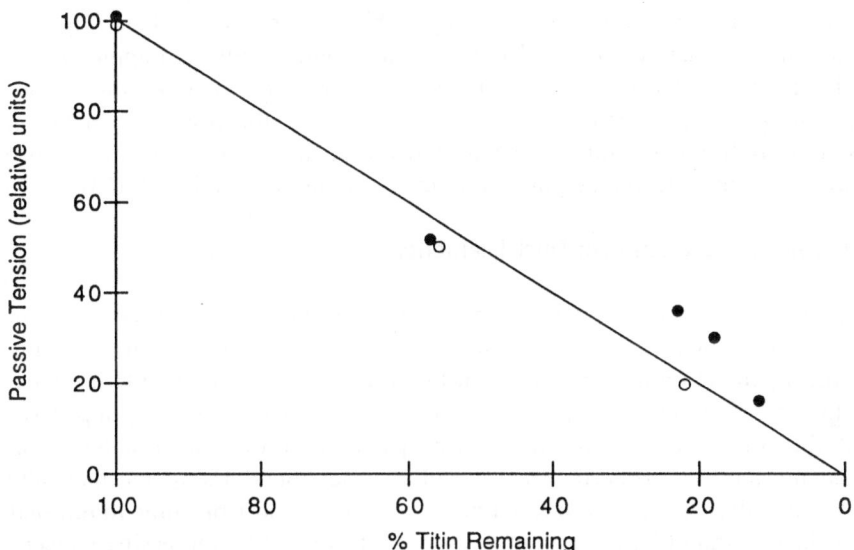

B. Passive Tension vs. Remaining Nebulin

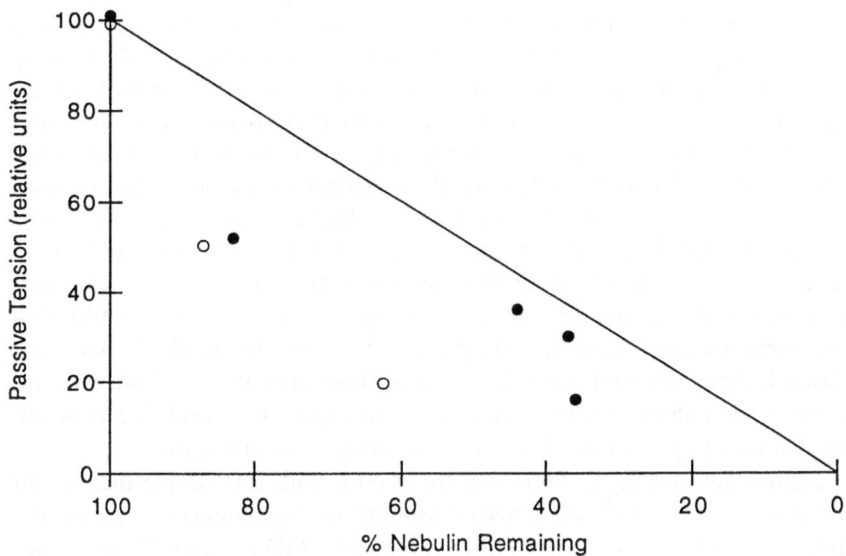

remaining as a function of intact titin or nebulin. Both techniques show a nearly one to one correspondence between the decrease in passive tension and the fragmentation of titin in single rabbit psoas muscle fibers (Fig. 6a). In contrast, passive tension decreases faster than nebulin is fragmented by either technique (Fig. 6b). The quantitative agreement between these studies, despite their distinct shortcomings, supports the original conclusion that titin fragmentation results in a decrease in resting tension because stretch of intact titin filaments is the source of that tension [5, 28, 108].

2 Selective Extraction of Thick Filaments

In addition to the strategy of directly studying the effects of fragmentation or removal of titin on muscle mechanics, valuable information regarding titin organization and function can be derived from experiments in which other myofibrillar components are selectively removed. Several groups have studied the effects of salt induced depolymerization of myosin filaments on the mechanical properties of relaxed muscle fibers. Using mechanically skinned fibers from frog skeletal muscle, Higuchi and Umazume found that as thick filament length was progressively decreased by increasing concentrations of KCl, passive tension decreased linearly with decreasing A-band length [22]. As A-band length approached zero, greater than 90% of the passive tension was eliminated at a sarcomere length of 4.0 μm. In contrast, 35% of the resting tension remained at sarcomere lengths greater than 5.0 μm, even after subsequent extraction of actin filaments with KI [22]. Several other investigators have also observed that a component of passive tension remains in extremely stretched skeletal and cardiac sarcomeres after extraction of thick and thin filaments [2, 12, 79, 96]. On the basis of morphological and biochemical analysis of the remaining structures, this residual force is generally attributed to intermediate filaments [99]. The physiological data derived from high salt extraction of myofilaments suggested a simple model in which KCl induces depolymerization of the thick filaments from their ends. As the thick filament length decreases, regions of titin that had originally been immobilized by interaction with the thick filaments are released. Once released from the A-band, these regions are elastic and increase the length of titin that can be stretched [22]. The result is a continuous decrease in passive tension with decreasing A-band length.

Some examples in the literature are at odds with this simple model. For example, Salviati et al. used pyrophosphate to depolymerize myosin filaments in skinned rabbit muscle fibers to one third their original length; they found little change in passive tension with this protocol, although subsequent removal of the residual myosin with 0.6 M NaCl greatly reduced passive tension [80]. Likewise, Granzier and Irving found that passive tension decreased far less than expected in cardiac myocytes when thick filament length was reduced by 60% with KCl [12].

The simple model used to interpret these results assumed that titin bound to the A-band is slack, and becomes elastic once the associated myosin is removed by depolymerization. These assumptions have been tested by using anti-titin antibodies to study titin epitope behavior following a salt-induced reduction in A-band length. These experiments demonstrate that titin epitopes that are normally inextensible (antibodies 3B9 and RT15 in Fig. 4) become elastic when the region of the A-band to which they are bound is removed [21, 96]. In addition, in frog skeletal muscle the region of titin associated with the A-band shortened upon myosin depolymerization,

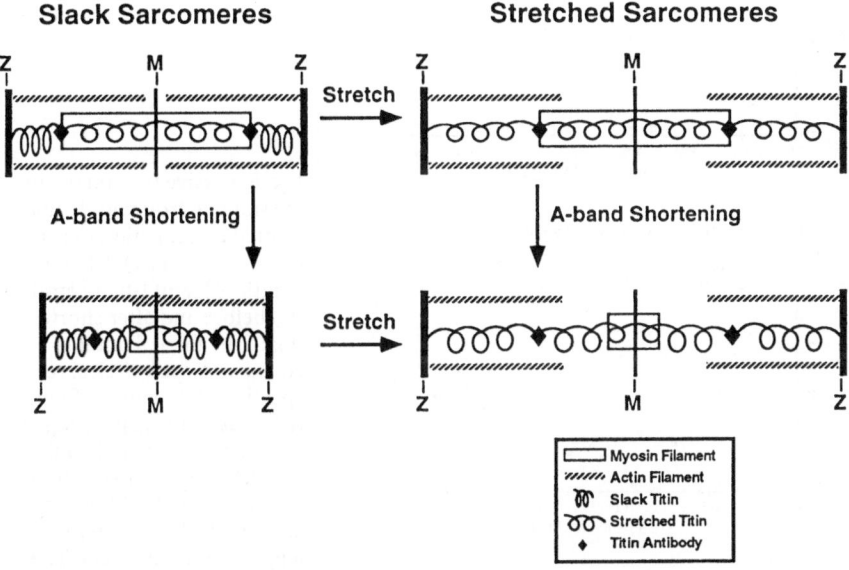

Fig. 7. Schematic model depicting the effects of thick filament shortening on titin elasticity and slack length. Titin is depicted as a spring that stretches between the Z-line and the M-line. The region of titin in the vicinity of the thick filament is stretched to twice its free slack length and immobilized by interactions with the thick filament. Elastic behavior is thus restricted to the region of titin between the thick filaments and the Z-line (upper left and right). When the thick filament is shortened by salt-induced depolymerization, part of the titin is freed from the thick filament and tends to shorten to its slack length, thus reducing the length of the slack sarcomere (lower left). The region of titin freed from the thick filament adds to the original elastic region of titin, creating a longer elastic element that elongates when the sarcomere is stretched (lower right). The model is based on Higuchi et al.'s Fig. 9 [21]

causing slack sarcomeres to shorten from 2.1 μm to 1.5 μm after the A-band length had decreased from 1.6 to 0.42 μm [21]. The Z-line to thick filament distance at slack length thus increased by only 0.3 μm, even though the length of the thick filaments had been decreased by 0.6 μm at each end. Although they did not directly observe the behavior of titin epitopes, Roos and Brady found that KCl induced depolymerization of thick filaments caused a large decrease in the slack sarcomere length of isolated cardiac myocytes

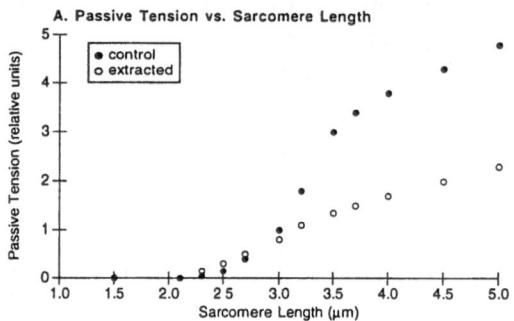

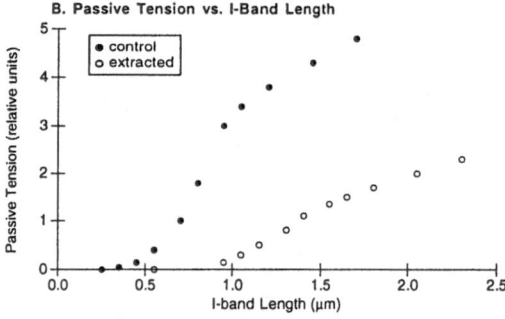

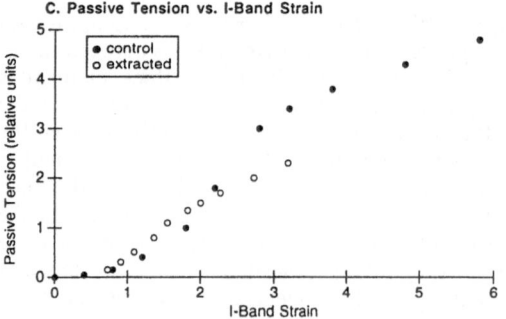

Fig. 8. Passive tension of single skinned frog muscle fibers plotted as a function of sarcomere length (**A**), I-band length (**B**), and I-band strain (**C**) before and after shortening of the thick filaments (filled and open circles, respectively). In control fibers, A-bands were 1.6 μm long and the slack sarcomere length was 2.1 μm; after extraction, A-bands were 0.42 μm long and the slack sarcomere length was 1.5 μm [21]. The data in (A) are replotted from Higuchi et al.'s Fig. 7 [21]. I-band length (I) was calculated as (L-A)/2, where L is sarcomere length and A is A-band length. I-band strain was calculated as I/I_0-1, where I_0 is the length of the I-band in the slack sarcomere. Note that I-band strain appears to be the main determinant of passive tension (C)

[79]. These results suggest that, in general, titin bound to the thick filaments is stretched relative to its unbound slack length in both skeletal and cardiac muscles (Fig. 7).

Figures 8 and 9 illustrate some features of a model based on these observations. Figure 8 shows the passive tension measured by Higuchi et al. before and after partial depolymerization of thick filaments in single frog skeletal muscle fibers [21]. The tension data is plotted as a function of sarcomere length (Fig. 8a), I-band length (defined as the thick filament to Z-line distance) (Fig. 8b), and I-band strain (Fig. 8c). Clearly I-band strain is a good predictor of passive tension. The results show that, once released from

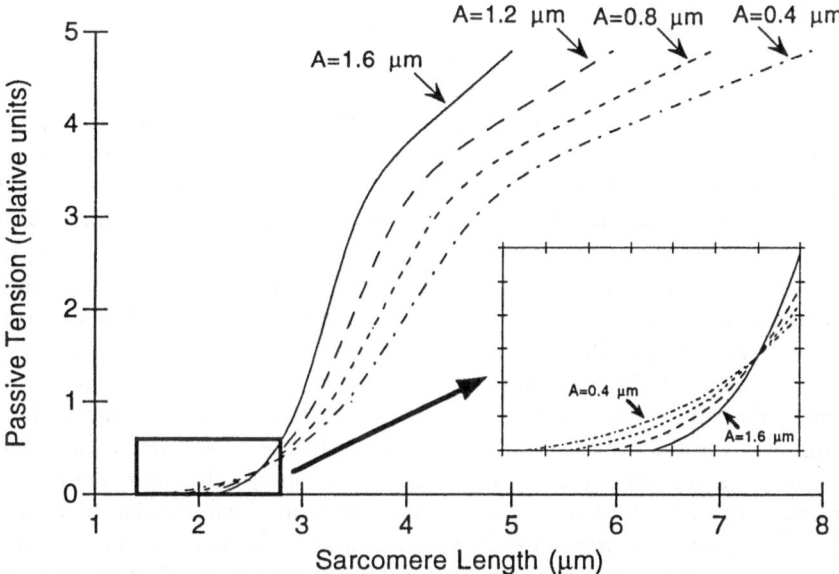

Fig. 9. Calculations of passive tension as a function of sarcomere length with varying A-band lengths. The curve for A = 1.6 μm corresponds to the data for single skinned frog muscle fibers measured by Higuchi et al. [21] and plotted as filled circles in Fig. 8a. This curve was converted to a relation between passive tension and I-band strain as in Fig. 8c. For each measured tension in Fig. 8a, the sarcomere length was calculated at each A-band length by assuming that the corresponding I-band strain was unchanged. The I-band slack length, I_0, was calculated by assuming that the titin is stretched to twice its unrestrained slack length in the region where it is bound to the thick filaments. In control fibers, the A-band length, A, is 1.6 μm and I_0 is 0.25 μm; then for any A-band length, $I_0 = (1.6-A)/4 + 0.25$. With the assumption that I-band strain determines passive tension, I/I_0 is the same at different A-band lengths when the passive tension is the same, so that the sarcomere length at which a particular tension is generated could easily be calculated for each value of A. For simplicity, the results are presented as smooth curves joining the calculated values

the thick filaments, the elastic modulus of A-band titin is similar to that of I-band titin. Figure 9 shows model calculations of passive tension-sarcomere length relations after A-bands have been depolymerized to various lengths. The model assumes that the dependence of passive tension on I-band strain remains constant (as in Fig. 8c) and that titin is uniformly stretched to twice its slack length in the region where it is bound to the thick filaments. As shown in the inset of Fig. 9, the slack sarcomere length decreases as A-band length decreases, so that passive tension at short sarcomere lengths is elevated relative to control fibers (A = 1.6 μm). At longer lengths, passive tension decreases with decreasing A-band length. However, the decrease in passive tension is less than if titin were unstrained when bound to the thick filaments in the A-band.

The model developed above can account for some previously unexplained findings. Figure 10 compares some results in the literature with calculated values based on this model. The passive tension-sarcomere length curves are shown for frog skeletal muscle before and after a reduction in A-band length from 1.6 μm to 0.42 μm (Fig. 10a); the slack length in unextracted fibers was 2.1 μm [21]. Similar curves are shown for rabbit skeletal muscle before and after a reduction in A-band length from 1.6 μm to 0.53 μm (Fig. 10b); in this case the slack length of unextracted fibers was 2.6 μm [80]. Note that a decrease in the length of the thick filaments leads to large decreases in passive tension at sarcomere lengths greater than 3.0 μm in frog muscle (Fig. 10a), while only small effects occur in rabbit muscle (Fig. 10b) (open and filled circles). These effects are well predicted by the model in which the titin is assumed to be stretched to twice its slack length when bound to the thick filament, and titin strain in the I-band determines the passive tension (dashed curves). Figure 10c shows the time course of changes in passive tension and A-band length in single cardiac myocytes during exposure to elevated KCl concentrations [12]. This exposure led to a rapid decrease in A-band length (crosses), followed by a gradual decrease in passive tension to approximately 60% of its control level (open circles). The model predicts a slightly lower level of passive tension at equilibrium than was observed (compare dashed lines in Fig. 10c), but most of this difference can be accounted for by the amount of tension attributed to intermediate filaments at this sarcomere length [12]. However, the slow time course of passive tension decrease relative to the decrease in A-band length is not predicted by the model (compare open and filled circles in Fig. 10c). These results suggest that titin released by depolymerization of the thick filaments may undergo slow changes in conformation from a stiff to a more elastic state. However, Higuchi and Umazume observed that the decrease in passive tension occurred simultaneously with thick filament depolymerization in

Fig. 10. Passive tension as a function of sarcomere length for single skinned fibers from frog semitendinosus muscle (**A**) and rabbit psoas muscle (**B**). Passive tensions are shown before and after shortening of the thick filaments (filled and open circles, respectively). The dashed curves are calculated according to the method in Fig. 9. (**C**) Time course of A-band shortening (x) and passive tension decrease (open circles) in isolated myocytes from rat hearts during exposure to high salt solution. Passive tension was measured at a sarcomere length of 2.7 μm. The changes in tension predicted according to the method in Fig. 9 are also shown (filled circles). The dashed lines correspond to the final levels of actual and calculated tensions after thick filament shortening has ended. Data in (**A**) are replotted from Fig. 7 of Higuchi et al. [21]; Data in (**B**) are replotted from Fig. 3a of Salviati et al. [80]. Data in (**C**) are replotted from Fig. 6 of Granzier and Irving [12]. For frog fibers (**A**), A-bands were initially 1.6 μm long and the slack sarcomere length was 2.1 μm; after extraction, the A-bands were 0.42 μm long and the slack sarcomere length was 1.5 μm [21]. For rabbit fibers (**B**), A-bands were initially 1.6 μm long and the slack sarcomere length was 2.6 μm; after extraction, the A-bands were 0.53 μm long [80]. For cardiac myocytes (**C**), A-bands were initially 1.6 μm long and the slack sarcomere length was 1.9 μm [12]

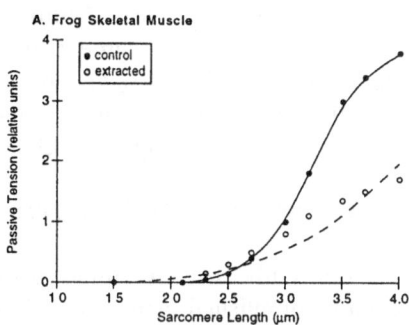

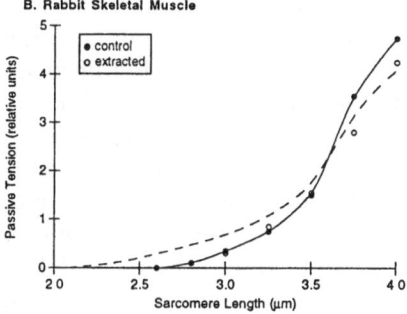

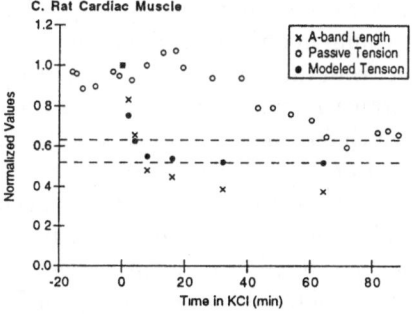

frog skeletal muscle [22]. The differences may be due to the widely different sarcomere lengths and I-band strains at which these time courses were measured in the two studies.

In summary, depolymerization of thick filaments by elevated salt results in the release of A-band titin. The segment of titin bound to the thick filament appears to be stretched to twice its slack length. Therefore, the recruitment of A-band titin to the I-band results in a decrease in slack sarcomere length. The released titin appears to have an elastic modulus that is similar to that of the segment of titin that was originally located in the I-band. The change in passive tension in response to thick filament depolymerization thus depends on sarcomere length, the extent of thick filament shortening, and the original slack length.

3 Selective Extraction of Thin Filaments and Z-Lines

More recently, selective removal of thin filaments from muscle fibers has been accomplished using the actin severing protein gelsolin. Funatsu et al. found that this procedure results in a small increase in resting tension in skinned rabbit psoas fibers, along with a small decrease in slack sarcomere length from 2.2 to 2.1 µm [5]. Granzier and Wang found a similar decrease in slack length upon thin filament removal in rabbit psoas fibers, but this was accompanied by a much larger increase in tension upon stretch of the relaxed fiber [13]. Results similar to Granzier and Wang's were obtained in isolated myofibrils from skeletal muscle by Yasuda et al. [106]. Although these authors observed a large increase in passive tension after removal of thin filaments from single rabbit psoas myofibrils, they found that myofibrils from cardiac muscle exhibited little change in passive tension after extraction of the region of actin filaments that were not overlapped with myosin filaments [34]. The different effects of thin filament extraction on passive tension in cardiac and skeletal muscle may be due to the presence of nebulin in only skeletal muscle; removal of actin may allow nebulin to interact with titin in the I-band, resulting in decreased elasticity, as suggested by Granzier and Wang [13]. This effect would not occur in cardiac muscle, which does not contain nebulin [8, 32, 35, 84, 100].

In rabbit cardiac muscle, the selective removal of thin filaments did not affect the elastic behavior of a titin epitope located in the I-band (SM1 in Fig. 4) [6]. However, recent studies from two laboratories have shown that thorough extraction of actin filaments from cardiac myofibrils increases the extensibility of titin in the region flanking the Z-line, as probed by an antibody that binds in this region (T12 in Fig. 4) [50, 52, 88]. This effect is accompanied by a significant decrease in passive stiffness [50, 52], along with a small decrease in slack sarcomere length caused by a shortening of the previously stiff region of titin flanking the Z-line [88]. The results show that the short region of titin immediately flanking the Z-lines is stiffened by interaction with actin filaments. Furthermore, when bound to the thin filaments this region of titin is stretched relative to its free slack length.

Finally, selective extraction of the Z-lines with calcium activated neutral protease or low salt did not result in movement of the T12 epitope flanking

the residual Z-lines [90]. Apparently, interaction of titin with the thin filaments maintains sarcomeric organization in the Z-line region, even after
extraction of most of the electron dense material in the Z-lines.

B Segmental Extension of Titin and the Mechanical Properties of Relaxed Muscle Fibers

1 Stress-Induced Mobilization of Previously Anchored Epitopes of Titin

As discussed above, titin epitopes that are fixed relative to the thick filaments can behave elastically once the thick filaments are depolymerized.
Several studies have shown that these epitopes can also be recruited to the
elastic region by high levels of mechanical stress. Itoh et al. observed that A-
band titin epitopes near the ends of the thick filaments (antibody 3B9 in Fig.
4) become misaligned and enter the I-band upon stretching frog skeletal
muscle sarcomeres from 3.5 to 4.0 μm [36]. Trombitas and colleagues obtained similar results with an antibody that binds to an epitope 50 nm outside of the A-band (antibody T11 in Fig. 4) [87, 90]. They found that this
epitope becomes elastic only at sarcomere lengths greater than 3.6 μm,
where the thick and thin filaments no longer overlap. They also found that
the region of titin adjacent to the Z-lines remains inelastic in sarcomeres
stretched to lengths up to 5.2 μm, but that upon stretch to 6 μm this region
finally moves away from the Z-line (antibody T12 in Fig. 4) [90]. Similar
results have been obtained in cardiac myofibrils by Linke et al., who observed a sudden extension of the A-band region of titin in sarcomeres
stretched to 3.4 μm (BD6 in Fig. 4) [49].

2 Segmental Extension of Titin and the Relation between Passive Tension and Sarcomere Length

As discussed above, sufficient mechanical strain can mobilize previously
fixed regions of titin, essentially stripping titin from its anchor points. This
process has important consequences for the mechanical properties of relaxed muscle fibers. Wang and colleagues have correlated the behavior of
titin epitopes with the passive mechanical properties of muscle fibers over a
wide range of sarcomere lengths [95, 96]. They found that a titin epitope
near the middle of the I-band (antibody RT13 in Fig. 4) behaved elastically
as the sarcomere was stretched, increasing its distance from both the Z-line
and the A-bands. However, in plots of antibody to Z-line spacings against
sarcomere lengths, a decrease in slope occurred in highly stretched sarcomeres. The length at which this change in titin epitope behavior occurred

corresponds to the length at which further stretch of the sarcomere does not result in an increase in passive tension. These effects are accompanied by a disordering of the sarcomere that is manifested as a ~15% decrease in A-band length, misalignment of the ends of the thick filaments, and a broadening of the transverse stripes produced by anti-titin antibodies [96]. Therefore, the mechanical property of sarcomere yield is correlated with a change in titin epitope translocation and sarcomere structure that is easily accounted for by the recruitment of A-band titin into the elastic I-band region [95, 96].

Figure 11 schematically shows how structural changes involving titin can account for various features of the relation between resting tension and sarcomere length [95, 96]. In this model, I-band titin behaves as a nonlinear elastic element. The slack length of this segment of titin along with the

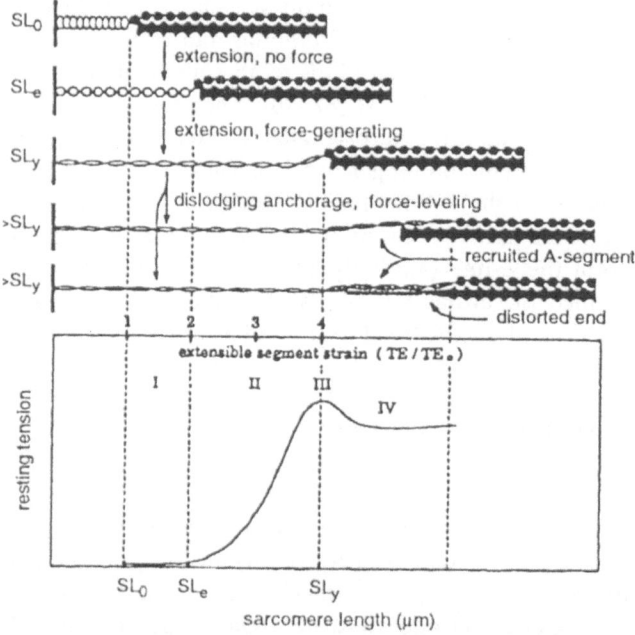

Fig. 11. Segmental extension of titin accounts for the relation between passive tension and sarcomere length. I-band titin is slack in sarcomeres at slack length, and is easily stretched in region I of the length tension curve. Stretch of titin in region II yields exponentially increasing tension, until the tension is so high that titin becomes dislodged from its binding sites on the thick filament. The sarcomere length at which this occurs is the yield point, SL_y. Further extension into region IV of the length-tension relation occurs without any significant increase in tension, as more titin is dislodged from the thick filament. This Fig. is reproduced from Wang et al. [95], with permission of the authors

length of the thick filaments determines the slack sarcomere length. The relative strain of titin in the I-band determines the passive tension. When the tension becomes high enough to rip titin from its anchor points in the A-band, sarcomere yield occurs.

One consequence of this model is that cycles of sarcomere stretch and release should produce reproducible tensions so long as the yield point is not reached. However, irreversible changes should be observed once yield has occurred and A-band titin has been recruited to the segment of titin in the I-band. Wang et al. observed these features in rabbit psoas fibers subjected to multiple cycles of stretch and release [96]. They found that the decrease in passive tension in post-yield sarcomeres could be quantitatively explained by an increase in length of the elastic I-band segment of titin, without any changes in the number or intrinsic elasticity of the titin filaments. Similar results have been observed in cardiac myofibrils [49].

Recruitment of A-band titin to the I-band by extreme stretch is accompanied by an increase in slack sarcomere length in both skeletal and cardiac muscle [49, 96]. Wang et al. suggested that this could be explained by assuming that titin is compressed when it is bound to the A-band; as an alternate explanation, they suggested that the region of titin that is recruited from the A-band may remain associated with fragmented pieces of thick filaments that prevent shortening of the released segment of titin to its intrinsic slack length [96]. Since A-band titin shortens when it is released by depolymerization of thick filaments [21], the latter explanation appears more likely. Fragmentation of thick filaments is also consistent with the decrease in A-band length and deformation of thick filaments observed in post-yield sarcomeres [96].

3 Dependence of Passive Tension on Titin Isoforms

On the basis of variations in electrophoretic mobility [1, 11, 12, 26, 32, 49, 89, 95, 100] and antibody reactivity [25], several investigators have demonstrated that different striated muscles of an animal can contain different isoforms of titin. Developmental changes in titin isoforms have also been documented [16, 107]. The variation in apparent molecular weight of titin is correlated with profound differences in the passive mechanical properties of relaxed fibers from different muscles.

In a study by Wang et al., the passive mechanical properties and titin isoforms of single muscle fibers from six different skeletal muscles of the rabbit were compared [95]. They found that several properties correlate with titin size. Mechanically, muscle fibers with larger titin isoforms start developing significant levels of passive tension at longer sarcomere lengths, and reach their elastic limit, or yield point, at longer sarcomere lengths (Fig. 11, SL_e and SL_y, respectively) [95]. Ultrastructurally, the elastic behavior of an epitope in the I-band region of titin (RT13 in Fig. 4) changes its behavior at a characteristic length corresponding to the yield point, presumeably because sarcomere yield is due to an increase in the extensible segment of titin caused by stripping titin from its anchor points in the A-band. This change in epitope translocation occurs at longer sarcomere lengths in fibers containing larger isoforms of titin, corresponding to the different lengths at which sarcomere yield occurs in these muscles [95]. The close correspondence between the mechanical yield point and changes in titin epitope translocation in the I-band has also been observed in isolated myofibrils from cardiac and skeletal muscle [51].

Several other groups have confirmed the basic finding that smaller titin isoforms in skeletal muscle correlate with higher resting tensions [13, 26, 51], even though the total amount of titin appears to be constant [26]. In addition, cardiac sarcomeres, which contain a much smaller isoform of titin than any of the vertebrate skeletal muscles examined (see Fig. 1), have a shorter slack length and develop high levels of passive tension at much shorter lengths than skeletal muscle sarcomeres [11, 12, 49, 51, 53, 89].

The differences in passive mechanical properties and titin epitope behavior between muscle cells containing different isoforms of titin are easily explained if it is assumed that the differences in size between the various isoforms are entirely accounted for by variations in the length of the normally extensible I-band region of titin [11, 12, 26, 95]. In this simple model, isoform variations in the length of the extensible segment of titin account for differences in the passive mechanical properties of relaxed muscle fibers. It should be noted that because the I-bands in slack sarcomeres are much shorter in cardiac than in skeletal muscles, the inextensible regions of I-band titin that are adjacent to the Z-lines and A-band take up a much larger percentage of the contour length of unstrained I-band titin than in skeletal muscle [11, 89]. Nevertheless, passive tension in cardiac myocytes is roughly equivalent to the passive tension in skeletal muscle fibers when the elastic regions of titin in the two muscle types are equally strained [11].

Variations in the length of the I-band region of titin have been confirmed by cDNA cloning of titin from various muscles [46]. These experiments show that the I-band region of titin contains tandem repeats of immunoglobulin-like (Ig-like) domains [46, 59, 81], as well as a region rich in proline (P), glutamate (E), valine (V), and lysine (K) residues, termed the PEVK domain [46]. Tissue-specific differences in titin are caused by alternative splicing, which leads to variations in the length of the PEVK domain as well as in the numbers of tandem Ig-like domains in the I-band region of titin [46].

C Molecular Basis of Titin Elasticity

1 Antibody Studies Revisited

Detailed models regarding the molecular basis of titin elasticity were originally based on a knowledge of the structure of individual titin domains in combination with studies of the thermal and chemical stability of these domains [4, 71, 75, 76, 77, 83]. Recently, the sequencing and expression of titin cDNAs has made possible a more direct approach using antibodies that specifically label known residues within the titin sequence. By observing the

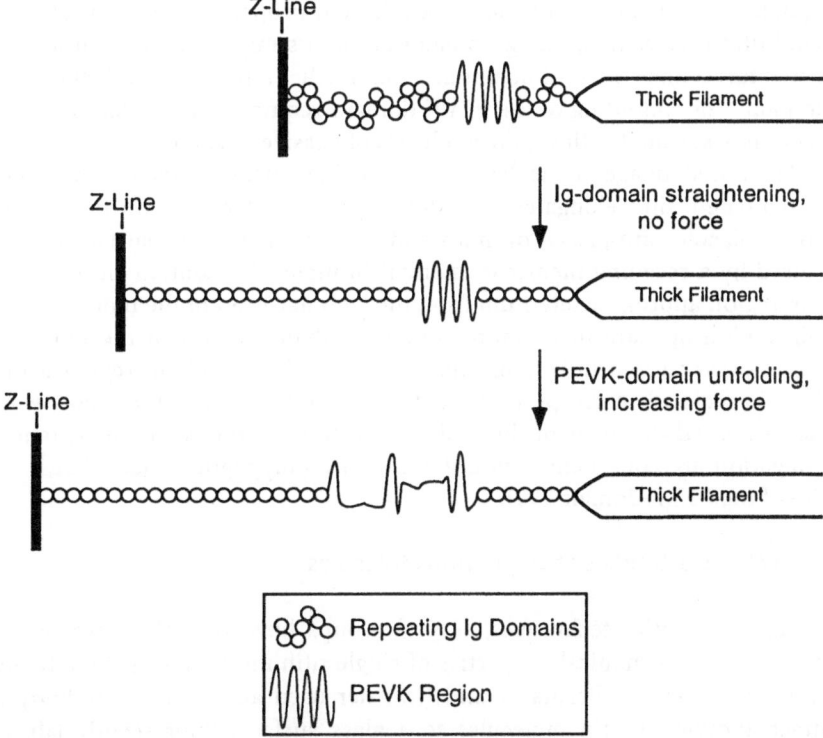

Fig. 12. Behavior of titin domains in the I-band during sarcomere stretch. When slack sarcomeres are stretched, tandem Ig domains are first straightened, with little increase in tension. Further extension leads to unfolding of the PEVK domain, which is associated with exponentially increasing tension. The model is based on the recent work of Linke et al. [51] and Gautel and Goulding [10]

behavior of these antibodies with sarcomere stretch, conclusions can be drawn regarding the elongation or unfolding of particular domains in various regions of the passive force-length relation. Using this approach, two groups have shown that with small amounts of stretch above slack length (region I in Fig. 11), the tandem Ig-domains between the Z-line and the PEVK region of titin elongate with little increase in passive tension (Fig. 12) [10, 51]. The data are consistent with a straightening of the Ig domains without unfolding. As significant passive tension develops (region II in Fig. 11), most of the change in I-band length occurs in the region from the edge of the A-band up to and including the PEVK domain (Fig. 12) [10, 51]. Assuming that the Ig-domains in between the A-band and the PEVK region behave the same as those between the Z-line and the PEVK region, these data suggest that generation of significant levels of passive tension is associated with unfolding of the PEVK domain [10, 51]. The titin molecule thus essentially contains two elements in series with different properties (Fig. 12). When I-band titin is stretched, the Ig-domains are first straightened with little associated force; once the Ig-domains are taut, further stretch unravels the PEVK domain. The unfolding of the PEVK domain requires substantial force, and hence is associated with significant levels of passive tension.

Shadowed images of isolated titin provide further evidence that, once straightened, titin elongates by unfolding of the PEVK region [92]. When titin molecules are placed on mica and straightened by the surface tension caused by a receding meniscus, many titin molecules contain an unusually thin region approximately 1 μm from the globular head of the molecule. The thin region appears to be more compliant than the rest of the molecule, varying several fold in length. The location of the compliant region within the titin molecule corresponds to that predicted for the PEVK domain, while the decreased diameter of the molecule in this region as well its extent of elongation are consistent with unfolding of a polypeptide chain that is the size of the PEVK domain [92].

2 Mechanical Studies of Single Titin Molecules

Using a variety of techniques, several groups have recently succeeded in studying the mechanical properties of single titin molecules by directly manipulating them. Kellermayer and Granzier used an anti-titin antibody to attach individual titin molecules to a glass surface; fluorescently labeled actin filaments were then bound to other points in the titin molecule and mechanically manipulated with a microneedle [42]. They found that single titin filaments could be stretched to at least four times their unextended length, while generating forces up to at least 100 piconewtons. More recently, two groups have used antibodies that bind near the ends of the molecule to tether individual titin filaments to beads that could be manipulated in an optical trap [43, 93]. These experiments show that the mechanical properties of the individual titin molecules are strikingly similar to those of relaxed muscle fibers: First, the force-extension curves of titin molecules exhibit the same general shape found in relaxed muscle fibers; second, stress relaxation is more pronounced after larger stretches than after shorter stretches; third, complete cycles of stretch and release exhibit increasing amounts of hysteresis as the degree of stretch is increased; and fourth, an apparent yield phenomenon occurs at very high extensions. The data from the optical trap experiments could be fit by entropic elasticity models, and were found to be consistent with straightening of a randomly coiled polypeptide at short extensions, followed by unfolding transitions at higher extensions, when forces exceeded 20–60 piconewtons; evidence for refolding was found only after the molecule was shortened to lengths at which tension was very low [43, 93].

Results similar to those observed using optical traps have been obtained by Rief et al.; these investigators used atomic force microscopy to study the mechanical properties of isolated titin and recombinant titin domains [78]. In these experiments, random segments of titin were immobilized between a gold surface and the tip of the atomic force microscope (AFM). The force measured during retraction of the AFM tip from the surface exhibited a region of variable length during which force steadily increased, followed by a repeating series of tension peaks followed by stress relaxation. The number of tension peaks varied between experiments, but occurred at a regular spacing of approximately 25 nm. Recombinant constructs containing four or eight titin Ig-domains exhibited similar behavior, except that the number of tension peaks was equal to or less than the number of Ig-domains present in the constructs. The tension peaks observed with both intact titin and the recombinant fragments started at approximately 150 piconewtons and increased in height with increasing extension, indicating that the Ig-domains were heterogeneous and were unfolded at varying tensions [78]. The observed tension peaks could be reproduced within a single specimen only after complete release of the molecule, indicating that refolding of the Ig-domains can only occur at very low forces.

In summary, the mechanical properties of isolated titin molecules are consistent with the model illustrated in Fig. 12, in which Ig-domains are straightened by low forces and the PEVK region extends at high forces. The mechanical experiments add the additional insights that stress relaxation and hysteresis are caused by the unfolding of domains at high forces, and that refolding occurs only when the molecule is released to very low force

levels. The unfolding transitions also give rise to an apparent yield phenomenon at very high extensions that is consistent with the unfolding of
titin Ig-domains [43, 78, 93].

As discussed in a previous section, yield in muscle fibers is associated
with recruitment of previously immobile regions of titin from the A-band to
the elastic I-band region (Fig. 11) [95, 96]. This detachment of titin from the
A-band may occur at forces below that needed to unfold the titin Ig-domains
in the I-band. A second possibility, suggested by Tskhovrebova et al. [93] , is
that unfolding of Ig-domains and/or fibronectin-like domains of titin in the
A-band occurs at high forces, and that this causes the detachment of titin
from the thick filaments. In the latter case, unfolding transitions would act
as feedback sensors that detect extreme levels of tension in the titin molecule; the unfolding transitions would thus signal the release of limited regions of titin from the thick filaments, with a concomitant decrease in passive tension.

IV Titin's Role in Myofilament Ordering

A Axial Centering of Thick Filaments

Simple models of the sarcomere that include only thick and thin filaments
and Z-lines include no basis for continuity between Z-lines when the myofibrils are relaxed. In these schemes, there is no structural link between successive Z-lines, and the A-bands are mechanically disconnected from the
rest of the sarcomere. These difficulties of two-filament models are ad-
dressed by the inclusion of titin as an elastic molecule that bridges the gap
between the Z-lines and the thick filaments, as well as by intermediate filaments that form longitudinal connections between the peripheries of successive Z-disks in the longitudinal direction [99]. Evidence for titin's involvement in axial ordering of thick filaments has been obtained by selective
destruction of titin. Digestion of titin by trypsin leads to sarcomere disorder;
thick filaments tend to lose their central position, and in some cases dislo-
cation of whole A-bands from the I-bands has been observed [20, 67]. Likewise, destruction of titin by ionizing radiation leads to axial misalignment of
thick filaments upon stretch, and this disorder is irreversibly increased by
activation [28].

In addition to the problem of centering thick filaments in relaxed sarcomeres, two-filament models of the sarcomere also suffer from instabilities
upon activation. This is because the amount of tension pulling on each half
of a thick filament is proportional to the fraction of its cross-bridge bearing
length that overlaps with thin filaments. Any initial imbalance would thus

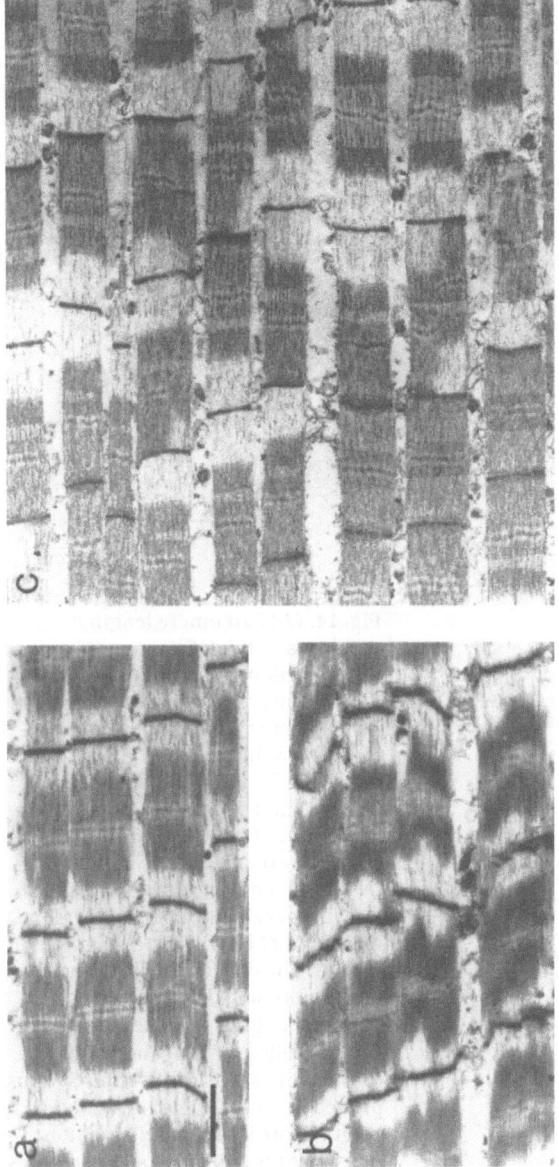

Fig. 13. Electron micrographs of rabbit psoas fibers fixed at rest (**a**), and after 1.7 (**b**) or 7.5 (**c**) minutes of isometric contraction. After prolonged activation, many sarcomeres contain asymetric sarcomeres, with the A-band adjacent to the Z-line (**c**). Calibration bar: 1 μm. Previously published in [30]

cause the thick filament to be pulled further away from the center of the sarcomere during contraction, until the thick filament moves completely to one end of the sarcomere. This behavior has been observed in skinned rabbit psoas fibers after prolonged activation, when many sarcomeres can be observed in which the thick filaments have moved from the center of the sarcomere (Fig. 13) [30]. Upon relaxation, all of the thick filaments return to the center of the sarcomere [30].

Careful examination of Fig. 13c reveals that thick filament movement depends on sarcomere length, and that significant movement is not observed

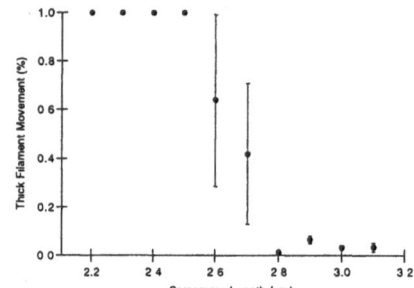

A. Thick Filament Movement vs. Length

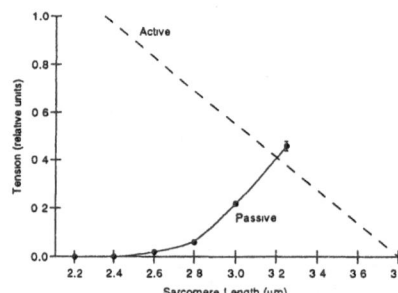

B. Tension vs. Length

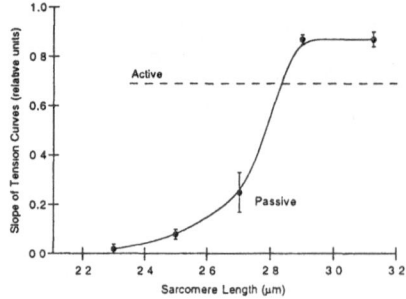

C. Tension Slopes vs. Length

Fig. 14. (A) Sarcomere length dependence of thick filament movement during prolonged activation. Thick filament movement is expressed as a percentage of its maximum movement from the center in a sarcomere of a given length. **(B)** Dependence of passive and active tensions on sarcomere length. Passive tensions (filled circles) were measured 5 minutes after stretch and normalized to the maximum calcium activated force measured at a sarcomere length of 2.4 μm. The dashed line is the active tension predicted on the basis of filament lengths. **(C)** The slopes of the passive (filled circles) and active (dashed line) tension-length relations plotted as a function of sarcomere length. Note that thick filament movement is abolished as sarcomeres are stretched to lengths where the slope of the passive tension-length relation exceeds the slope of the active tension-length relation (compare A and C). Previously published in [30]

in the longest sarcomeres present in the field. Figure 14a shows that thick filaments move all the way to the Z-line in sarcomeres between 2.2 and 2.5 μm in length, while thick filament movement is not observed in sarcomeres longer than 2.8 μm. These observations can be understood in terms of a model that includes the titin filaments as elastic elements linking the thick filaments to the Z-lines. In its simplest formulation, this model predicts that thick filament movement will be abolished when a certain amount of thick filament movement leads to more tension due to stretch of the titin filament resisting the movement than is gained by increasing the number of cross-bridges that can interact with actin on the opposite side of the sarcomere [30, 31] (Fig. 15). This means that thick filament movement will not occur in sarcomeres stretched to lengths where the passive tension-length relation is steeper than the active tension-length relation (Fig. 14b, c), provided that the passive tension is due to stretch of the titin filaments. Comparison of Fig. 14a with 14c shows that the experimental observations from single psoas muscle fibers are in quantitative agreement with the predictions of the model. Further support for this model of titin's effect on thick filament movement comes from a comparison of muscle fibers that have different isoforms of titin; in rabbit soleus fibers, significant passive tension is produced only at longer sarcomere lengths than in rabbit psoas fibers [26, 51, 95], and thick filament movement is abolished at higher sarcomere lengths [26]. Finally, the elastic behavior of an epitope of titin in the I-band (SM1 in Fig. 4) showed that titin lengthens on one side of the sarcomere but shortens on the opposite side as thick filaments move away from the center of the sarcomere, in agreement with the schematic model shown in Fig. 15 [29].

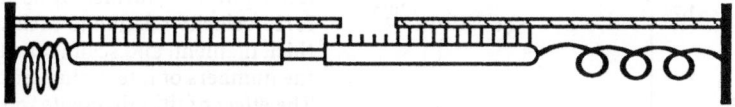

Fig. 15. Schematic model of an activated sarcomere in which the thick filament is moving to the left, resulting in shortening of titin on the left side of the sarcomere and stretch of titin on the right side of the sarcomere. Thick filament movement is resisted by the cross-bridges and the titin on the right side of the sarcomere. If the sum of these resistive tensions equals the sum of the cross-bridge and titin forces on the left side of the thick filament, thick filament movement will be halted. Note that the stretched titin filament on the right side of the sarcomere transmits tension to the Z-line, and helps keep tension output high even though the number of cross-bridges interacting with actin on the right side of the sarcomere is reduced

The rate of thick filament movement during isometric contraction, as well as the accompanying tension output, can be modeled by applying the relation between tension and velocity to each side of the thick filament; the effect of titin filaments on this movement can be included by assuming that passive tension is accounted for by stretch of the elastic region of titin in the I-band [31].The instability giving rise to thick filament movement away from the center of the sarcomere upon activation manifests itself slowly. In rabbit psoas sarcomeres between 2.4 and 2.5 µm long, significant movement is not detected in the first 80 seconds of isometric contraction, during which thick filaments move away from the center of the sarcomere at a very slow rate (Fig. 16a) [31]. This initial lag phase is not affected by the titin filaments, because they are not stretched to lengths where they produce significant tension. However, titin slows the thick filaments as they approach the Z-line (Fig. 16a), and helps keep tension from decreasing (Fig. 16b).

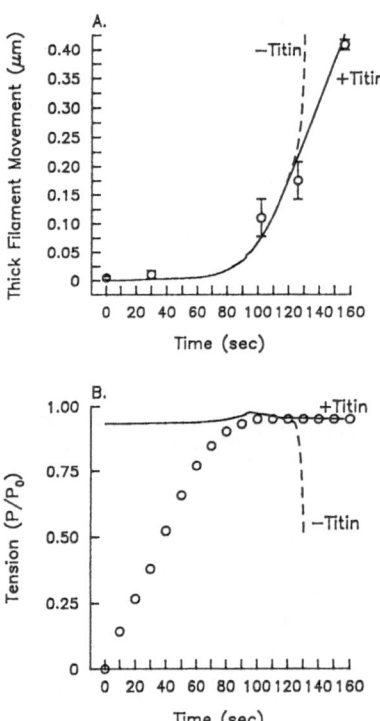

Fig. 16. Time course of thick filament movement (A) and isometric tension (B) in skinned rabbit psoas sarcomeres between 2.4 and 2.5 µm in length. Small bundles were activated with saturating levels of calcium at time zero. Continuous curves were calculated from models incorporating (solid curves) or excluding (dashed curves) titin filaments. Note that a long lag phase precedes significant movement (A), and that titin prevents tension output from decreasing (B) by effectively slowing the motion as thick filaments approach the Z-line (A). The calculations were performed by applying the force-velocity relation to each side of the thick filament, and scaling the forces by the numbers of interacting cross-bridges. The effect of titin filaments was included by adding the appropriate force to each side of the thick filament, assuming that all of the tension measured in relaxed sarcomeres is due to stretch of the I-band region of titin. The calculations assume an initial thick filament displacement of 0.09 nm from the center of the sarcomere. Previously published in [31]

The model described in the preceding paragraph presupposed a small initial imbalance in the sarcomere caused by imperfect centering of the thick filaments. The observed lag before significant thick filament movement occurs could be fit by assuming an initial thick filament displacement of 0.09 nm from the center of the sarcomere (Fig. 16a) [31]. However, even perfectly centered thick filaments can become unstable and move from the center of the sarcomere due to stochastic effects caused by the independent generation of tension by individual cross-bridges on each side of the thick filament. Modeling studies by Pate and Cooke showed that these stochastic properties can lead to a time course of thick filament movement with the same shape as shown in Fig. 16a, but that the lag phase is on average much shorter at longer sarcomere lengths due to the smaller number of interacting cross-bridges on each side of the thick filament [70]. These authors suggest, however, that in an actual sarcomere stability would be increased by lateral coupling of adjacent thick filaments, effectively increasing the number of mechanically linked cross-bridges interacting on each side of the sarcomere. Therefore, the modeled and observed time course of thick filament movement indicate that thick filaments should remain reasonably centered in sarcomeres during isometric contractions *in vivo* [31, 70], which are typically limited to at most a few seconds in duration [14, 15, 104].

B Radial Ordering of the Myofilament Lattice

In addition to axial ordering of the thick and thin filaments, sarcomeres are highly ordered in the radial direction. One of the first indications that axial and radial forces are coupled in relaxed muscle fibers was the discovery that the radial spacing of the myofilaments decreased linearly with passive tension as the sarcomere is stretched [23]. The decrease in lattice spacing with stretch was reduced when titin was digested by mild trypsin treatment; however, the relation between lattice spacing and resting tension was unchanged [19]. These data strongly suggested that titin filaments form lateral linkages that position the thick filaments in the radial direction (Fig. 17).

Additional evidence for titin's involvement in radial ordering comes from observing the effect of fragmenting titin by ionizing radiation. Podolsky et al. observed that the (1,0) and (1,1) equatorial x-ray intensities obtained from skinned muscle fibers were initially unchanged after titin destruction [74]. However, these x-ray intensities progressively decreased when the filament lattice was repeatedly perturbed by cycling between neutral and low pH. Modeling of the data showed that these results could be explained by progressively increasing disorder in the distribution of the thick filaments about their average position in the lattice [74]. The data sup-

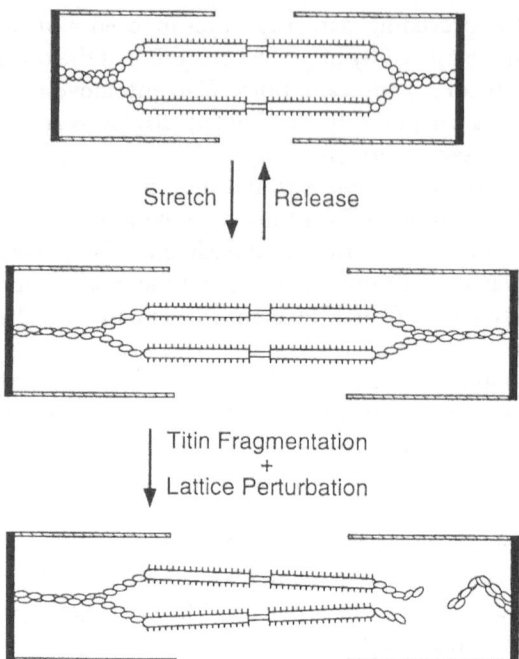

Fig. 17. Schematic model illustrating a possible mechanism for the coupling of axial and radial forces in the relaxed sarcomere. The model shows the bundling of titin emanating from the ends of two thick filaments in the I-band. This arrangement adds a radial component to the force produced when titin is stretched, causing a decrease in the spacing between thick filaments [23]. This arrangement also helps maintain the lateral order of the thick filaments, which is destabilized after titin is destroyed by radiation [74]

port the model proposed by Higuchi and Umazume in which titin forms elastic links between thick filaments in the radial direction, as well as linking thick filaments to the Z-line in the axial direction (Fig. 17) [23].

Funatsu et al. obtained morphological support for this model by electron microscopy of deep-etch, rotary shadowed cardiac muscle [6]. Since cardiac muscle is devoid of nebulin, removal of thin filaments by gelsolin allowed these authors to clearly observe the morphology of titin filaments in the I-band. They found that titin filaments emanating from the tips of several thick filaments were laterally bundled at the N2-line, an electron dense line sometimes observed between the thick filaments and Z-lines. Titin filaments are thus organized in a way that allows them to maintain both the lateral and axial order of the thick filaments.

V Effects on Actomyosin Interactions

Since titin extends axially throughout the entire sarcomere, it is natural to consider whether it affects actomyosin interactions apart from its effects on axial and radial order of the myofilaments. Titin destruction either by ionizing radiation [28] or by trypsin digestion [20] results in decreased levels of isometric tension upon activation. These effects can be explained by axial disorder of the thick filaments after titin breakage, leading to decreased thick and thin filament overlap on one side of the thick filament. However, the decrease in active tension approximately parallels the decrease in passive tension after titin is fragmented by ionizing radiation [28], whereas active tension declines slower than passive tension when titin is proteolyzed by trypsin [20]. As suggested by Higuchi [20], the differences may be due to fragmentation of titin at random points along its length by ionizing radiation, compared to proteolytic cleavage at a defined point in the I-band region of titin by trypsin treatment. One possibility is that the extent of the axial disorder of the thick filaments prior to activation may be more severe after random fragmentation of titin than after the defined scission caused by trypsin; since the initial positions of the thick filaments in the sarcomere determine the lag before active tension drops in the absence of intact titin (Fig. 16), this could account for the different results obtained by the two methods.

Although sarcomere disorder was likely to explain the decrease in active tension after fragmentation of titin, the evidence could not rule out an effect of titin on actomyosin interaction. In contrast, two recent papers report that titin and recombinant fragments of titin slow the velocity of actin filament sliding in an *in vitro* motility assay [41, 48]. These studies showed that titin binds actin filaments and slows their velocity by tethering them to the nitrocellulose surface, creating a load against which actomyosin interactions must work. The binding is stronger at higher calcium concentrations, suggesting a regulatory function [41].

However, two lines of evidence suggest that titin does not actively modulate the production of active tension *in vivo*. First, the calcium sensitivity of active tension production is unchanged after digestion of titin by trypsin; in addition, the velocity of unloaded shortening is unchanged [20]. Second, the elastic behavior of an epitope of titin in the I-band (SM1 in Fig. 4) is unaffected by calcium ions or cross-bridge activity [29]. These results suggest that titin functions as a truly passive element in the sarcomere, and that interactions with actin filaments are weak and readily reversible.

VI Summary: the Role of Titin in the Physiology of Muscle

Over the range of sarcomere lengths found *in vivo*, most of the passive tension produced by relaxed intact skeletal muscle originates in the myofibrils [56]. Linke et al. reached similar conclusions regarding cardiac muscle [53], although Granzier and Irving observed a significant contribution from intercellular collagen fibers [12]. The discrepancy may be due to a difference in experimental protocols, or to varying amounts of collagen in different regions of the heart. However, both groups agree that the myofibrils contribute a large portion of the relaxed myocardium's resistance to stretch. The experiments discussed in the preceding sections in turn provide several lines of evidence that the passive mechanical properties of the myofibrils are determined by the elastic properties of titin: (1) Titin forms elastic links between the thick filaments and the Z-lines; the passive tension progressively decreases in parallel with breakage of these links. (2) Release of titin from its anchor points results in a decrease in passive tension. (3) Passive tension varies with titin isoforms. (4) Passive tension is determined by the strain imposed on the elastic region of titin; to a first approximation, this relation remains constant when the slack length of the elastic region of titin is varied by depolymerization of thick filaments, by release of titin from thick filaments by high levels of stress, or by titin isoform variations. (5) The mechanical properties of isolated titin molecules are similar to those of relaxed muscle fibers, and the sizes of these forces are consistent with the amount of titin present in muscle fibers. (6) The dependence of thick filament movement on sarcomere length during isometric contraction, the time course of the movement, as well as the tension output during the movement can be modeled by assuming that all of the passive tension originates in the titin filaments.

The properties of titin endow muscle tissues with passive mechanical characterisitics that have functional consequences. The length of the titin helps set the slack length of the muscle [11, 21, 95, 96], and provides a restoring force that causes the sarcomeres to relengthen after shortening [11, 18]. The presence of two types of domains arranged in series in the I-band region of titin contributes to the nonlinear elasticity of striated muscle [10, 51]. This allows lengthening of muscle fibers over a limited region with very little tension, while lengthening above this region requires exponentially increasing amounts of tension. Ease of lengthening over a limited range is advantageous in certain situations. For example, opposing muscle groups are anatomically arranged so as to power movement of limbs in opposite directions; one group of muscles contracts during normal motion, while the opposing group lengthens. Keeping resistance to passive stretch at a minimal level allows voluntary movements to occur efficiently. On the other hand, excessive lengthening of striated muscles can lead to irreversible damage; when sarcomeres are released after being stretched to lengths where the thick and thin filaments no longer overlap, the thin filaments do not return to their original positions in the filament lattice, and active force generation is reduced [24, 54]. Therefore, high levels of passive tension above the physiological range of sarcomere lengths help keep the muscle from being stretched to lengths where its ability to contract would be impaired. The yield associated with release of titin from the thick filaments may help prevent breakage of titin and rupturing of the sarcomere under conditions of extraordinary strain [96].

These principals can also be applied to cardiac muscle, which has the smallest isoform of titin found in vertebrate striated muscle to date [11, 12, 49, 89, 100]. The shorter titin isoform sets the slack length of cardiac myocytes on the ascending limb of the length-tension curve; this allows the heart to adapt to increased filling with a stronger contraction, a phenomenon commonly referred to as the Frank-Starling law of the heart [37, 69]. The nonlinear elasticity of titin would not impair filling of the heart over a limited range, but would provide enough diastolic tone to resist overfilling of the heart. These concepts are consistent with the finding that dilated failing hearts exhibit decreased amounts of titin as well as disorganization of titin [17, 39, 66].

As discussed above, skeletal and cardiac muscles contain different titin isoforms that appear to be optimized for the different physiological demands encountered by these tissues. The passive mechanical properties of different skeletal muscles from the same organism also appear to be adjusted by variations in titin isoforms [13, 26, 51, 95]. Although in some cases variations in titin isoforms appear to correlate with different physiological fiber types, fast-twitch and slow-twitch fibers that coexist in the same muscle have the same passive mechanical properties [26, 27]. Therefore, titin isoform composition appears to be related to the anatomical location of the muscle fiber. However, the precise role of the fine-tuning of the passive mechanical properties of different skeletal muscles remains unclear. A systematic comparison of the working range of various muscles *in vivo* along with a determination of their titin isoforms may shed light on this topic.

In addition to its influence on the passive mechanical properties of muscles, titin is necessary to maintain myofilament order in both the axial and radial directions. Disruption of this order by titin fragmentation impairs the ability of muscle fibers to generate active tension [20, 28]. Titin provides a force to position the thick filaments at the center of the sarcomere in relaxed muscle, and to keep tension from dropping due to thick filament movement during prolonged contractions [31]. However, the properties of the cross-bridges allow for only a very small movement of the thick filaments from their central position in the sarcomere during contractions of physiological duration [31, 70]. Therefore, the primary ordering function of the titin filaments *in vivo* is to keep the thick filaments centered during passive stretch and to prevent sarcomere asymmetry from accumulating over several contractions by recentering the thick filaments each time the muscle is relaxed [31].

References

1.	Akster HA, Granzier HLM, Focant B (1989) Differences in I band structure, sarcomere extensibility, and electrophoresis of titin between two muscle fiber types of the perch (Perca fluviatilis L.). J Ultrastruct Mol Struct Res 102:109–121
2.	Brady AJ, Farnsworth SP (1986) Cardiac myocyte stiffness following extraction with detergent and high salt solutions. Am J Physiol 250:h932–943
3.	Carlsen F, Knappeis GG, Buchthal F (1961) Ultrastructure of the resting and contracted striated muscle fiber at different degrees of stretch. J Biophys Biochem Cytol 11:95–117
4.	Erickson HP (1994) Reversible unfolding of fibronectin type III and immunoglobulin domains provides the structural basis for stretch and elasticity of titin and fibronectin. Proc Natl Acad Sci USA 91:10114–10118
5.	Funatsu T, Higuchi H, Ishiwata S (1990) Elastic filaments in skeletal muscle revealed by selective removal of thin filaments with plasma gelsolin. J Cell Biol 110:53–62
6.	Funatsu T, Kono E, Higuchi H, Kimura S, Ishiwata S, Yoshioka T, Maruyama K, Tsukita S (1993) Elastic filaments in situ in cardiac muscle: deep-etch replica analysis in combination with selective removal of actin and myosin filaments. J Cell Biol 120:711–724
7.	Furst DO, Nave R, Osborn M, Weber K (1989) Repetitive titin epitopes with a 42 nm spacing coincide in relative position with known A band striations also identified by major myosin-associated proteins. An immunoelectron-microscopical study on myofibrils. J Cell Sci 94:119–125
8.	Furst DO, Osborn M, Nave R, Weber K (1988) The organization of titin filaments in the half-sarcomere revealed by monoclonal antibodies in immunoelectron microscopy: a map of ten nonrepetitive epitopes starting at the Z line extends close to the M line. J Cell Biol 106:1563–1572
9.	Gassner D (1986) Myofibrillar interaction of blot immunoaffinity-purified antibodies against native titin as studied by direct immunofluorescence and immunogold staining. Eur J Cell Biol 40:176–184
10.	Gautel M, Goulding D (1996) A molecular map of titin/connectin elasticity reveals two different mechanisms acting in series. FEBS Lett 385:11–14
11.	Granzier H, Helmes M, Trombitas K (1996) Nonuniform elasticity of titin in cardiac myocytes: a study using immunoelectron microscopy and cellular mechanics. Biophys J 70:430–442

12. Granzier HL, Irving TC (1995) Passive tension in cardiac muscle: contribution of collagen, titin, microtubules, and intermediate filaments. Biophys J 68:1027–1044

13. Granzier HL, Wang K (1993) Passive tension and stiffness of vertebrate skeletal and insect flight muscles: the contribution of weak cross-bridges and elastic filaments. Biophys J 65:2141–2159

14. Grimby L, Hannerz J (1977) Firing rate and recruitment order of toe extensor motor units in different modes of voluntary conraction. J Physiol (Lond) 264:865–879

15. Hannerz J (1974) Discharge properties of motor units in relation to recruitment order in voluntary contraction. Acta Physiol Scand 91:374–385

16. Hattori A, Ishii T, Tatsumi R, Takahashi K (1995) Changes in the molecular types of connectin and nebulin during development of chicken skeletal muscle. Biochim Biophys Acta 1244:179–184

17. Hein S, Scholz D, Fujitani N, Rennollet H, Brand T, Friedl A, Schaper J (1994) Altered expression of titin and contractile proteins in failing human myocardium. J Mol Cell Cardiol 26:1291–1306

18. Helmes M, Trombitas K, Granzier H (1996) Titin develops restoring force in rat cardiac myocytes. Circ Res 79:619–626

19. Higuchi H (1987) Lattice swelling with the selective digestion of elastic components in single-skinned fibers of frog muscle. Biophys J 52:29–32

20. Higuchi H (1992) Changes in contractile properties with selective digestion of connectin (titin) in skinned fibers of frog skeletal muscle. J·Biochem (Tokyo) 111:291–295

21. Higuchi H, Suzuki T, Kimura S, Yoshioka T, Maruyama K, Umazume Y (1992) Localization and elasticity of connectin (titin) filaments in skinned frog muscle fibres subjected to partial depolymerization of thick filaments. J Muscle Res Cell Motil 13:285–294

22. Higuchi H, Umazume Y (1985) Localization of the parallel elastic components in frog skinned muscle fibers studied by the dissociation of the A- and I-bands. Biophys J 48:137–147

23. Higuchi H, Umazume Y (1986) Lattice shrinkage with increasing resting tension in stretched, single skinned fibers of frog muscle. Biophys J 50:385–389

24. Higuchi H, Yoshioka T, Maruyama K (1988) Positioning of actin filaments and tension generation in skinned muscle fibres released after stretch beyond overlap of the actin and myosin filaments. J Muscle Res Cell Motil 9:491–498

25. Hill C, Weber K (1986) Monoclonal antibodies distinguish titins from heart and skeletal muscle. J Cell Biol 102:1099–1108

26. Horowits R (1992) Passive force generation and titin isoforms in mammalian skeletal muscle. Biophys J 61:392–398

27. Horowits R, Dalakas MC, Podolsky RJ (1990) Single skinned muscle fibers in Duchenne muscular dystrophy generate normal force. Ann Neurol 27:636–641

28. Horowits R, Kempner ES, Bisher ME, Podolsky RJ (1986) A physiological role for titin and nebulin in skeletal muscle. Nature 323:160–164

29. Horowits R, Maruyama K, Podolsky RJ (1989) Elastic behavior of connectin filaments during thick filament movement in activated skeletal muscle. J Cell Biol 109:2169–2176

30. Horowits R, Podolsky RJ (1987) The positional stability of thick filaments in activated skeletal muscle depends on sarcomere length: evidence for the role of titin filaments. J Cell Biol 105:2217–2223

31. Horowits R, Podolsky RJ (1988) Thick filament movement and isometric tension in activated skeletal muscle. Biophys J 54:165–171
32. Hu DH, Kimura S, Maruyama K (1986) Sodium dodecyl sulfate gel electrophoresis studies of connectin-like high molecular weight proteins of various types of vertebrate and invertebrate muscles. J Biochem (Tokyo) 99:1485–1492
33. Huxley AF, Peachey LD (1961) The maximum length for contraction in vertebrate striated muscle. J Physiol (Lond) 156:150–165
34. Ishiwata S, Yasuda K, Shindo Y, Fujita H (1996) Microscopic analysis of the elastic properties of connectin/titin and nebulin in myofibrils. Adv Biophys 33:135–142
35. Itoh Y, Matsuura T, Kimura S, Maruyama K (1988) Absence of nebulin in cardiac muscles of the chicken embryo. Biomed Res 9:331–333
36. Itoh Y, Suzuki T, Kimura S, Ohashi K, Higuchi H, Sawada H, Shimizu T, Shibata M, Maruyama K (1988) Extensible and less-extensible domains of connectin filaments in stretched vertebrate skeletal muscle sarcomeres as detected by immunofluorescence and immunoelectron microscopy using monoclonal antibodies. J Biochem (Tokyo) 104:504–508
37. Jewell BR (1977) A reexamination of the influence of muscle length on myocardial performance. Circ Res 40:221–230
38. Jin JP (1995) Cloned rat cardiac titin class I and class II motifs. Expression, purification, characterization, and interaction with F-actin. J Biol Chem 270:6908–6916
39. Kawaguchi N, Fujitani N, Schaper J, Onishi S (1995) Pathological changes of myocardial cytoskeleton in cardiomyopathic hamster. Mol Cell Biochem 144:75–79
40. Kawamura Y, Kume H, Itoh Y, Ohtsuka S, Kimura S, Maruyama K (1995) Localization of three fragments of connectin in chicken breast muscle sarcomeres. J Biochem (Tokyo) 117:201–207
41. Kellermayer MS, Granzier HL (1996) Calcium-dependent inhibition of in vitro thin-filament motility by native titin. FEBS Lett 380:281–286
42. Kellermayer MS, Granzier HL (1996) Elastic properties of single titin molecules made visible through fluorescent F-actin binding. Biochem Biophys Res Commun 221:491–497
43. Kellermayer MSZ, Smith SB, Granzier HL, Bustamante C (1997) Folding-unfolding transitions in single titin molecules characterized with laser tweezers. Science 276:1112–1116
44. Kempner ES (1988) Molecular size determination of enzymes by radiation inactivation. Adv Enzymol Relat Areas Mol Biol 61:107–147
45. Labeit S, Gautel M, Lakey A, Trinick J (1992) Towards a molecular understanding of titin. Embo J 11:1711–1716
46. Labeit S, Kolmerer B (1995) Titins: giant proteins in charge of muscle ultrastructure and elasticity. Science 270:293–296
47. Labeit S, Kolmerer B, Linke WA (1997) The giant protein titin. Emerging roles in physiology and pathophysiology. Circ Res 80:290–294
48. Li Q, Jin JP, Granzier HL (1995) The effect of genetically expressed cardiac titin fragments on in vitro actin motility. Biophys J 69:1508–1518
49. Linke WA, Bartoo ML, Ivemeyer M, Pollack GH (1996) Limits of titin extension in single cardiac myofibrils. J Muscle Res Cell Motil 17:425–438

50. Linke WA, Ivemeyer M, Labeit S, Hinssen H, Ruegg JC, Gautel M (1997) Actin-titin interaction in cardiac myofibrils: probing a physiological role. Biophys J 73:905–919
51. Linke WA, Ivemeyer M, Olivieri N, Kolmerer B, Ruegg JC, Labeit S (1996) Towards a molecular understanding of the elasticity of titin. J Mol Biol 261:62–71
52. Linke WA, Ivemeyer M, Ruegg JC, Gautel M (1997) A physiological role for actin-titin interaction in cardiac myofibrils. Biophys J 72:A389
53. Linke WA, Popov VI, Pollack GH (1994) Passive and active tension in single cardiac myofibrils. Biophys J 67:782–792
54. Locker RH, Daines GJ, Leet NG (1976) Histology of highly-stretched beef muscle. III. Abnormal contraction patterns in ox muscle, produced by over-stretching during prerigor blending. J Ultrastruct Res 55:173–181
55. Locker RH, Leet NG (1975) Histology of highly-stretched beef muscle. I. The fine structure of grossly stretched single fibers. J Ultrastruct Res 52:64–75
56. Magid A, Law DJ (1985) Myofibrils bear most of the resting tension in frog skeletal muscle. Science 230:1280–1282
57. Magid A, Ting-Beall HP, Carvell M, Kontis T, Lucaveche C (1984) Connecting filaments, core filaments, and side-struts: a proposal to add three new load-bearing structures to the sliding filament model. Adv Exp Med Biol 170:307–328
58. Maruyama K (1986) Connectin, an elastic filamentous protein of striated muscle. Int Rev Cytol 104:81–114
59. Maruyama K, Endo T, Kume H, Kawamura Y, Kanzawa N, Nakauchi Y, Kimura S, Kawashima S, Maruyama K (1993) A novel domain sequence of connectin localized at the I band of skeletal muscle sarcomeres: homology to neurofilament subunits. Biochem Biophys Res Commun 194:1288–1291
60. Maruyama K, Kimura M, Kimura S, Ohashi K, Suzuki K, Katunuma N (1981) Connectin, an elastic protein of muscle. Effects of proteolytic enzymes in situ. J Biochem (Tokyo) 89:711–715
61. Maruyama K, Kimura S, Ohashi K, Kuwano Y (1981) Connectin, an elastic protein of muscle. Identification of titin with connectin. J Biochem (Tokyo) 89:701–709
62. Maruyama K, Kimura S, Yoshidomi H, Sawada H, Kikuchi M (1984) Molecular size and shape of beta-connectin, an elastic protein of striated muscle. J Biochem (Tokyo) 95:1423–1433
63. Maruyama K, Matsubara S, Natori R, Nonomura Y, Kimura S (1977) Connectin, an elastic protein of muscle. Characterization and Function. J Biochem (Tokyo) 82:317–337
64. Maruyama K, Sawada H, Kimura S, Ohashi K, Higuchi H, Umazume Y (1984) Connectin filaments in stretched skinned fibers of frog skeletal muscle. J Cell Biol 99:1391–1397
65. Maruyama K, Yoshioka T, Higuchi H, Ohashi K, Kimura S, Natori R (1985) Connectin filaments link thick filaments and Z lines in frog skeletal muscle as revealed by immunoelectron microscopy. J Cell Biol 101:2167–2172
66. Morano I, Hadicke K, Grom S, Koch A, Schwinger RH, Bohm M, Bartel S, Erdmann E, Krause EG (1994) Titin, myosin light chains and C-protein in the developing and failing human heart. J Mol Cell Cardiol 26:361–368

67. Natori R, Umazume Y, Natori R (1980) The elastic structure of sarcomere: the relation of connectin filaments with thick and thin filaments. Jikeikai Med J 27:83–97

68. Nave R, Furst DO, Weber K (1989) Visualization of the polarity of isolated titin molecules: a single globular head on a long thin rod as the M band anchoring domain? J Cell Biol 109:2177–2187

69. Noble MI (1978) The Frank-Starling curve. Clin Sci Mol Med 54:1–7

70. Pate E, Cooke R (1991) Simulation of stochastic processes in motile crossbridge systems. J Muscle Res Cell Motil 12:376–393

71. Pfuhl M, Gautel M, Politou AS, Joseph C, Pastore A (1995) Secondary structure determination by NMR spectroscopy of an immunoglobulin-like domain from the giant muscle protein titin. J Biomol NMR 6:48–58

72. Pierobon BS, Betto R, Salviati G (1989) The organization of titin (connectin) and nebulin in the sarcomeres: an immunocytolocalization study. J Muscle Res Cell Motil 10:446–456

73. Pierobon BS, Biral D, Betto R, Salviati G (1992) Immunoelectron microscopic epitope locations of titin in rabbit heart muscle. J Muscle Res Cell Motil 13:35–38

74. Podolsky RJ, Horowits R, Tanaka H (1991) Ordering mechanisms in striated muscle fibers. In: Ozawa E, Masaki T, Nabeshima Y (eds) Frontiers in muscle research. Elsevier Science Publishers, New York, NY

75. Politou AS, Gautel M, Improta S, Vangelista L, Pastore A (1996) The elastic I-band region of titin is assembled in a "modular" fashion by weakly interacting Ig-like domains. J Mol Biol 255:604–616

76. Politou AS, Gautel M, Pfuhl M, Labeit S, Pastore A (1994) Immunoglobulin-type domains of titin: same fold, different stability? Biochemistry 33:4730–4737

77. Politou AS, Thomas DJ, Pastore A (1995) The folding and stability of titin immunoglobulin-like modules, with implications for the mechanism of elasticity. Biophys J 69:2601–2610

78. Rief M, Gautel M, Oesterhelt F, Fernandez JM, Gaub HE (1997) Reversible unfolding of individual titin immunoglobulin domains by AFM. Science 276:1109–1112

79. Roos KP, Brady AJ (1989) Stiffness and shortening changes in myofilament-extracted rat cardiac myocytes. Am J Physiol 256:H539–551

80. Salviati G, Betto R, Ceoldo S, Pierobon BS (1990) Morphological and functional characterization of the endosarcomeric elastic filament. Am J Physiol 259:c144–149

81. Sebestyen MG, Wolff JA, Greaser ML (1995) Characterization of a 5.4 kb cDNA fragment from the Z-line region of rabbit cardiac titin reveals phosphorylation sites for proline-directed kinases. J Cell Sci 108:3029–3037

82. Sjostrand FS (1962) The connections between A- and I-band filaments in striated frog muscle. J Ultrastruct Res 7:225–246

83. Soteriou A, Clarke A, Martin S, Trinick J (1993) Titin folding energy and elasticity. Proc R Soc Lond B Biol Sci 254:83–86

84. Stedman H, Browning K, Oliver N, Oronzi-Scott M, Fischbeck K, Sarkar S, Sylvester J, Schmickel R, Wang K (1988) Nebulin cDNAs detect a 25-kilobase transcript in skeletal muscle and localize to human chromosome 2. Genomics 2:1–7

85. Trinick J, Knight P, Whiting A (1984) Purification and properties of native titin. J Mol Biol 180:331–356
86. Trinick JA (1981) End-filaments: a new structural element of vertebrate skeletal muscle thick filaments. J Mol Biol 151:309–314
87. Trombitas K, Baatsen PH, Kellermayer MS, Pollack GH (1991) Nature and origin of gap filaments in striated muscle. J Cell Sci 100:809–814
88. Trombitas K, Granzier H (1997) Actin-titin interaction in the I-band of rat cardiac myocytes. Biophys J 72:A276
89. Trombitas K, Jin JP, Granzier H (1995) The mechanically active domain of titin in cardiac muscle. Circ Res 77:856–861
90. Trombitas K, Pollack GH (1993) Elastic properties of the titin filament in the Z-line region of vertebrate striated muscle. J Muscle Res Cell Motil 14:416–422
91. Trombitas K, Pollack GH, Wright J, Wang K (1993) Elastic properties of titin filaments demonstrated using a freeze-break technique. Cell Motil Cytoskeleton 24:274–283
92. Tskhovrebova L, Trinick J (1997) Direct visualization of extensibility in isolated titin molecules. J Mol Biol 265:100–106
93. Tskhovrebova L, Trinick J, Sleep JA, Simmons RM (1997) Elasticity and unfolding of single molecules of the giant muscle protein titin. Nature 387:308–312
94. Wang K (1985) Sarcomere-associated cytoskeletal lattices in striated muscle. Review and hypothesis. Cell Muscle Motil 6:315–369
95. Wang K, McCarter R, Wright J, Beverly J, Ramirez MR (1991) Regulation of skeletal muscle stiffness and elasticity by titin isoforms: a test of the segmental extension model of resting tension. Proc Natl Acad Sci USA 88:7101–7105
96. Wang K, McCarter R, Wright J, Beverly J, Ramirez MR (1993) Viscoelasticity of the sarcomere matrix of skeletal muscles. The titin-myosin composite filament is a dual-stage molecular spring. Biophys J 64:1161–1177
97. Wang K, McClure J, Tu A (1979) Titin: major myofibrillar components of striated muscle. Proc Natl Acad Sci USA 76:3698–3702
98. Wang K, Ramirez MR, Palter D (1984) Titin is an extraordinarily long, flexible, and slender myofibrillar protein. Proc Natl Acad Sci USA 81:3685–3689
99. Wang K, Ramirez-Mitchell R (1983) A network of transverse and longitudinal intermediate filaments is associated with sarcomeres of adult vertebrate skeletal muscle. J Cell Biol 96:562–570
100. Wang K, Wright J (1988) Architecture of the sarcomere matrix of skeletal muscle: immunoelectron microscopic evidence that suggests a set of parallel inextensible nebulin filaments anchored at the Z line. J Cell Biol 107:2199–2212
101. Wang K, Wright J, Ramirez-Mitchell R (1984) Architecture of the titin/nebulin containing cytoskeletal lattice of the striated muscle sarcomere – evidence of elastic and inelastic domains of the bipolar filaments. J Cell Biol 99:435a
102. Wang SM, Greaser ML (1985) Immunocytochemical studies using a monoclonal antibody to bovine cardiac titin on intact and extracted myofibrils. J Muscle Res Cell Motil 6:293–312
103. Wang SM, Sun MC, Jeng CJ (1991) Location of the C-terminus of titin at the Z-line region in the sarcomere. Biochem Biophys Res Commun 176:189–193
104. Warmolts JR, Engel WK (1972) Open-biopsy electromyography. I. Correlation of motor unit behavior with histochemical muscle fiber type in human limb muscle. Arch Neurol 27:512–517

105. Whiting A, Wardale J, Trinick J (1989) Does titin regulate the length of muscle thick filaments? J Mol Biol 205:263–268
106. Yasuda K, Anazawa T, Ishiwata S (1995) Microscopic analysis of the elastic properties of nebulin in skeletal myofibrils. Biophys J 68:598–608
107. Yoshidomi H, Ohashi K, Maruyama K (1985) Changes in the molecular size of connectin, an elastic protein, in chicken skeletal muscle during embryonic and neonatal development. Biomed Res 6:207–212
108. Yoshioka T, Higuchi H, Kimura S, Ohashi K, Umazume Y, Maruyama K (1986) Effects of mild trypsin treatment on the passive tension generation and connectin splitting in stretched skinned fibers from frog skeletal muscle. Biomed Res 7:181–186

Control of Sarcomeric Assembly: The Flow of Information on Titin

M. Gautel[1], A. Mues and Paul Young

European Molecular Biology Laboratory, Postfach 102209,
69012 Heidelberg, Germany

Contents

1 Introduction: Spatial and Temporal Events in Myofibrillogenesis

Complex macromolecular structures are assembled following defined pathways, where the controlled transitions of intermediates from one distinct conformation to another allows their controlled integration into the nascent biological structure. Most macromolecular systems, like viruses or flagella, rely on the self-assembly properties of their protein subunits which are controlled in various ways. The sarcomere of striated muscle is a complex macromolecular assembly, built of many self-interacting proteins which are organised in a highly specific way into filamentous and anchoring structures (Fig. 1). A number of the sarcomeric proteins, like actin and myosin, can self-assemble *in-vitro* into synthetic filaments that closely resemble the native structures, but lack their narrow length distribution. This is an indication that control elements involved in filament assembly are either lost or inactivated during conventional protein preparations. In contrast, Maw and Rowe (1986) reported that myosin filaments, disassembled under high-salt conditions by *in-situ* dialysis of whole myofibrils, could be reconstituted to regain their uniform length distribution.

During myofibrillogenesis in vivo, the time point of appearance, the intracellular localisation, and activation state of the sarcomeric components must be accordingly controlled, presumably in a fixed sequence of events. Spatial and temporal relay of information is thus a key element of sar-

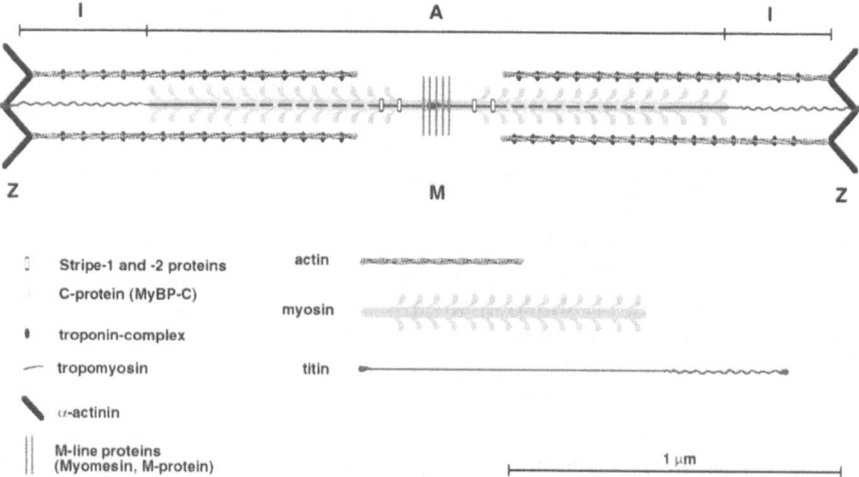

Fig. 1. A Schematic representation of the sarcomere showing the principal protein components of the Z-disk, I band, A band, M band

comeric assembly as well as of the function of the assembled structure. For this control, specific signalling to localised proteins are required.

The giant protein titin (Wang et al. 1979), described as connectin (Maruyama et al. 1977) is emerging as a major component that defines the positions of other sarcomeric proteins by providing them with specific, spatially defined binding sites (reviewed in Trinick, 1994 and 1996; Maruyama, 1997). Bridging the I-band and A-band (Maruyama et al. 1985; Wang and Wright, 1988; Whiting et al. 1989) and effectively the entire distance from Z-disk to M-band (Fürst et al. 1988; Vinkemeyer et al. 1993), titin is the sole component that can sample an entire half-sarcomere as a molecular entity, and that can thus relay information -spatial or temporal- to other proteins over large distances. Understanding of this flow of information along the titin molecule is beginning to emerge.

The titin molecule is constructed in a modular fashion from more than 280 immunoglobulin- and fibronectin-3 like domains (Labeit and Kolmerer, 1995a), and specific sequences in the Z-disk, I-band and M-band (Maruyama et al. 1993; Gautel et al. 1993a; Maruyama et al. 1994; Sebestyén et al. 1995; Labeit and Kolmerer, 1995a). The anchoring regions in the Z-disk and M-band contain contain specific modules and regions involved in the relay of information (Gautel et al. 1993a; Sebestyén et al. 1995; Gautel et al. 1996a; Young et al. 1998).

2 The Z-Disk Region of Titin

The assembly of the Z-disk is the starting point of myofibrillogenesis (reviewed in Vigoreaux, 1994; Fürst and Gautel, 1995). Titin appears as one of the first sarcomeric proteins during myofibrillogenesis and is detected in dot-like aggregates which colocalize with α–actinin in differentiating myoblasts (Tokuyasu and Maher 1987a, b; Wang et al. 1988; Fürst et al. 1989; Handel et al. 1989; Komiyama et al. 1990; Lin et al. 1994). These dots align on stress-fibres, and are organised into Z-disks during the progression of myofibrillogenesis (Dlugosz et al. 1984; Terai et al. 1989; Van der Ven et al. 1993, Dabiri et al. 1997). In good agreement with the proposed ruler function, titin/connectin is one of the first proteins to appear in the Z-disk in developing striated muscle. The integration of titin into nascent sarcomeres begins with the assembly of the aminoterminus into the Z-disk and proceeds towards the C-terminal M-band region (Fürst et al. 1989; van der Loop et al. 1996).

The identification of sequence insertions in Z-disk titin which can be phosphorylated in vitro by SP-specific protein serine/threonine kinases of the extracellular signal regulated kinase (ERK) family (Sebestyén et al. 1995;

Gautel et al. 1996) as well as by similar kinase activities from developing muscle extracts suggested that the titin N-terminus may be involved in controlling myofibril assembly. Since the anchorage of titin in the Z-disk is important in ensuring its ability to provide an elastic connection between Z-disks and thick filaments (Fürst et al. 1988; Itoh et al. 1988; Whiting et al. 1989; Wang et al. 1993), the structural elements of titin involved in this assembly are of particular interest.

The sarcomeric Z-disk of striated muscles is a greatly variable structure that adapts morphologically to the mechanical needs of the respective fibre type (Rowe 1973; Vigoreaux, 1994). Actin filaments from adjacent sarcomeres are anchored in the Z-disk by overlapping with four filaments from the opposite sarcomere, forming a square lattice. This lattice is cross-connected in a zig-zag pattern by Z-filaments, which are assumed to consist of α–actinin (See Squire, 1981; Yamaguchi et al. 1985; Luther 1991 and references therein). The number of α–actinin crosslinks and hence the thickness of Z-disks is greatly variable, presumably to adapt the Z-disk structure to the level of mechanical strain (Rowe, 1973; Vigoreaux, 1994). Within a given muscle fibre, however, their number is precisely regulated (reviewed in Squire, 1981).

2.1 Controlling the Variable Assembly of the Central Z-Disk Region: Titin and Nebulin in the Z-Disk

How is the variable assembly of the Z-disk controlled? In an expansion of the molecular ruler concept proposed for the assembly of the thick filament (Whiting et al. 1989), a ruler-like protein, containing variable numbers of specific binding sites for other Z-disk components, could control the number of α–actinin molecules and their spacing. The giant proteins titin and the actin-binding nebulin are obvious candidates to act as a molecular blueprint (see review by Trinick, 1994). A 60-120 kDa N-terminal region of titin (total molecular weight; 3 MDa!), is localised within the Z-disk (Yajima et al. 1996; Gautel et al. 1996a). This central Z-disk region is constructed from four immunoglobulin-like domains (Yajima et al. 1996; Gautel et al. 1996a), phosphorylation sites and variable copies of a specific protein motif, the 45-residue Z-repeat (Gautel et al. 1996a) (Fig. 2).

By comparison, the nebulin molecule is anchored in the Z-disk with the C-terminal region (Chen et al. 1993; Labeit and Kolmerer, 1995b; Wang et al. 1996), presumably largely by the interactions with F-actin mediated by α–helical actin-binding repeats (Pfuhl et al. 1994, Labeit and Kolmerer, 1995b; Pfuhl et al. 1996, Wang et al. 1996). Similar to titin, the actin-binding repeats of nebulin are found in different length variants in the Z-disk region. In this

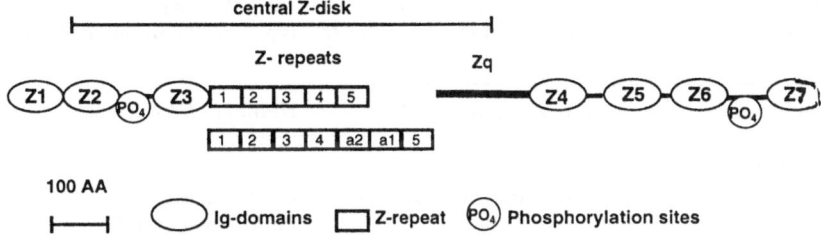

Fig. 2. Domain structure of the Z disk region of cardiac titin. Immunoglobulin domains are labelled Z1-Z4. The positions of the differentially expressed Z repeats and of the phosphorylated sequence insertions (the first one called Zis1, the second Zis2 and so forth as in Labeit and Kolmerer 1995) are shown. Domain nomenclature according to Labeit and Kolmerer, 1995 and Gautel et al. 1996a

region, the fetal isoform reported by Wang et al. (1996) differs in length from the adult isoform of Labeit and Kolmerer (1995b). A C-terminal Src-homology 3 domain (SH3) and putative phosphorylation motifs could participate in the control of Z-disk assembly (Labeit and Kolmerer, 1995b; Wang et al. 1996). A short nebulin like protein, nebulette, shows similar features and seems to represent a cardiac-specific Z-disk ruler (Moncman and Wang, 1995).

In thin Z-disks with a single Z-filament in projection, like chicken *M. Pectoralis major* (Fürst et al. 1988; Yajima et al. 1996), the N-terminal region of titin cannot be discriminated in immunoelectron microscopy from the Z-disk centre. Studies in thick cardiac Z-disks suggested a central position with little or no overlap (Fürst et al. 1988; Gautel et al. 1996a). However, if titin were to control the number of Z-filaments, one would expect some degree of titin overlap in the central Z-disk corresponding at least partly to the degree of thin filament overlap. In agreement with this, a more highly resolved study using improved techniques has localised both the titin N terminus and the nebulin SH3 domain to the edge of the central Z disk (Young et al. 1998).

The Z-repeats are differentially spliced in the central Z-disk region of titin (Gautel et al. 1996a; Ohtsuka et al. 1997a; Sorimachi et al. 1997, Peckham et al. 1997). This observation begs the question of whether the number of titin Z-repeats and the Z-disk morphology are correlated? From the data available such a correlation is beginning to emerge, see Table 1. Thus muscle types with thicker Z disks express titin isoforms with more Z repeats. In particular, the number of Z repeats correlates to the number Z filaments. How, then, is the information for Z-disk assembly encoded in the titin molecule translated into the actual structure?

Table 1. Summary of the available data which demonstrate the correlation betweeen the number of Z filaments observed in a given muscle type and the number of titin Z repeats expressed in that muscle

Muscle type	Number of Z-filaments	Number of Z repeats
Mammalian		
psoas	2–4 (rabbit) Rowe 1973	4–7 (human) Peckham et al. 1997 4–7 (rabbit) Sorimachi et al. 1997
heart	3–4 (mammals) Yamaguchi et al. 1985	5–7 (human) Gautel et al. 1996
slow fibre	3–4 (rabbit) Rowe 1973	5–7 (rabbit) Sorimachi et al. 1997
Non mammalian		
fast white	1 (fish) Luther 1991	2 (chicken) Ohtsuka et al. 1997b

2.2 Two Types of Interactions of Z-Disk Titin with α–Actinin

Since nebulin is proposed to act as the thin-filament ruler, we might expect that titin mediates interactions with proteins other than F-actin. The presumed role of titin as an organiser of the sarcomeric Z-disk was substantiated by reports that native titin could interact with α–actinin (Jeng and Wang, 1992). This was confirmed by Turnacioglou et al (1996). They identified the Z-disk protein Zeugmatin as a presumed breakdown product of titin and demonstrated that a short segment of recombinant Z-disk titin (or "zeugmatin") would bind to α–actinin as well as to vinculin. More specifically, recombinant fragments of Z-disk titin bind to α–actinin in a region of about 63 kDa (Turnacioglou et al. 1996; Ohtsuka et al. 1997b). Genetic interaction searches with the two-hybrid system have since identified a binding site for Z-disk titin in the C-terminal domain of α–actinin (Ohtsuka et al. 1997a). This domain is homologous to calmodulin, (CaM) but contains a non-functional EF-hand (Blanchard et al. 1989). This interaction was initially localised to a subset of the titin Z-repeats (Ohtsuka et al. 1997a; Sorimachi et al. 1997). Titin Z-repeats form two subgroups: the two invariant flanking repeats are most closely related to each other, and the differentially spliced central repeats form a distinct subgroup (Gautel et al. 1996a). The interaction studies in yeast two-hybrid assays suggested that only the flanking pair of repeats bind to α–actinin (Ohtsuka et al. 1997a; Sorimachi et al. 1997; Young et al. 1998). Biochemical analysis of recombinant central Z-repeats has now revealed that central type of repeats can interact with the

CaM-like domain of α–actinin, similar to the flanking repeats (Young et al. 1998). The titin Z-repeats therefore represent multiple binding modules, which are plausibly responsible for the consecutive arrangement of variable numbers of α–actinin molecules as Z-filaments, and hence for Z-disk thickness (Gautel et al. 1996a).

In addition to the Z-repeats, recent work has revealed that another interaction site for α–actinin exists on titin in a region denoted as Zq. This 70 residue motif is localised in a single copy at the Z-disk edge and interacts with the two central spectrin-like repeats of the α–actinin rod (Young et al. 1998). This might explain why in transfection experiments in myogenic cells α–actinin, lacking the C-terminal domain, is also sorted to the sarcomeric Z-disk (Schultheiss et al. 1992), but causes Z-disk misassembly during the progression of myofibrillogenesis and the formation of Z-rods reminiscent of those formed in nemaline myopathy (Schultheiss et al. 1992). The unique interaction of titin and α–actinin at the Z-disk edge therefore possibly represents a termination motif for α–actinin incorporation, and perturbation of this mechanism would be expected to result in Z-disk anomalies. The dominant-negative phenotype observed by Schultheiss et al. (1992) could thus be explained by the competition of the titin termination motif by an assembly-incompetent mutant molecule. It is clear, therefore, that sorting of α–actinin to the Z-disk is not solely controlled by the interaction of titin Z-repeats and the α–actinin CaM-like domain, but that a second binding site between the titin Zq motif and the α–actinin rod is also crucial.

The way in which the titin/α–actinin interactions are regulated is still unclear. Calcium would appear not to play a major role since muscle-type α–actinins are defective in their Calcium-binding properties (Blanchard et al. 1989. Possible other mechanisms could involve phosphorylation of titin or α–actinin, or additional regulatory molecules. The close proximity of the two types of α–actinin binding sites in titin are highlighted by a sequence comparison in Fig. 3. Note that several consensus substrate sites for serine threonine kinases (Pearson and Kemp, 1991) are present in this region (underlined). Also, the high degree of cross-species conservation of the hydrophobic binding site in the C-terminal Z-repeat (zr5) and of the 70 residue region in the titin Zq motif is evident. The close proximity of the two distinct binding sites suggests that for α-actinin molecules at the Z-disk periphery, interaction sites include the rod domain as well as the two opposed C-terminal domains which could bind to zr5, and zr1 from opposite titin filaments. Since the rod-binding site on titin is present in only one copy per half-sarcomere, the further centrally sorted α–actinin molecules can interact only via their C-terminal domains with the central type of titin Z-

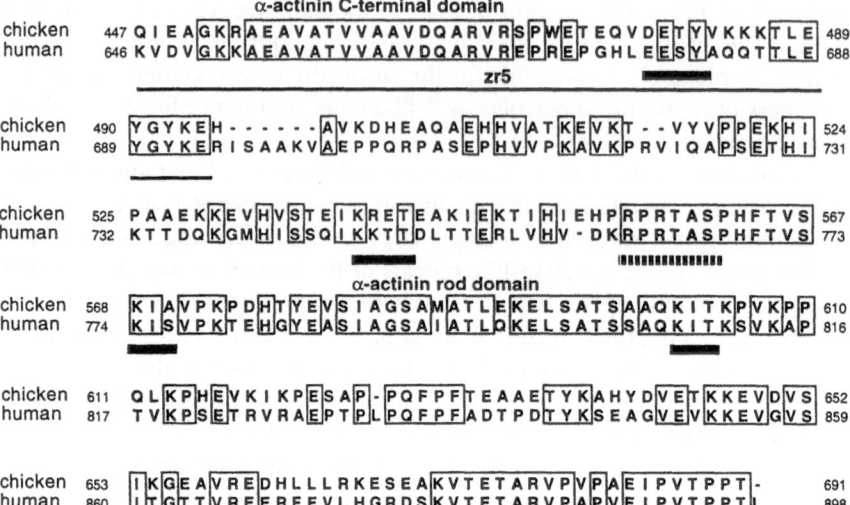

Fig. 3. A multiple-sequence alignment demonstrates the high cross-species conservation of the two principal sorting signals for α-actinin between chicken titin (EMBL D83390) and human titin (EMBL X90568) The Z-repeat zr5 is shown with the binding site defined by Ohtsuka et al. (1997b) shaded in grey. Note the totally conserved cluster of hydrophobic residues which are the hallmark of Z-repeats. This binding site is separated by about 70 residues of lowly conserved sequence from the rod-binding domain in Zq (residues in human titin) which again is highly conserved between chicken and human titin. Putative serine-threonine kinase phosphorylation sites are underlined with hatched bars. A single predicted tyrosine-kinase site in zr5 is underlined with a solid bar. Identical residues are boxed

repeat (Young et al. 1998). Regulation of these binding sites would seem to be crucial in Z disk assembly but has yet to be demonstrated.

2.3 Phosphorylation Motifs and Regulatory Protein Interactions in the Z-Disk

For the assembly of the Z-disk, two mechanisms have to be spatially and temporally co-ordinated: the capping of actin filaments, probably by the capZ protein (Schafer et al. 1995, 1996), and the crosslinking of the overlapping filaments by α–actinin. The molecular ruler proteins of the Z-disk, titin and nebulin/nebulette, are therefore expected to provide specific attachment sites for these proteins in a concerted way, as well as for factors possibly involved in the control of capping activity and crosslinking.

Actinin-associated LIM protein (ALP) from muscle was recently shown to bind to the α–actinin rod (Xia et al. 1997) by its PDZ domain. The binding

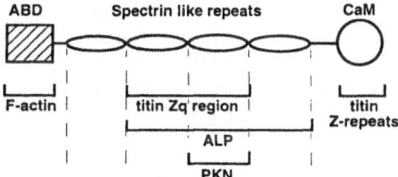

Fig. 4. The domain structure of α-actinin. The binding sites of known protein ligands of the skeletal isoform are shown. In the Z disk anti-parallel α-actinin dimers cross link actin filaments. CaM = Calmodulin like C terminal domain; ABD = Actin binding domain PKN = Protein kinase N; ALP = Actinin associated LIM protein. Structural as well as regulatory proteins share overlapping binding sites. See text for references

site of ALP overlaps with the titin binding site on the α–actinin rod; however, the central two spectrin-like repeats do not suffice for ALP interaction. Therefore, those α–actinin molecules not bound to titin will have free binding sites for ALP. Since muscle LIM proteins (MLP) have been implicated in muscle differentiation and cytoarchitectural organisation, (Arber et al. 1997; Pomiès et al. 1997), a more direct involvement of the titin/α–actinin/actin complex and additional proteins in the control of myofibrillogenesis is possible. This is underlined by the recent observation that the serine-threonine kinase PKN, which is activated by the small GTPase Rho, binds to the third spectrin-like repeat as well (Mukai et al. 1997). Rho-like GTPases are involved in the organization of the actin cytoskeleton (Paterson et al. 1990; Ridley and Hall, 1992; Narumiya et al. 1997), such as stress-fibre and focal adhesion reorganization. It seems to emerge, therefore, that the binding sites on the α–actinin rod co-ordinate multiple protein-interactions relevant for the control of Z-disk assembly (Fig. 4).

A possible key for the understanding of the control of Z-disk assembly resides in two phosphorylation motifs located at the edge of the central Z-disk and in the Z-disk periphery (Z-disk comb; Gautel et al. 1996a; Young et al. 1998). In both regions, multiple copies of serine-proline rich motifs are inserted between the immunoglobulin domains of titin (Sebestyén et al. 1995; Labeit and Kolmerer, 1995a). The most striking motif, consisting of several copies of the amino acid motif SPIRM, is localised outside the central Z-disk in a region termed Zis5. In-vitro phosphorylation assays demonstrated that recombinant titin fragments from this region can be readily phosphorylated by cdc2 kinase as well as by ERK2 (Sebestyén et al. 1995). The SPIRM motifs are, however, an evolutionary rather young feature of the titin molecule as suggested by the observation that the copy number of the motif varies highly between avian and mammalian titins (Fig. 5). The regulatory function of this motif may therefore not be a species-conserved feature,

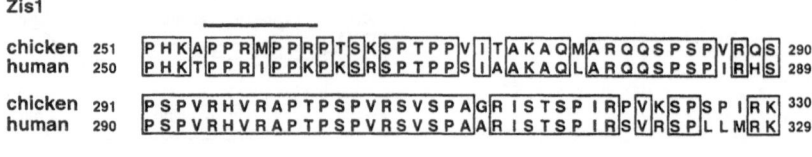

Fig. 5. Alignment of the phosphorylation sites in the interdomain insertion Zis1 and Zis5 of the Z disk region of titin from different species. The putative SH3 domain binding site is shown (bar). (Human titin: EMBL X90568; chicken titin/connectin: D88390; rabbit titin: U28657). Note the variable copy number of SPARM phosphorylation sites between mammalian and avian titin. Identical residues are boxed; phopshorylation sites are shaded in grey

similar to the situation in myomesin, where the presence of the putative phosphorylation motif in the KQSTAS-repeats is also species-dependent (Vinkemeier et al. 1993; Bantle et al. 1996).

A second interdomain-insertion of titin, localised between the second N-terminal Ig domain Z2 and the putative Ig-module Z3 is a similar target for SP-directed protein kinases of the ERK and cdk families (Gautel et al. 1996a). Recent ultrastructural investigations have localised this region to the central Z-disk and placed it in close vicinity to the SH3-domain of nebulin at the barbed end of the thin filament (Gautel et al. 1996a; Young et al. 1998). SH3 domains interact with proline rich sequences (Musacchio et al. 1994). These interactions are involved in the coupling of receptor tyrosine kinases to their effector molecules (Egan et al. 1993) as well as in the control of cytoskeletal organisation by several actin binding proteins like yeast ABP1p or mammalian cortactin (Drubin et al. 1990; Wu and Parsons, 1993). The solution structure of the nebulin SH3 domain was recently solved (Politou et al. 1997), and the structural features of the ligand binding site suggest that a consensus sequence PPXXPPK will represent the preferred ligand. This sequence is represented in the interdomain insertion Zis1 of Z-disk titin (TPPRIPPKKP) close to several consensus phosphorylation sites for ERK 2, and it was hence suggested that the nebulin SH3 domain could interact in a phosphorylation-controlled manner with the titin N-terminus (Gautel et al. 1996a; Politou et al. 1997). However, other SH3-containing proteins are localised to the Z-disk like the recently described ArgBP2 (Wang et al. 1997), which contains three SH3 domains. Future cell biological and biochemical investigations will elucidate the complex interactions of SH3 containing proteins at the Z-disk.

2.4 Titin, α–Actinin and Nebulin: a Molecular Model of the Sarcomeric Z-Disk

The wealth of biochemical and ultrastructural data that has emerged recently allows to correlate ultrastructure and molecular interactions in a model for the sarcomeric Z-disk (Young et al. 1998). The spacing of Z-filaments (α–actinin) in longitudinal projections of Z-disks is about 38 nm (Yamaguchi et al. 1985; Luther 1991). However, the overlap of actin filaments is highly variable, ranging from about 10 nm to over 100 nm (Yamaguchi et al. 1985; Luther 1991). Correspondingly, the number of crosslinking Z-filaments ranges between one and four (see Table 1). In reconstructions of Z disks the spacing of orthogonal Z filaments has been determined to be 15–20 nm in fish (Luther, 1991) and 16+/–9 nm in mammalian Z-disks (Schroeter et al. 1996). In nemaline rods, which are presumably binary complexes of actin and α–actinin, the spacing is about 19 nm, (Morris and Squire, 1990). Allowing for the standard deviation of these measurements, pairs of Z-filaments are therefore placed at angles of 80–90° roughly every 17–18 nm. This is about a quarter turn of the F-actin long-pitch helix for each Z-filament (Luther 1991). The interactions of titin Z-repeats and successive pairs of α–actinin molecules must roughly allow for this spacing and also, to some degree, follow the helical structure of the actin filaments. This helix has a variable half-repeat spacing of between 35 to 38 nm (Egelman et al. 1982). Single titin Z-repeats depending on their secondary structure, could span a length of up to 17 nm (Young et al. 1998). This length agrees well with the orthogonal spacing of Z-filaments in both mammalian and non-mammalian Z-disks. The successive arrangement of α–actinin molecules into Z-filaments can therefore be readily explained by successive pairs of titin binding sites spaced about 17 nm apart. The helical path of these binding sites would be indirectly enforced by the interactions of the complex with F-actin.

These ultrastructural and biochemical observations can be summarised in a molecular model for the layout of titin, actin and α–actinin in the central Z-disk as depicted in Fig. 6 (Young et al. 1998). The path titin takes as it enters the central Z-disk is constrained by the protein interactions with α–actinin as well as by the ultrastructural position of the N-terminus (Fig. 7) and of Zq and of Z4 immediately at the outer edge of the central Z-disk

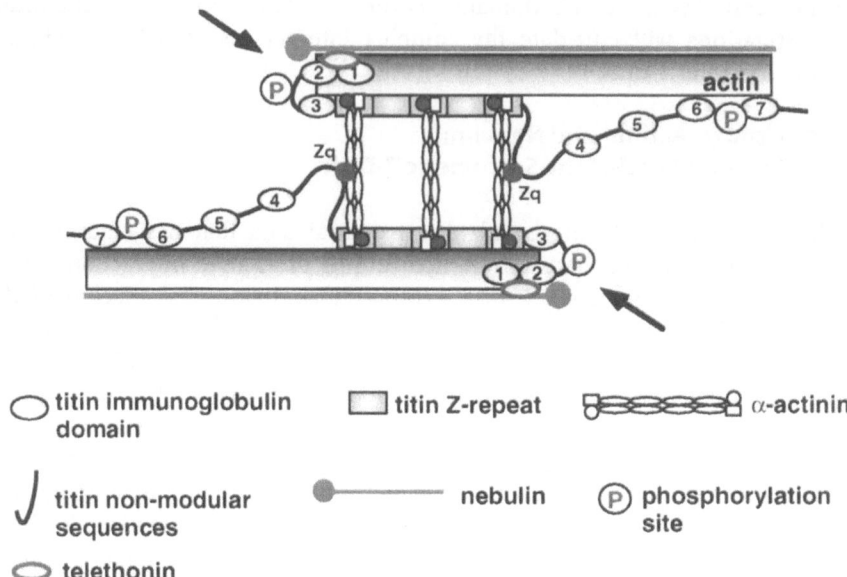

○ titin immunoglobulin   ☐ titin Z-repeat   ▱▱▱▱ α-actinin
  domain

∫ titin non-modular   ●———— nebulin   Ⓟ phosphorylation
  sequences   site

○ telethonin

Fig. 6. The most likely path of titin into the Z-disk runs parallel to the thin filament
for most of its length. Established protein interaction sites are highlighted in red.
Titin immunoglobulin domains are numbered accoding to Labeit and Kolmerer
(1995). Along the outmost pair of α-actinin molecules, titin is linked to the α-actinin
rod and to the CaM-like domain. In the central Z-disk, titin Z-repeats coordinate
successive α-actinin molecules. The nebulin SH3 domain (circle) and the proline-
rich N-terminus of titin are in close proximity and could interact; proteins binding
to Z1Z2 or Z3 of titin or to the C-terminal region of nebulin are likely to be involved
in the control of barbed-end capping. The position of regulatory SP-rich phosphorylation sites (Gautel et al. 1996) is shown. Arrows point to the presumed signalling
complex at the barbed end of the thin filament, where additional proteins bind
(telethonin and cap-Z). Only every second α-actinin molecule is shown in the central Z-disk, as they would appear in longitudinal sections of the lattice. Since the
spacing of thin filaments and concomitantly the angle of the Z-filaments is variable
(Yamaguchi et al. 1985), the α-actinin links are shown schematically at right angles.
The position of the titin N-terminus is arbitrarily shown with some flexibility in the
linker segment. (Modified after Young et al. 1998)

(Gautel et al. 1996a; Yajima et al. 1996). In this model titin binds firstly to the
centre of the α–actinin rod and then runs along the outermost α–actinin
molecule for half its length. The next interaction site is localised in zr5,
where titin will bind to the C-terminal domain of α–actinin (Fig. 3). At this
point, the titin molecule must change the orientation of its path to be linked
to the successive α–actinin molecules now interacting with the Z-repeats
and will therefore run parallel to the actin filaments in the Z-repeat region.

The ultrastructural analysis now places the position of the N-terminal domains of titin at the edge of the actin overlap region, with molecules entering from two sarcomere halves crossing over the Z-disk centre by about 30 nm. Thus the two C-terminal domains of α–actinin dimers could interact with the antiparallel Z-repeats, similar to the way in which the two actin-binding domains interact with antiparallel actin filaments. Only at the Z-disk edge will α–actinin molecules be bound additionally by the interaction of their rod domains with the Zq motif, which is present only once per titin molecule. The maximal number of Z-filaments predicted by this model would be the highest integer of half the number of Z-repeats which is well in agreement with the data in Table 1.

The most plausible path of titin places the titin N-terminus close to the pointed end of the thin filament, and the previously defined regulatory phosphorylation sites (Gautel et al. 1996a) could therefore be directly involved in the control of thin filament capping. This concept is supported by the observation that the position of the regulatory SH3 domain of nebulin colocalises with the titin N-terminus (Fig. 6). This supports the previously proposed interaction between the titin N-terminal region and the nebulin SH3 domain (Gautel et al. 1996a; Politou et al. 1997), while the nebulin actin-binding repeats are proposed to span into the Z-disk (Labeit and Kolmerer, 1995b). Titin interactions of Z1Z2 with telethonin (Valle et al. 1997), a novel muscle protein, are conformation-dependent (Mues et al. 1998) and suggest complex regulatory events during the assembly of the Z-disk.

3 The I-Band Region of Titin

3.1 Differential Titin Expression in the Cardiac Conduction System

An important physiological function of titin is to act as the elastic link between thick and thin filaments (Horowits et al. 1986). Titin is therefore uniquely positioned to sense mechanical strain and to relay this information to other regions of the sarcomere, possibly over distances of hundreds of nanometres. The molecular cloning and sequencing of titin/connectin (Labeit and Kolmerer, 1995a) revealed the complete primary structure of the giant molecule and demonstrated a surprising complexity in the I-band region. Notably, elements of very different predicted secondary structure were found, spliced together in apparently numerous permutations (Maruyama et al. 1994; Labeit and Kolmerer, 1995a). Two major variants, specifying cardiac or skeletal-specific isoforms, were denoted as the N2-isoforms of titin (not to be confused with the morphological N2-line; Ben-

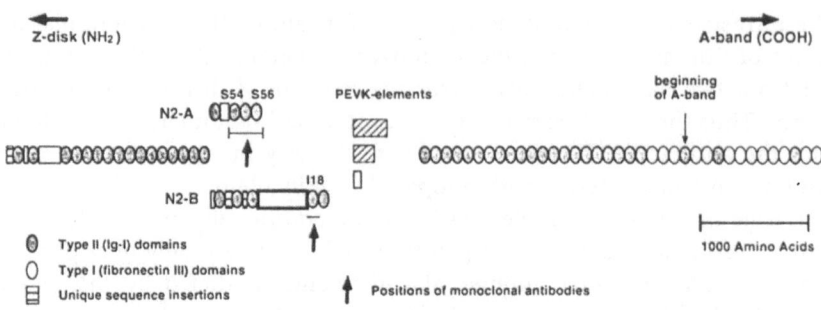

Fig. 7. Domain patterns of the two known splice variants in the cardiac titin I-band (modified from Labeit and Kolmerer 1995). The positions of the monoclonal antibodies are marked by arrows (Modified from Gautel et al. 1996b). The cardiac-specific N2-B isoform contains a large non-modular sequence insertion N-terminal to the I18 epitope monitored in Fig. 9

nett et al. 1997). These regions separate stretches of Ig-domains arranged in tandem from a region rich in proline, glutamate, valine and lysine and hence termed PEVK (Labeit and Kolmerer, 1995a). The PEVK region provides the major spring of the elastic region of titin (Gautel and Goulding, 1996; Linke et al. 1996).

Monoclonal antibodies against the titin N2 isoforms (positions on the molecule marked in Fig. 7) showed in indirect immunofluorescence studies that both isoforms are co-expressed in all cardiac muscle compartments (Gautel et al. 1996b). This co-expression was found to remain constant throughout embryogenesis, and the cardiac isoform of titin was never found expressed in skeletal muscle. This is in contrast to a number of other sarcomeric proteins, like β-myosin, which are also expressed transiently in developing fast skeletal muscle. This argues that both N2 isoforms are constitutive components of cardiac titin, and that the N2-B isoform fulfils functions specific entirely to cardiac muscle. A very specific function is suggested by the observation that fibres of the cardiac conducting system (atrioventricular node and His' bundle) appear to express the two N2 isoforms at different levels with the apparent amount of N2-A exceeding that of N2-B (Gautel et al. 1996b; Fig. 8). Since the conducting system is not a force-generating element of cardiac muscle, the requirements of the supramolecular architecture of its titin fibrils may be different. The strictly linked co-expression of the cardiac N2 isoforms of titin irrespective of developmental stage suggests that these may be involved in functions not directly related to the actual elastic behaviour of the cardiac I-band. The laboratory of K. Maruyama has proposed a model of the cardiac I-band where a junction-line joins titin filaments diverging towards the Z-disk (Funatsu et al. 1993;

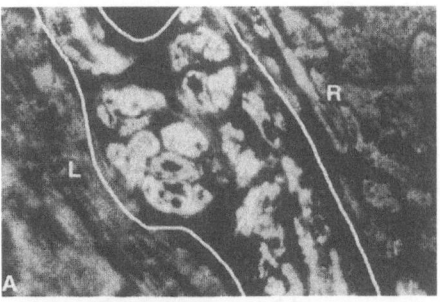

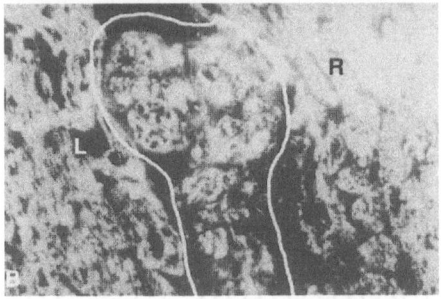

Fig. 8. Inidirect immunofluorescence showing the expression of N2-A and N2-B isoforms in cardiac muscle (modified after Gautel et al. 1996b). Monoclonal antibodies against the N2-A isoform (A) preferentially stain the conducting system (atrioventricular node and His-Bundle, encircled), whereas the N2-B isoform (B) is expressed similarly in both the ventricular tissues (left: L; right: R) as well as the conducting system. Titin isoforms are therefore expressed differentially in the signalling cells of the heart

Maruyama 1994). Sequence analysis and secondary structure predictions suggest the N2 regions as possible candidates for intermolecular titin-titin interactions, as the N2-A isoform contains intermediate-filament like sequences in avian and human titin (Maruyama et al. 1994). However, analysis of the extensibility of the cardiac isoform of titin revealed that this insertion must contain an extensible region of high stiffness, as its extension becomes significant only after the other structural elements in the I-band (Ig-domains and a sequence denoted as PEVK) have been extended (Gautel et al. 1996b). Since this happens only beyond the physiological working range (Gautel et al. 1996b; Fig. 9), it is conceivable that the N2-B insertion senses mechanical strain and communicates this information. Cardiac titin could thus be involved in the adaptation of active force production (Frank-Starling Mechanism). Such a role could explain the differential expression between ventricular myocardium and the conducting system.

A

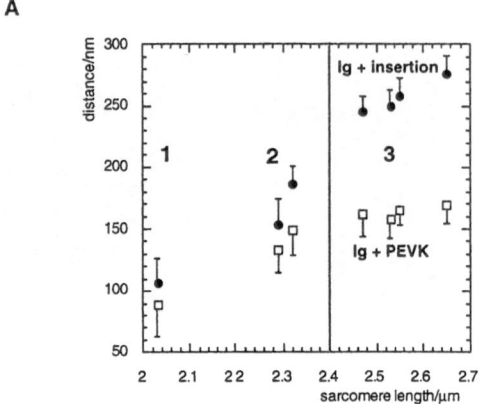

B

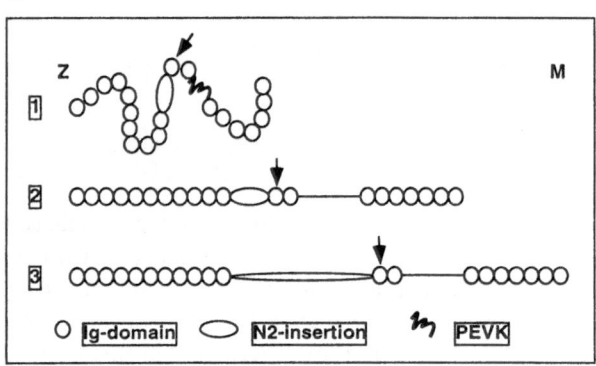

Fig. 9. **A.** Extension of the cardiac titin isoform in intact sarcomeres. The relation of the position of the I18-epitope to sarcomere length in cardiac papillary muscle is plotted. Error bars: mean standard deviation. Closed circles: distance between the I18-epitope and the Z-disk edge. Open squares: distance between the I18 epitope and the A-band. Note the greater extensibility of the I18-Z-segment especially at higher sarcomere lengths, starting at the upper end of the physiological working range (vertical line). (Modified after Gautel et al. 1996b). **B.** Three phases of cardiac titin extension can be interpreted from this extension diagram of the molecule: 1. The collapsed Ig-domain chain and the PEVK segment are straightened at short sarcomere lengths. 2. The Ig-domain chain is straightened out and the PEVK element is extended at the upper end of the physiological working range. 3. Beginnning at the upper end of the physiological working range, the curve is consistent with the unique insertion in N2-B being extended (although domain unfolding cannot be ruled out). This conformational change could integrate and signal information on the mechanical strain level to other regions of the sarcomere

4 The A-Band Region of Titin

4.1 Titin at the Thick Filament End: the Myosin Termination Motif

The thick filaments of striated muscle are assembled by the polymerization of myosin molecules into highly ordered filaments of precisely tailored length. It is now generally accepted that titin is involved in the control of the myosin-filament length; yet, the mechanisms of this control are still largely unknown.

Native titin can aggregate myosin rods (Maruyama et al. 1985) and is tightly bound to the thick filament under physiological conditions. Due to this association with the thick filament, the function of titin as a molecular ruler for the assembly of the myosin filament was proposed (Whiting et al. 1989; reviewed by Trinick, 1994). The binding of titin to myosin occurs mainly in the C-terminal region of myosin (light meromyosin, LMM; Labeit et al. 1992; Soteriou et al. 1993) and could be mapped to a short peptide in the LMM-region (Houmeida et al. 1995).

The cloning of human A-band titin (Labeit et al. 1990; Labeit et al. 1992; Labeit and Kolmerer, 1995a) showed that titin is mostly constructed in a modular fashion by repetitive patterns of immunoglobulin-like domains (Ig) and fibronectin-3 like modules (fn3) (Labeit et al. 1990). The fn3-domains of titin/connectin are found predominantly in the proposed A-band region of the molecule (Labeit and Kolmerer, 1995a). In this region, the form these patterns take is generally that of groups of two or three fn3-domains preceded by an Ig-domain. The arrangement of these groups into super-repeats of seven or 11 domains is probably correlated to the different structure of the thick filament in the C-and D zones and the sorting of the associated myosin-binding proteins (reviewed in Gautel, 1996c). This pattern is broken only in two places; the first is at the end of the A-band region near the C-terminus where the pattern is more diverse and is followed by the MLCK-like catalytic kinase domain. It was suggested that this transition correlates to the change of parallel to antiparallel myosin packing (Labeit et al. 1992). The second is towards the Z-disk near the end of the A-band, where the first two groups are a followed by a unique stretch of six fn3-domains. The striking feature of these groups is that there are 49 of them, correlating closely to the number of levels of myosin heads seen in one half of the thick filament (Bennett and Gautel, 1996). This suggests that the fn3-domains are specific for the interaction of titin with myosin. This hypothesis is supported by the observations on binding of recombinant multi-module titin fragments to myosin. Also, the binding affinity to myosin increases with the number of fn3 domains (Labeit et al. 1992). In contrast, Ig domains are found over the entire length of the titin molecule (Labeit and Kolmerer, 1995a) and may therefore be responsible for the interactions with ligands other than myosin. Labeit and Kolmerer (1995a) have suggested the first of the fn3 domain groups to be outside the A-band into the I-band, separated from the first putative A-band domain, A1, by the group of six fn3 domains. A more precise understanding of how the myosin-binding domains and the termination of the thick filament correlate was recently achieved (Bennett and Gautel, 1996).

Previous ultrastructural studies of striated muscle have demonstrated a characteristic structure at the end of the thick filament. Here, the last two levels of crossbridges are separated from the contiguous array of myosin-heads in the A-band by a gap of 15–20 nm (Craig and Offer, 1976; Craig 1977; Sjöström and Squire, 1977). The axial position of titin antibodies in the putative "junction" region between the thick filament and the I-band could be investigated by immunoelectron microscopy (Bennett and Gautel, 1996). These results showed that several epitopes are localised very close to the A/I junction and allow a correlation of the packing of the myosin filament to the domain pattern of titin (Fig. 10). Immunoelectron microscopy now revealed

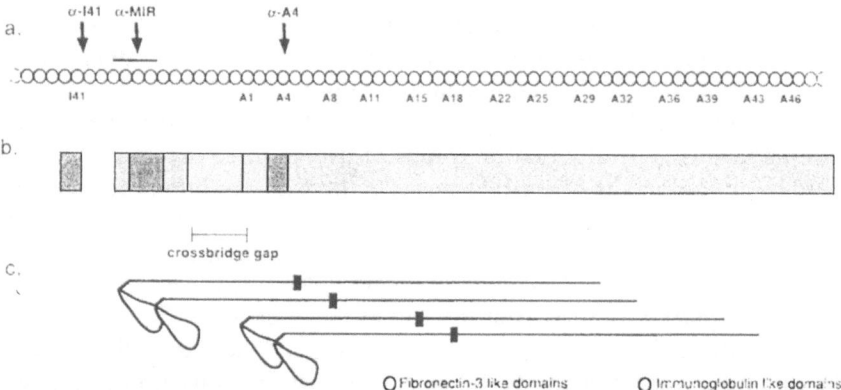

Fig. 10. The correlation of titin domain patterns in the A-band and the arrangement of myosin molecules at the thick-filament end (modified after Bennett and Gautel, 1996). The ultrastructural positions of the sequence-specific antibodies (arrows in a) correlate closely to a unique structure at the thick filament end, the crossbridge gap (shaded in light grey in b), and to a unique stretch of six fibronectin-3 (fn3) domains on titin. The I41-epitope, labelling the last of the tandem-ig domains, is just outside the A-band. The ultrastructural features of the thick filament and the position of antibody label (shaded in dark grey) are shown in b. The altered packing of myosin molecules (c), leading to the crossbridge gap and to thick-filament termination, is therefore correlated to a unique arrangement of the myosin-binding domain type of titin. This mechanism is likely to be involved in the control of thick-filament length by titin

that one of these epitopes, MIR (Gautel et al. 1993b), is within the A-band in the region of the last two levels of myosin heads, just outside the 'cross-bridge-gap' described by Craig and Offer (1976). In agreement with the sequence assignment of Labeit and Kolmerer (1995a), α-A4 labels a position in the A-band just towards the M-line from the crossbridge-gap. The epitope for MIR epitope is preceded by only three fn3 domains. Therefore, all fn3 domains are closely associated to the thick filament, supporting the notion that they may indeed be involved in the binding of myosin to titin. Since the binding site on LMM identified by Houmeida et al. (1995) is significantly away from the myosin heads, other interactions between titin and myosin must come into play at the thick filament end (Bennett and Gautel, 1996). There are two other features of the sequence which further reinforce the idea that the fn3 domains are associated with myosin: these groups are all of two or three domains except for the stretch of six near the MIR epitope, suggesting that this might be related to the missing crossbridge region. The unusual and single block of six fn3 domains exactly correlates to the only gap in the crossbridges of myosin at the thick-filament end. It was therefore proposed that this motif is involved in the perturbed packing of myosin molecules towards the end of the thick filament in the sense of a termination motif (Bennett and Gautel 1996). This could be an important mechanism in the proposed role of titin as a molecular ruler for determining the lengths of thick filaments and keeping the last crossbridge in axial register.

4.2 Myosin-Binding Protein C as a Sensor for cAMP- and Ca^{2+} Signalling

Myosin filaments in striated muscle are assembled by the polymerization of myosin in association with a family of myosin-binding proteins (MyBP-C and MyBP-H) and titin. These associated proteins (Offer et al. 1973) all share the molecular building plan of the immunoglobulin superfamily (Einheber and Fischman, 1990; Fürst et al. 1992; Vaughan et al. 1993; Kasahara et al. 1994; Gautel et al. 1995; Yasuda et al. 1995; Fig. 11A). While titin is thought to fulfil the role of a molecular ruler, the MyBP-C might play a role as a spatially defined regulatory protein (Craig and Offer, 1976b; Jeacocke and England, 1980; Hartzell and Titus, 1992; Hartzell and Glass, 1984; Hartzell 1985; Garvey et al. 1988; Schlender and Bean, 1991; Gautel et al. 1995; Weisberg and Winegrad, 1996). Myosin, MyBP-C and titin form a stable ternary complex where MyBP-C is arranged regularly in 9 of the 11 thick-filament stripes (Offer et al. 1973; Dennis et al. 1984; Bennett et al. 1986; Okagaki et al. 1993).

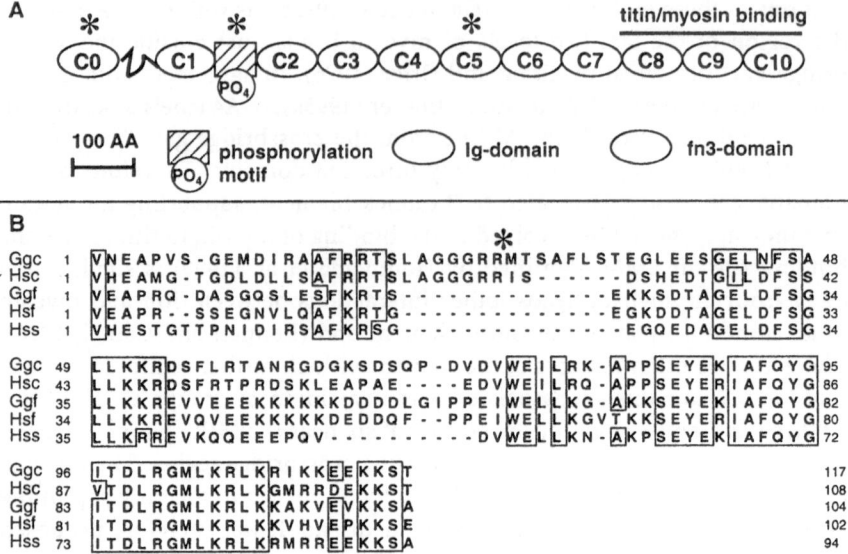

Fig. 11. A The domain structure of the cardiac myosin-binding protein C (c-MyBP-C) with the cardiac-specific regions highlighted by asterisks. the polar molecule bears a signal transduction module at the N-terminus. For further details, see review by Bennett et al. in this issue. **B** multiple sequence alignment of the human and chicken slow and fast skeletal and cardiac isoforms of MyBP-C. The phosphorylation sites for cAMP-dependent protein kinase are shaded in grey; the cardiac-specific insertion contains an additional site (asterisk). Note the striking conservation between isoforms and species which argues for a crucial control function of this module. Ggc: chicken cardiac (); Hsc: human cardiac (); Ggf: chicken fast (); Hsf: human fast (); Hss: human slow ()

Mutations in the gene of the cardiac isoform of MyBP-C (cMyBP-C) are the cause of the chromosome-11 associated form of familial hypertrophic cardiomyopathy (FHC). Most mutations identified so far lead to truncated proteins lacking the thick-filament binding region of the cMyBP-C (Bonne et al. 1996; Watkins et al. 1996; Rottbauer et al. 1997; Carrier et al. 1997). Protein analysis of cardiac biopsies from a patient with a cMyBP-C associated FHC indicated that the truncated protein is not present in detectable amounts (Rottbauer et al. 1997). A possible explanation for the cellular events in cMyBP-C associated FHC could therefore be a haplophenotype caused by an altered protein stoichiometry. This would suggest that the missing allele cannot be complemented by any of the other known MyBP-C isoforms.

Skeletal muscle contains at least two isoforms of MyBP-C, expressed predominantly either in slow or fast muscle fibres (Yamamoto and Moos, 1983; Dennis et al. 1984; Bennett et al. 1986). However, these isoforms can be coexpressed in single myofibrils in varying stoichiometries leading to diverse arrangements of the characteristic sarcomeric stripes (Bennett et al. 1986). Skeletal muscle therefore has a great potential to adapt in a flexible way to alterations in MyBP-C isoform stoichiometry by the modification of coexpression ratios.

Immunological and in-situ hybridisation studies in mouse and man demonstrate a successive expression of the skeletal-muscle specific isoforms in an ordered way (Gautel et al. 1998). This sequential appearance resembles the conserved programme of myogenesis also observed, although on a shorter time-scale, in cultured avian skeletal muscle cells (Sutherland et al. 1993; Lin et al. 1994). However, it was found that the cardiac isoform of MyBP-C is not expressed in skeletal muscle throughout development (Gautel et al. 1998). This observation is in contrast to the situation in avians where a splice-variant of cardiac MyBP-C is transiently expressed in embryonic skeletal muscle (Obinata et al. 1984; Yasuda et al. 1995). The embryonic splice variant of cardiac MyBP-C expressed in avian skeletal muscles lacks the cardiac-specific phosphorylation site (Yasuda et al. 1995). It was shown that the deletion of this phosphorylation loop in the human cardiac isoform reduces sensitivity to cAMP-dependent protein kinase phosphorylation (Gautel et al. 1995). Although avians appear to lack a specific embryonic isoform of skeletal MyBP-C, their embryonic skeletal splice variant of the cardiac MyBP-C gene is therefore similar in its regulatory properties to the genuine skeletal isoforms. This cardiac-specific site is also the substrate site for an associated Ca^{2+}/Calmodulin activated kinase (Fig. 11B). The cardiac MyBP-C is therefore a titin- and myosin-associated sensor for cAMP- and Ca^{2+}/Calmodulin- activated protein kinase signalling.

In mammalian cardiac muscle, the cardiac MyBP-C transcript and protein are detected early on and the expression coincides with that of sarcomeric myosin heavy chain in the earliest detectable stages (Gautel et al. 1998). The expression of cardiac MyBP-C is therefore tightly restricted to cardiac muscle which does not express the slow or fast skeletal isoforms of MyBP-C under normal circumstances. Unlike in skeletal muscle, which expresses up to three isoforms of MyBP-C in addition to MyBP-H, a trans-complementation of isoforms therefore seems to be impossible in cardiac muscle (Gautel et al. 1998). This could plausibly explain why a dominant-negative phenotype is observed in cardiac muscle upon mutations in the cardiac MyBP-C gene, while no skeletal myopathies have been associated with a member of the MyBP-C family.

5 The M-Band Region of Titin

The sarcomeric M-band region of titin provides a similarly complex blue-print for the integration of titin into one of the antipodal anchoring planes as does the N-terminal region in the Z-disk. Particularly, this region contains a serine/threonine kinase domain, substrate sites for developmentally regulated kinases as well as regulated binding sites for other M-band proteins like myomesin. We can thus assume that the M-band plays crucial roles in signal transduction during myofibrillogenesis as well as in the differentiated sarcomere.

5.1 Titin Kinase: Worms do it Like Humans?

Recently, a growing family of giant modular proteins specific for the striated muscles of vertebrates and invertebrates have been identified and partially sequenced. Sequencing of the vertebrate titins has demonstrated the presence of a catalytic kinase domain near their carboxyterminus which is homologous to the myosin light chain kinase (MLCK) family (Labeit et al. 1992; Sebestyén et al. 1996). A catalytic kinase domain is also found in the nematode mini-titin twitchin (Benian et al. 1989) as well as in the so-called 'mini-titins' of insects and molluscs. This kinase domain is generally flanked by Ig- or fn3 domains (Benian et al. 1989; Labeit et al. 1992, Heierhorst et al. 1994; Ayme-Southgate et al. 1995). Sequence comparison and evolutionary analysis of titin, twitchin and MLCKs have suggested, however, that the titin kinase may have acquired functions distinct from MLCK during the evolution of cross-striated muscle (Higgins et al. 1994) despite a strong overall homology in the molecular architecture (Fig. 12). The molecular sites involved in the regulation of enzymatic activity might well be arranged simi-

--▶

Fig. 12. Multiple-sequence alignment of titin kinase (Hs-Titin; EMBL X 90568) with the human MLCK from hypocampus (Hc-hypoc; X85337), rabbit skeletal muscle MLCK (Oc-skel; J05194), human death-associated protein kinase (Hs-Dap; X76104) and C. elegans twitchin (Cel-Twitchin; X15423). Identical residues are boxed. Note that titin kinase displays unique exchanges of catalytic residues (arrowheads). The autoinhibitory domain (beginning at the horizontal arrow) contains binding sites for calmodulin (CaM; marked for the skeletal MLCK) or S100A1 (marked for twitchin kinase). The CaM-binding site of titin kinase is positioned similarly as the S100A1 binding site of twitchin; structural investigation of titin kinase will reveal if structural similarities also exist

ATP

```
Oc-skel       34  I VELRTGNVSSEFSMNSKEALGGGKFGAVCTCTEKSTGLK   73
Hs-hypoc     529  TVTINTEQKVSDF-YDIEERLGSGKFGDCFRLVEKKTRKV  567
Hs-Dap         1  -MTVFRQENVDDY-YDTGEELGSGQFAVVKKCREKSTGLQ   38
Cel-Twitchin 5145 PVEIKHDDHVLDH-YDIHEELGTGAFGVVHRVTERATGNN  5183
Hs-Titin    24726 ASHSSTKELYEKYM--IAEDLGRGEFGIVHRCVETSSKKT 24763

Oc-skel       74  LAAKVIKKQTPK------DKEMVMLEIEVMNQLNHRNLIQ  107
Hs-hypoc     568  WAGKFFKAYSAK------EKENIRQEISIMNCLHHPKLVQ  561
Hs-Dap        39  YPAKFIKKRRTKSSRRGVSREDIEREVSILKEIQHPNVIT   78
Cel-Twitchin 5184 FAAKFVMT-----PHESDKETVRKEIQTMSVLRHPTLVN  5217
Hs-Titin    24764 YMAKFVKVKGT------DQVLVKKEISILNIARHRNILH  24796

Oc-skel      108  LYAAIETPHEIVLFMEYIEGGELFERIVDEDYHLTEVDTM  147
Hs-hypoc     562  CVDAFEEKANIVMVLEIVSGGELFERIIDEDFELTERECT  601
Hs-Dap        79  LHEVYENKTDVILILELVAGGELFD-FLAEKESLTEEEAT  117
Cel-Twitchin 5218 LHDAFEDDNEMVMIYEFMSGGELFEKVADEHNKMSEDEAV  5257
Hs-Titin    24797 LHESFESMEELVMIFEFISGLDIFERINTSAFELNEREIV 24836

Oc-skel      148  VFVRQICDGILFMHKMRVLHLDLKPENILCV--NTTGHLV  185
Hs-hypoc     602  KYMRQISEGVEYIHKQGIVHLDLKPENIMCV--NKTGTRI  639
Hs-Dap       118  EFLKQILNGVYYLHSLQIAHFDLKPENIMLLDRNVPKPRI  157
Cel-Twitchin 5258 EYMRQVCKGLCHMHENNYVHLDLKPENIMFT--TKRSNEL  5295
Hs-Titin    24837 SYVHQVCEALQFLHSHNIGHFDIRPENIIY--QTRRSSTI 24874

Oc-skel      186  KIIDFGLARRYNPNEKLKVNFGTPEFLSPEVVNYDQISDK  225
Hs-hypoc     634  KLIDFGLPRRLENAGSLKVLFGTPEFVAPEVINYEPIRYA  673
Hs-Dap       158  KIIDFG-------NEFKNIFGTPEFVAPEVVNYEPLGLE  189
Cel-Twitchin 5296 KLIDFGLTAHLDPKQSVKVTTGTAEFAAPEVAEGKPVGYY  5335
Hs-Titin    24875 KIIEFGQARQLKPGDNFRLLFTAPEYYAPEVHQHDVVSTA 24914

Oc-skel      226  TDMWSLGVITYMLLSGLSPFLGDDDTETLNNVLSGNWYFD  265
Hs-hypoc     674  TDMWSIRVICYILVSGPFPFMGDNDNETLANVTSATWDFD  713
Hs-Dap       190  ADMWSIIGVITYILLSGASPFLGDTKQETLANVSAVNYEFE  229
Cel-Twitchin 5336 TDMWSVGVLSYILLSGLSPFGGENDDETLRNVKSCDWNMD  5375
Hs-Titin    24915 TDMWSLGTLVYVLLSGINPFLAETNQQIIENIMNAEYTFD 24954

Oc-skel      266  EETFEAVSDEAKDFVSNLLIVKEQGARMSAAQCLAHPWLN  -    304
Hs-hypoc     714  DEAFDEISDDAKDFISNLLKKDMKNRLDLAQCLQHPWLM  -    752
Hs-Dap       230  DEYFSNTSALAKDFIRRLLVKDPKKRMTIQDSLQHPWLK  -    268
Cel-Twitchin 5376 DSAFSGISEDGKDFIRKLLLADPNTRMTIHQALEHPWLTP  5415
Hs-Titin    24955 EEAFKEISIEAMDFVDRLLVKERKSRMTASEALQHPWLKQ 24994
```

S100A1

CaM

```
Oc-skel      305  -NLAEKAKRCNRRLKSQILLKKYLMKRRWKKNFIAVSAAN  343
Hs-hypoc     753  ----KDTKNMEAKKLSKDRMKKYMARRKWQKTGNAVRAIG  788
Hs-Dap       269  PKDTQQALSRKASAVNMEKFKKFAARKKWKQSVRLISLCQ  308
Cel-Twitchin 5416 GNAPGRDSQIPSSRYTKIRDSIKTKYDAWPEPLPPL---G  5452
Hs-Titin    24995 KIERVSTKVIRTLKHRRYYHTLIKKDLNMVVSA------A 25028

Oc-skel      344  RFKKISSSGALMALGV-----------  359
Hs-hypoc     789  RLSSMAMISGLSGRKSSTGSPTSP-----  812
Hs-Dap       309  RLSRSFLSRS----------------  318
Cel-Twitchin 5453 RISNYSSLRKHRPQEYSIRDAFWDRSEAQ  5481
Hs-Titin    25029 RISCGGAIRSQKGVSVAKVKVASIEIG--  25064
```

larly despite a divergence in function. Also, titin kinase was shown to have a
fixed position at the M-band edge (Obermann et al. 1996) – in contrast to the
A-band localisation of twitchin kinase (Vibert et al. 1993). Such differential
localisations make identical substrate specificities unlikely (Obermann et al.
1996). On the other hand, the insect twitchin-like protein projectin is very
differently localised in synchronous and asynchronous muscle in Drosophila, yet bears the same catalytic kinase domain (Ayme-Southgate et al.
1995). Although the subsarcomeric localisation of giant muscle protein
kinases seems to be important for their function, it may not give the ultimate clue for understanding their physiological roles. The function of titin
kinase may be relevant during myofibrillogenesis and not in the assembled
structure, and the position of the kinase domain could therefore be quite
different in assembling myofibrils.

For the twitchin kinases from *C. elegans* and *A. california*, the constitutively active forms of these enzymes phosphorylate myosin-light chain peptides (Lei et al. 1994; Heierhorst et al. 1995; Heierhorst et al. 1996a). More
strikingly, the mollusc enzyme from Aplysia could be shown to be activated
by mammalian $S100A1_2$ (Heierhorst et al. 1996b). S100 proteins are EF-hand
containing, calcium binding proteins which form homo- or heterodimers
(Schäfer et al. 1996). In muscle, they had been implied in the regulation of F-
actin capping proteins (Schafer et al. 1996), but a role in protein kinase signalling was unexpected. However, whereas Aplysia twitchin kinase is activated over 1000-fold by mammalian $S100A1_2$, the *Caenorhabditis* enzyme
shows only weak stimulation (Heierhorst et al. 1996a). For mammalian titin
kinase, no stimulation of MLCK activity by $S100A1_2$ could be observed (M.
Gautel, unpublished observation). It is likely that the substrate of titin
kinase will be distinctly different from myosin light chains and possibly a
diffusible one (Obermann et al. 1996). The binding site for $S100A1_2$ in
twitchin kinase is at the C-terminus of the MLCK-like catalytic domain. An
amphipathic α-helix presumably forms several charge interactions and hydrophobic contacts crucial for the recognition of $S100A1_2$ (Kobe et al. 1996).
This region, however, is only weakly conserved between titin and twitchin
kinases (Fig. 12). The same amphipathic helix can also bind to calmodulin,
however without activating the enzyme (Gautel et al. 1995; Heierhorst et al.
1996a). In analogy to the genuine MLCK (Kennelly et al. 1987), the C-
terminal domain contains a 'pseudosubstrate sequence' which interacts with
the substrate binding site to block enzyme activity. The structural basis for
this intrasteric regulation was partly elucidated by the crystal structures of
the twitchin kinase domains from *C. elegans* (Hu et al. 1994; no coordinates
available) and *A. california* (Kobe et al. 1996). Extensive contacts between
the inhibitory domain and the substrate binding cleft block both the access

of ATP and the peptide substrate (Kobe et al. 1996). It is plausible that titin kinase needs additional factors for its full activation, like the phosphorylation of specific residues involved in autoinhibition.

The coordination of ATP in the protein kinases involves an extensive network of salt bridges and hydrogen bonds that tightly coordinate the γ-phosphate moiety and the acceptor serine hydroxyl group (Lowe et al. 1997). In the catalytic loop between beta-strands 6 and 7, a conserved lysine residue (Lys168 in cAMP-dependent protein kinase, cAPK) coordinates the γ-phosphate in relation to the serine acceptor hydroxyl. Of the MLCK-like family and protein serine/threonine kinases in general, titin kinase is the only enzyme where this residue is replaced by arginine (Fig. 12). Since the pK_a of this side chain is much higher, the electrostatic attraction to the γ-phosphate is expected to be significantly increased. The catalysis rate of titin kinase could be different from that of other kinases. In addition to this remarkable difference, the aspartate preceding the activation loop in the tightly conserved DFG motif (Johnson et al. 1996; Asp184 in cAPK) is converted to glutamate in titin kinase. Titin kinase is the only protein kinase which displays this exchange. Since this residue coordinates the Mg^{2+} ion which again locks the β- and γ-phosphates into place (Bossemeyer et al. 1993; Lowe et al. 1997), the increase in side chain length by one methylen group and the concomitant decrease in pK_a of the carboxyl group is expected to influence catalysis rate and possibly substrate specificity. The two bulky side chains are also expected to influence the overall topology of the catalytic site in titin kinase.

Whereas significant structural insight into intrasteric regulation of the giant muscle protein kinases has come from the studies of the invertebrate homologues, our understanding of the functions of titin kinase will be improved significantly only when the genuine substrate(s) and regulators will be found, and when the structural implications of the specific amino acid exchanges mentioned above are elucidated.

5.2 Phosphorylation Motifs in the M-Band

Major reorganisation of the cytoskeleton occurs during myofibrillogenesis in differentiating muscle cells, as well as during myofibril breakdown induced, for example, by regenerating stimuli, acidosis, tumour necrosis factor or glucocorticoid stimulation.

Phosphorylation motifs possibly involved in relaying such signals to the sarcomeric cytoskeleton are typically extended insertions between the immunoglobulin domains, rich in proline, serine and basic residues (Figs. 5 and 13). In the sarcomeric Z-disk, two phosphorylation repeats can sense

Mis4
                                    *
TSPPRVKSPEPRVKSPEAVKSPKRVKSPEPSHPKAVSPTETKPTPIEKVQHLPVSA

      Titin KSP   PRVKSPEPR
                PRVKSPEAV
                EAVKSPKRV
                KRVKSPEPS

      NFH KSP   VEVKSPEKA
                KETKSPVKE

cdk5 consensus KHHKSPKHR

Fig. 13. Phosphorylation sites of M band titin in Mis4. Note the 4 repeats of the sequence VKSP (in bold). The phosphorylated serine is highlighted by an asterisk. This sequence is compared to the similar neurofilament H (NFH) substrate sequences (Shetty et al. 1993) and to the cdk5 consensus phosphorylation motif determined by screening a combinatorial library (Songyang et al. 1996). Despite a high homology of the titin VKSP repeats to the neurofilament cdk5 substrate sites, the +3 position (highlighted) does not fit the cdk5 consensus. Mis4 also contains two SPXR sites, which are consensus sites for cdk2 and ERK (Songyang et al. 1994)

kinase signals (see above). In the M-line, a similar inter-domain linker contains 4 repeats of the sequence VKSP (Gautel et al. 1993b; Fig. 13). This motif does not resemble the SPXR motifs in the Z-disk, suggesting that the kinase activities found in differentiating myoblasts which can phosphorylate this region are different from those acting on the Z-disk. KSP-motifs are the major phosphorylation sites for the brain-specific cyclin-dependent kinase cdk5 in neurofilaments (Lee et al. 1988; Shetty et al. 1993; Songyang et al. 1996). The kinase activities from differentiating myocytes which can phosphorylate titin KSP-motifs *in-vitro* also share functional characteristics with cyclin-dependent kinases (Gautel et al. 1993b). The titin KSP-motifs lack, however, the conserved basic residue 3 residues C-terminal to the serine site which is important for substrate recognition by cdk5 (Songyang et al. 1996) (Fig. 13). The identity of the muscle kinase phosphorylating the titin KSPs, and the functional significance of the phosphorylation *in-vivo*, are unknown to date.

The functional significance of phosphorylation of inter-domain linkers is highlighted in the case of myomesin, a protein of the M-line which binds tightly to the C-terminal region of titin, and crosslinks titin with myosin (Obermann et al. 1995; Obermann et al. 1996; Obermann et al. 1997). The interaction between titin and myomesin is mediated by a single titin immunoglobulin domain and three of the fibronectin domains of myomesin. A short linker peptide between two of the domains contains the phosphorylation motif KARLKSRPSAP. The highlighted serine is a target site for cAPK–like activities in muscle. Phosphorylation of myomesin by these activities was shown to block the interaction with the titin immunoglobulin domain M4 (Obermann et al. 1997). This is the first example where the interaction of sarcomeric proteins has been demonstrated to be regulated by interdomain linker phosphorylation. It appears likely that this will be a general mechanism acting both on titin and its associated proteins.

6 Kinases Acting on the Cytoskeleton in Muscle Differentiation

Titin is strongly phosphorylated *in-vivo* (Sommerville and Wang, 1987), similar to nebulin. Biochemical experiments show that in vitro, muscle extracts can phosphorylate titin at both the N- and C-terminal regions, and that the titin fragments containing the multiple phosphorylation repeats are good substrates for purified cdc2- and ERK2 kinases. However, the endogenous kinases involved in titin phosphorylation have not yet been identified directly. Since cdc2 kinase is involved in cell-cycle control, it is unlikely to mediate a differentiation signal after mitotic arrest. However, differentiation in neuronal and myogenic cells is promoted by the brain proline specific kinase cdk5, a member of the cyclin regulated kinases (Lew et al. 1992; Lew et al. 1994; Tsai et al. 1994; Lazaro et al. 1997).

Myogenic differentiation is mediated also by signal transduction pathways that result in the activation of MAP/ERK kinases (Thorburn and Thorburn, 1994; Thorburn et al. 1997). In cardiomyocytes, possible pathways that lead to MAP/ERK activation appears to be via tyrosine-kinase signalling by the insulin-like growth factor-1 (IGF-1) receptor (Ito et al. 1993; Donath et al. 1994; Harder et al. 1996). Other upstream pathways involve G-protein coupled receptor signalling, which is proposed to be involved in adrenergic stimulation of cardiac hypertrophy via the Ras-pathway (LaMorte et al. 1994; Thorburn and Thorburn, 1994; reviewed by Schaub et al. 1997). It is conceivable that the titin multiphosphorylation repeats actually sense phosphorylation signals from several convergent pathways linked to G-protein coupled tyrosine-kinase receptors by coupling of Ras with Raf-1 kinase (Vojtek et al. 1993; Ramocki et al. 1997) or Rac/Rho with Rho-associated kinase (Vojtek and Cooper, 1995; Coso et al. 1995; Thorburn et al. 1997; Matsui et al. 1996; reveiwde by Hefti et al. 1997). Interestingly, Rho signalling can occur directly to the Z-disk by the Rho-activated protein kinase N (PKN; Watanabe et al. 1996) which can bind to alpha-actinin (Mukai et al. 1997), and possibly by myotonic dystrophy kinase, which shows homologies to Rho-activated kinases (Ishizaki et al. 1996).

Conversely, however, the activation of v-Src leads to a rapid disassembly of myofibrillar structures starting with the disassembly of the Z-disk and the I-Z-I complex (Castellani et al. 1995; Castellani et al. 1996), as well as transcriptional repression of muscle specific genes including myogenic factor genes (Falcone et al. 1991; Yoon and Boettiger, 1994). Activated v-Src is associated with constitutively activated MAP kinases (Cowley et al. 1994). In the actin cytoskeleton, major direct substrates are cortactin and tensin (Thomas et al. 1995), and activation of v-Src indeed leads to phosphotyrosine incorporation into the remnants of the I-Z-I complex (Castellani et al. 1995). Similar phenotypic effects have been reported for the phorbol-ester mediated activation of protein kinase C (PKC), which equally leads to a sequential disassembly of sarcomeric structures, starting at the Z-line (Croop et al. 1982; Dlugosz et al. 1983; Lin et al. 1987; Lin et al. 1989; Choi et al. 1991). These effects seem to be mediated by the α-PKC and the Ca^{2+}-independent ε- and δ-PKC isoforms in myocardium (Puceat et al. 1994). PKC activation induced by α-adrenergic stimulation was reported to result in hypertrophic remodelling of the sarcomere (Chien et al. 1991). However, activation of PKC by phorbol esters in a background of activated v-Src abolishes Src-induced I-Z-I breakdown, suggesting that the overstimulation of MAP-kinase activity downstream of v-Src and the substrates of PKC diverge in different pathways (Castellani et al. 1995). Parallel processing is a feature common of ERK/MAP kinases (Cano and Mahadevan, 1995). It is conceivable that simultaneous activation of several pathways involving also phosphorylation of PKC substrate sites inactivate specific elements in the MAP kinase pathway (Johnson and Vaillancourt, 1994).

What are the actual ERK/MAP kinases involved in controlling myogenic differentiation? Surprisingly, very little is known about this important question. Although the ubiquitous ERK2 was shown to phosphorylate titin in-vitro (Sebestyén et al. 1995; Gautel et al. 1996a), muscle-specific kinases with a role in myogenic differentiation are more likely to be the in-vivo kinases phosphorylating titin. A novel ERK, ERK6 (also known as SAPK3; Lechner et al. 1996; Goedert et al. 1997) was described as a kinase promoting the fusion of C2C12 cells (Lechner et al. 1996). A mutant enzyme with a dysfunctional activation loop exerts a dominant-negative effect on myoblast fusion. However, the upstream regulation and the cellular substrates of this kinase in muscle have not yet been reported. In-vitro assays with brain tau protein (Goedert et al. 1997) confirms that ERK6/SAPK3 phosphorylates serine-proline rich cytoskeletal substrates, but the specificity for sarcomeric substrates remains to be demonstrated. Also, it is not known whether ERK6/SAPK3 selectively promotes myoblast fusion or also stimulates myofibrillogenesis. Combinatorial approaches demonstrated that the promiscuity of the ERK/MAP kinases is rather high (Songyang et al. 1994; Songyang et al. 1996). The sequence comparisons in Figs. 5 and 13 show that the titin

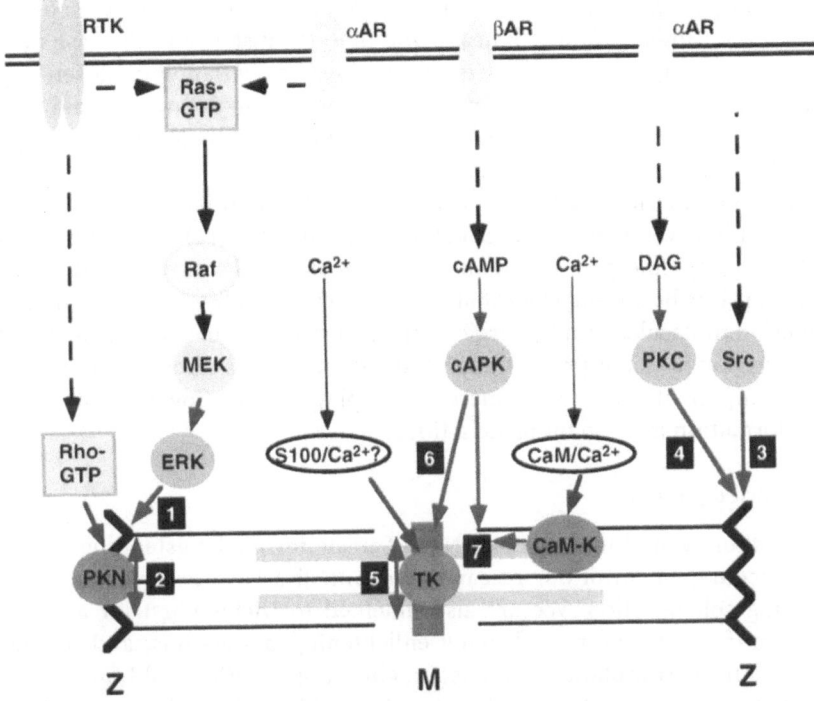

Fig. 14. Schematic representation of the emerging signalling pathways which converge on the sarcomere. In most cases, the exact links of the pathways or the cellular substrates of the kinases are unknown. Three protein serine/threonine kinases are known to localize to titin ligands of titin: the alpha-actinin associated PKN at the Z-disk, the Ca2+/Calmodulin activated protein kinase (CaM-K) associated with myosin-binding protein C (MyBP-C) in the A-band of heart muscle, and titin kinase (TK) at the M-band. Signalling via receptor tyrosine kinases (RTK), G-protein- or phospholipase-coupled α-adrenergic receptors (α-AR), or adenylcyclase-coupled β-adrenergic receptors (β-AR) results in activation of kinases acting on the sarcomeric cytoskeleton: ERK-phosphorylation of sites in Z-disk titin (1), activation of PKN by Rho-GTP, and phosphorylation of unidentified substrates in the Z-disk by activated Src kinase (3), Protein kinase N (PKN; 2) or Protein kinase C (PKC; 4). Cyclic-AMP dependent protein kinase phosphorylates MyBP-C (7) and myomesin in the M-line (6). The substrate and activation of titin kinase (TK, 5) remain to be identified. Red arrows show phosphorylation on substrates interacting with the sarcomere. Kinases are shaded dark, G-proteins light. Second messengers are Calcium^{2+} ions, cyclic adenosine monophosphate (cAMP) and diacylglycerol (DAG). Signalling pathways which are incomplete are marked with dashed arrows. See text for references

multiphosphorylation sites display key residues for consensus substrate sites for a number of these kinases, and that, for instance in the case of the M-line interdomain insertion Mis5, these sites are overlapping. Closely adjacent recognition sites for different signal transduction pathways could result in a very complex 'computation' of signals on the titin molecule, resulting in greatly varied cellular responses.

The sarcomeric cytoskeleton consisting of titin and its ligands is directly involved in several signal transduction pathways (Fig. 14). These links have only recently emerged and many components need still to be identified; however, it is becoming clear that receptor activation by growth factors, cell adhesion molecules or adrenergic agonists not only signals to the nucleus to alter gene expression programmes, but communicates directly with specialized components of the sarcomere. A key player in the flow and integration of information is the giant protein titin.

Acknowledgements

The authors would like to thank Matti Saraste for his longstanding support and encouragement of our research and Annalisa Pastore for a stimulating ongoing collaboration. We are also indebted to Dieter Fürst for a trustful and fruitful joint effort and many enlightening discussions, and to Klaus Weber for his stimulating and encouraging support. We would furthermore like to thank all our other collaborators for making this work successful.

The work of our group would have been impossible without funding support by the Deutsche Forschungsgemeinschaft and the European Union.

References

Arber S, Hunter J, Ross J, Hongo M, Sansig G, Borg J, Perriard J-C, Chien K and Caroni P (1997). MLP-deficient mice exhibit a disruption of cardiac cytoarchitectural organization, dilated cardiomyopathy, and heart failure. Cell 88:393–403

Ayme-Southgate A, Southgate R, Saide J, Benian GM and Pardue ML (1995). Both synchronous and asynchronous muscle isoforms of projectin (the Drosophila bent locus product) contain functional kinase domains. J Cell Biol 128:393–403

Bantle S, Keller S, Haussmann I, Auerbach D, Perriard E, Mühlebach S and Perriard J-C (1996). Tissue-specific isoforms of chicken myomesin are generated by alternative splicing. J Biol Chem 271:19042–19052

Benian GM, Kiff JE, Neckelmann N, Moerman DG and Waterston RH (1989). Sequence of an unusually large protein implicated in the regulation of myosin activity in C. elegans. Nature 342:45–50

Bennett PM and Gautel M (1996). Titin domain patterns correlate with the axial disposition of myosin at the end of the thick filament. J Mol Biol 259:896–903

Bennett PM, Hodkin TE and Hawkins C (1997). Evidence that the tandem Ig domains near the End of the muscle thick filament form an inelastic part of the I-band titin. J Struct Biol 120:93–104

Bennett P, Craig R, Starr R and Offer G (1986). The ultrastuctural localization of C-protein, X-protein and H-protein in rabbit muscle. J Muscle Res Cell Motil 7:550–567

Blanchard A, Ohanian V and Critchley D (1989). The structure and function of alpha-actinin. J Muscle Res Cell Motil 10:280–289

Bonne G, Carrier L, Bercovici J, Cruaud C, Richard P, Hainque B, Gautel M, Labeit S, James M, Weissenbach J, Vosberg H-P, Fiszman M, Komajda M and Schwartz K (1995). Cardiac myosin binding protein-C gene splice aceptor site mutation is associated with familial hypertrophic cardiomyopathy. Nature Genetics 11:438–440

Bossemeyer D, Engh RA, Kinzel V, Ponstingl H and Huber H (1993). Phosphotransferase and substrate binding mechanism of the cAMP-dependent protein kinase catalytic subunit from porcine heart as deduced from the 2.0 Å structure of the complex with Mn^{2+} adenylyl imidodiphosphate and inhibitor peptide PKI(5–24). EMBO J 12:849–859

Cano E and Mahadevan LC (1995). Parallel signal processing among mammalian MAPKs. TIBS 20:117–122

Carrier L, Bonne G, Bährend E, Yu B, Richard P, Niel F, Hainque B, Cruaud C, Gary F, Labeit S, Bouhour J-B, Dubourg O, Desnors M, Hagege AA, Trent RJ, Konajda M, Fiszman M and Schwartz K (1997). Organization sequence of human cardiac myosin binding protein C gene (MYBPC3) Identification of mutations predicted to produce truncated proteins in familial hypertrophic cardiomyopathy. Circulation Res 80:427–434

Castellani L, Reedy MC, Gauzzi MC, Provenzano C, Alema S and Falcone G (1995). Maintenance of the differentiated state in skeletal muscle: activation of v-Src disrupts sarcomeres in quail myotubes. J Cell Biol 130:871–885

Castellani L, Reedy M, Airey JA, Gallo R, Ciotti MT, Falcone G and Alema S (1996). Remodeling of cytoskeleton and triads following activation of v-Src tyrosine kinase in quail myotubes. J Cell Sci 109:1335–1346

Chen M-J G, Shih C-L and Wang K (1993). Nebulin as an actin zipper. J Biol Chem 268:20327–20334

Chien KR, Knowlton KU, Zhu H and Chien S (1991). Regulation of cardiac gene expression during myocardial growth and hypertrophy: molecular studies of an adaptive physiologic response. FASEB J 5:3037–3046

Choi JK, Holtzer S, Chacko SA, Lin ZX, Hoffman RK and Holtzer H (1991). Phorbol esters selectively and reversibly inhibit a subset of myofibrillar genes responsible for the ongoing differentiation program of chick skeletal myotubes. Mol Cell Biol 11:4473–4482

Coso OA, Chiariello M, C, Y. J, Teramoto H, Crespo P, Xu N, Miki T and Gutkind J S (1995). The small GTP-binding proteins Rac1 and cdc42 regulate the activity of the JNK/SAPK signaling pathway. Cell 81:1137–1146

Cowley S, Paterson H, Kemp B and Marshall C (1994). Activation of MAP kinase kinase is necessary and sufficient for PC12 differentiation and for transformation of NIH 3T3 cells. Cell 77: 841–852

Craig R and Offer G (1976a). Axial arrangement of crossbridges in thick filaments of vertebrate skeletal muscle. J Mol Biol 102:325–332

Craig R and Offer G (1976b). The localization of C-protein in rabbit skeletal muscle. Proc R Soc Lond B 192:451–461

Craig R (1977). Structure of A-segments from frog and rabbit skeletal muscle. J Mol Biol 109:69–81

Croop J, Dubyak G, Toyama Y, Dlugosz A, Scarpa A and Holtzer H (1982). Effects of 12-O-tetradecanoyl-phorbol-13-acetate on myofibril integrity and Ca^{2+} content in developing myotubes. Developmental Biology 89:460–474

Dabiri GA, Turnacioglou KK, Sanger JM and Sanger JW (1997). Myofibrillogenesis visualized in living embryonic cardiomyocytes. Proc Natl Acad Sci USA 94:9493–9498

Dennis JE, Shimzu T, Reinach FC and Fischman DA (1984). Localization of C-protein isoforms in chicken skeletal muscle: ultrastructural detection using monoclonal antibodies. J Cell Biol 98:1514–1522

Dlugosz AA, Tapscott SJ and Holtzer H (1983). Effects of phorbol 12-myristate 13-acetate on the determination program of chick skeletal myoblasts. Cancer Res 43:2780–2789

Dlugosz AA, Antin PB, Nachmias VT and Holtzer H (1984). The relationship between stress fiber-like structures and nascent myofibrils in cultured cardiac myocytes. J Cell Biol 99:2268–2278

Donath MY, Zapf J, Eppenberger-Eberhardt M, Froesch ER and Eppenberger HM (1994). Insulin-Like Growth Factor I Stimulates Myofibril Development and Decreases Smooth Muscle alpha-Actin of Adult Cardiomyocytes. Proc Natl Acad Sci USA 91:1686–1690

Drubin DG, Mulholland J, Zhu ZM and Botstein D (1990). Homology of a yeast actin-binding protein to signal transduction proteins and myosin-I. Nature 343:288–290

Egan SE, Giddings BW, Brooks MW, Buday L, Sizeland AM and Weinberg RA (1993). Association of Sos Ras exchange protein with Grb2 is implicated in tyrosine kinase signal transduction and transformation. Nature 363:45–51

Egelman EH, Francis N and DeRosier DJ (1982). F-actin is a helix with a random variable twist. Nature 298:131–135

Einheber S and Fischman DA (1990). Isolation and characterization of a cDNA clone encoding avian skeletal muscle C-protein: an intracellular member of the immunoglobulin superfamily. Proceedings of the National Academy of Sciences of the USA 87:2157–2161

Falcone G, Alemà S and Tatò F (1991). Transcription of muscle-specific genes is repressed by reactivation pp60^{v-src} in postmitotic myotubes. Mol Cell Biol 11:3331–3338

Funatsu T, Higuchi H and Ishiwata S (1990). Elastic filaments in skeletal muscle revealed by selective removal of thin filaments with plasma gelsolin. J Cell Biol 110:53–62

Fürst DO, Osborn M, Nave R and Weber K (1988). The organization of titin filaments in the half-sarcomere revealed by monoclonal antibodies in immunoelectron microscopy; a map of ten non-repetitive epitopes starting at the Z line extends close to the M line. J Cell Biol 106:1563–1572

Fürst DO, Osborn M and Weber K (1989). Myogenesis in the mouse embryo: differential onset of expression of myogenic proteins and the involvement of titin in myofibril assembly. J Cell Biol 109:517–527

Fürst DO, Vinkemeier U and Weber K (1992). Mammalian skeletal muscle C-protein: purification from bovine muscle, binding to titin and the characterization of a full length human cDNA. J Cell Sci 102:769–778

Fürst DO and Gautel M (1995). The anatomy of a molecular giant: How the sarcomere cytoskeleton is assembled from immunoglobulin superfamily molecules. J Mol Cell Cardiol 27:951–960

Garvey JL, Kranias EG and Solaro RJ (1988). Phosphorylation of C-protein,troponin-I and phospholamban in isolated rabbit hearts. Biochem J 249:709–714

Gautel M, Leonard K and Labeit S (1993a). Phosphorylation of KSP motifs in the C-terminal region of titin in differentiating myoblasts. EMBO J 12:3827–3834

Gautel M, Lakey A, Barlow DP, Holmes Z, Scales S, Leonard K, Labeit S, Mygland A, Gilhus NE and Aarli JA (1993b). Titin antibodies in myasthenia gravis: identification of a major immunogenic region of titin. Neurology 43:1581–1585

Gautel M, Zuffardi O, Freiburg A and Labeit S (1995). Phosphorylation switches specific for the cardiac isoform of myosin binding protein-C: a modulator of cardiac contraction? EMBO J 14:1952–1960

Gautel M, Castiglione Morelli M, Pfuhl M, Motta A and Pastore A (1995). A calmodulin-binding sequence in the C-terminus of human cardiac titin kinase. Eur J Biochem 230:752–759

Gautel M, Goulding D, Bullard B, Weber K and Fürst DO (1996a). The central Z-disk region of titin is assembled from a novel repeat in variable copy numbers. J Cell Sci 109:2747–2754

Gautel M, Lehtonen E and Pietruschka F (1996b). Assembly of the cardiac I-band region of titin/connectin: expression of the cardiac-specific regions and their relation to the elastic segments. J Muscle Res Cell Motil 17:449–461

Gautel M (1996c). The super-repeats of titin/connectin and their interactions: glimpses at sarcomeric assembly. Adv Biophys 33:27–37

Gautel M and Goulding D (1996). A molecular map of titin/connectin elasticity reveals two different mechanisms acting in series. FEBS lett 385:11–14

Gautel M, Fürst DO, Cocco A and Schiaffino S (1998). Isoform transitions of the myosin-binding protein C family in developing human and mouse muscles suggests a lack of isoform transcomplementation in cardiac muscle. Circulation Res (in press)

Goedert M, Hasegawa M, Jakes R, Lawler S, Cuenda A and Cohen P (1997). Phosphorylation of microtubule-associated protein tau by stress-activated protein kinases. FEBS Lett 409:57–62

Handel SE, Wang S-M, Greaser ML, Schultz E, Bulinski JC and Lessard JL (1989). Skeletal muscle myofibrillogenesis as revealed with monoclonal antibody to titin in combination with detection of the α- and γ-isoforms of actin. Dev Biol 132:35–44

Harder BA, Schaub MC, Eppenberger HM and Eppenberger-Eberhardt M (1996). Influence of fibroblast growth factor (bFGF) and insulin like growth factor (IGF-I) on cytoskeletal and contractile structures and on atrial natriuretic factor (ANF) expression in adult rat ventricular cardiomyocytes in culture. J Mol Cell Cardiol 28:19–31

Hartzell HC and Titus L (1982). Effects of cholinergic and adrenergic agonists on phosphorylation of 165,000-dalton myofibrillar protein in intact cardiac muscle. J Biol Chem 257:2111–2120

Hartzell HC and Glass DB (1984). Phosphorylation of purified cardiac muscle C-protein by purified cAMP-dependent and endogenous Ca$_2$-calmodulin-dependent protein kinases. J Biol Chem 259:15587–15596

Hartzell HC (1985). Effects of phosphorylated and unphosphorylated C-protein on cardiac actomyosin ATPase. J Mol Biol 186:185–195

Hefti MA, Harder BA, Eppenberger HM and Schaub MC (1997). Signaling pathways in cardiac myocyte hypertrophy. J Mol Cell. Cardiol 29:2873–2892

Heierhorst J, Probst WC, Vilim FS, Buku A and Weiss KR (1994). Autophosphorylation of molluscan twitchin and interaction of its kinase domain with calcium/calmodulin. J Biol Chem 269:21086–21093

Heierhorst J, Probst WC, Kohanski RA, Buku A and Weiss KR (1995). Phosphorylation of myosin regulatory light chains by the molluscan twitchin kinase. Eur J Biochem 233:426–431

Heierhorst J, Kobe B, Feil S, Parker MW, Benian GM, Weiss KR and Kemp B (1996a). Ca^{2+}/S100 regulation of giant protein kinases. Nature 380:636–639

Heierhorst J, Tang X, Lei J, Probst WC, Weiss KR, Kemp BE and Benian GB (1996b). Substrate specificity and inhibitor sensitivity of Ca^{2+}/S100-dependent protein kinases. Eur J Biochem. 242:454–459

Higgins D, Labeit S, Gautel M and Gibson T (1993). The evolution of titin and related giant muscle proteins. J Mol Evolution 38:395–404

Horowits R, Kempner ES, Bisher ME and Podolsky RJ (1986). A physiological role for titin and nebulin in skeletal muscle. Nature 323:160–164

Houmeida A, Holt J, Tskhovrebova L and Trinick J (1995). Studies of the interaction between titin and myosin. J Cell Biol 131:1471–1481

Hu SH, Parker MW, Lei JY, Wilce MCW, M BG and E KB (1994). Insights into autoregulation from the crystal structure of twitchin kinase. Nature 369:581–584

Ishizaki T, Maekawa M, Fujisawa K, Okawa K, Iwamatu A, Fujita A, Watanabe N, Saito Y, Kakizuka A, Morii N and Narumiya S (1996). The small GTP-binding protein Rho binds to and activates a 160 kDa Ser/Thr protein kinase homologous to myotonic dystrophy kinase. EMBO J 15:1885–1893

Ito H, Hiroe M, Hirata Y, Tsujino M, Adachi S, Shichiri M, Koike A, Nogami A and Marumo F (1993). Insulin-like growth factor-I induces hypertrophy with enhanced expression of muscle specific genes in cultured rat cardiomyocytes. Circ 87:1715–1721

Itoh Y, Suzuki T, Kimura S, Ohashi K, Higuchi H, Sawada H, Shimizu T, Shibata M and Maruyama K (1988). Extensible and less extensible domains of connectin filaments in stretched vertebrate skeletal muscle sarcomeres as detected by immunofluorescence and immunoelectron microscopy using monoclonal antibodies. J Biochem 104:504–508

Jeacocke S and England P (1980). Phosphorylation of a myofibrillar protein of Mr 150000 in perfused rat heart, and the tentative identification of this as C-protein. FEBS Lett 122:129–132

Jeng CJ and Wang SM (1992). Interaction Between Titin and a-Actinin. Biomedi Res 13:197–202

Johnson GL and Vaillancourt RR (1994). Sequential protein kinase reactions controlling cell growth and differentiation. Curr Opin Cell Biol 6:230–238

Johnson LN, Noble MEM and Owen DJ (1996). Active and inactive protein kinases: structural basis for regulation. Cell 85:149–158

Kasahara HMI, Sugiyama T, Kido N, Hayashi H, Saito H, Tsukita S and Kato N (1994). Autoimmune myocarditis induced in mice by cardiac C-protein. Cloning of complementary DNA encoding murine cardiac C-protein and partial characterization of the antigenic peptides. J Clin Invest 94:1026–1036

Kennelly PJ, Edelman AM, Blumenthal DK and Krebs EG (1987). Rabbit skeletal muscle myosin light chain kinase. The calmodulin binding domain as a potential active site-directed inhibitory domain. J Biol Chem 262:11958–11963

Kobe B, Heierhorst J, Feil SC, Parker MW, Benian GB, Weiss KR and Kemp BE (1996). Giant protein kinases: domain interactions and structural basis of autoregulation. EMBO J 15:6810–6821

Komiyama M, Maruyama K and Shimada Y (1990). Assembly of connectin (titin) in relation to myosin and a-actinin in cultured cardiac myocytes. J Muscle Res Cell Motil 11:419–428

Labeit S, Barlow DP, Gautel M, Gibson T, Holt J, Hsieh C-L, Francke U, Leonard K, Wardale J, Whiting A and Trinick J (1990). A regular pattern of two types of 100-residue motif in the sequence of titin. Nature 345:273–276

Labeit S, Gautel M, Lakey A and Trinick J (1992). Towards a molecular understanding of titin. EMBO J 11:1711–1716

Labeit S and Kolmerer B (1995a). Titins: giant proteins in charge of muscle ultrastructure and elasticity. Science 270:293–296

Labeit S and Kolmerer B (1995b). The complete primary structure of human nebulin and its correlation to muscle structure. J Mol Biol 248:308–315

LaMorte J, Thorburn J, Absher D, Spiegel A, Heller Browns J, Chien KR, Feramisco JR and Knowlton KU (1994). Gq- and Ras-dependent pathways mediate hypertrophy of neonatal rat ventricular myocytes following a1-adrenergic stimulation. J Biol Chem 269:13490–14496

Lazaro JB, Kitzmann M, Poul MA, Vandromme M, Lamb NJ, Fernandez A (1997). Cyclin dependent kinase 5, cdk5, is a positive regulator of myogenesis in mouse C2 cells. J Cel Sci 110:1251–1260

Lechner C, Zahalk a. M, Giot J, Moller N and Ullrich A (1996). ERK6, a mitogen-activated protein kinase involved in C2C12 myoblast differentiation. Proc Natl Acad Sci USA 93:4355–4359

Lee VM-Y, Otvos L, Carden J, Hollosi M, Dietzschold B and Lazzarini RA (1988). Identification of the major multiphosphorylation site in mammalian neurofilaments. Proc Natl Acad Sci USA 85:1998–2002

Lei J, Tang X, Chamber TC, Pohl J and Benian GM (1994). The protein kinase domain of twitchin has protein kinase activity and an autoinhibitory domain. J Biol Chem 269:21078–21085

Lew J, Winkfein RJ, Paudel HK and Wang JH (1992). Brain proline-directed protein kinase is a neurofilament kinase which displays high sequence homology to p34^{cdc2}. J Biol Chem 267:25922–25926

Lew J, Huang, Q-Q, Qi Z, Winkfein RJ, Ebersold R, Hunt T and Wang JH (1994). A brain-specific activator of cyclin-dependent protein kinase 5. Nature 371:423–426

Lin ZX, Eshelman JR, Forry S, Duran S, Lessard JL and Holtzer H (1987). Sequential disassembly of myofibrils induced by myristate acetate in cultured myotubes. J Cell Biol 105:1365–1376

Lin ZX, Eshleman J, Grund C, Fischman DA, Masaki T, Franke WW and Holtzer H (1989). Differential response of myofibrillar and cytoskeletal proteins in cells treated with phorbol myristate acetate. J Cell Biol 108:1079–1091

Lin ZX, Lu MH, Schultheiss T, Choi J, Holtzer S, Dilullo C, Fischman DA and Holtzer H (1994). Sequential appearance of muscle-specific proteins in myoblasts as a function of time after cell division: Evidence for a conserved myoblast differentiation program in skeletal muscle. Cell Motil Cytoskeleton 29:1–19

Linke WA, Ivemeyer M, Olivieri N, Kolmerer B, Rüegg JC and Labeit S. (1996). Towards a molecular understanding of the elasticity of titin. J Mol Biol 261:62–71

Lowe ED, Noble MEM, Skamnaki VTS, Oikonomakos NG, Owen DJ and Johnson LN (1997). The crystal structure of a phosphorylase kinase peptide substrate complex: kinase substrate recognition. EMBO J 16:6646–6658

Luther PK (1991). Three-dimensional reconstruction of a simple Z-band in fish muscle. J Cell Biol 113:1043–1055

Maruyama K, Matsubara S, Natori R, Nonomura Y, Kimura S, Ohashi K, Murakami F, Handa S and Eguchi G (1977). Connectin, an elastic protein of muscle: characterization and function. J Biochem 82:317–337

Maruyama K, Sawada H, Kimura S, Ohashi K, Higuchi H and Umazume Y (1984). Connectin filaments in stretched skinned fibers of frog skeletal muscle. J Cell Biol 99:1391–1397

Maruyama K, Yoshioka T, Higuchi H, Ohashi K, Kimura S and Natori R (1985). Connectin filaments link thick filaments and Z lines in frog skeletal muscle as revealed by immunoelectron microscopy. J Cell Biol 101:2167–2172

Maruyama K, Endo T, Kume H, Kawamura Y, Kanzawa N, Nakauchi Y, Kimura S, Kawashima S and Maruyama K (1993). A novel domain sequence of connectin localized at the I band of skeletal muscle sarcomeres: homology to neurofilament subunits. Biochem Biophys Res Comm 194:1288–1291

Maruyama K, Endo T, Kume H, Kawamura Y, Kanzawa N, Kimura S, Kawashima S and Maruyama K (1994). A partial connectin cDNA encoding a novel type of RSP motifs isolated from embryonic skeletal muscle. J Biochem 115:147–149

Maruyama K (1997). Connectin/titin, giant elastic protein of muscle. FASEB J 11:341–345

Matsui T, Amano M, Yamamoto T, Chihara K, Nakafuka M, Ito M, Nakano T, Okawa K, Iwamatsu A and Kaibuchi K (1996). Rho-associated kinase, a novel serine/threonine kinase, as a putative target for small GTP binding protein Rho EMBO J 15:2208–2216

Maw MC and Rowe AJ (1986). The reconstitution of myosin filaments in rabbit psoas muscle from solubilized myosin. J Muscle Res Cell Motil 7:97–109

Moncman CL and Wang K (1995). Nebulette: a 107 kD nebulin-like protein in cardiac muscle. Cell Motil Cytoskel 32:205–225

Morris EP and Squire J (1990). The three-dimensional structure of the nemaline rod Z-band. J Cell Biol 111:2961–2978

Mues A, van der Ven PF, Young P, Fürst DO, Gautel M (1998). Two immunoglobin-like domains of the Z-disc portion of titin interact in a conformation-dependent way with telethonin. FEBS Lett 428:111–114

Mukai H, Toshimori M, Shibata H, Takanaga H, Kitagawa M, Miyahara M, Shimakawa M and Y O (1997). Interaction of PKN with alpha-actinin. J Biol Chem 272:4740–4746

Musacchio A, Saraste M and Wilmanns M (1994). High-resolution crystal structures of tyrosine kinase SH3 domains complexed with proline-rich peptides. Nat Struct Biol 1:546–551

Nakagawa O, Fujisawa K, Ishizaki T, Saito Y, Nakao K and Narumiya S (1996). ROCK-I and ROCK-II, two isoforms of Rho-associated coiled-coil forming protein serine/threonine kinase in mice. FEBS Lett 392:189–193

Narumiya S, Ishizaki T and Watanabe N (1997). Rho effectors and reorganization of actin cytoskeleton. FEBS Lett 410:68–72

Nave R, Fürst D, Vinkemeier U and Weber K (1991). Purification and physical properties of nematode mini-titins and their relation to twitchin. J Cell Sci 98:491–496.

Obermann W, Plessmann U, Weber K and Fürst DO (1995). Purification and bio-chemical characterization of myomesin, a myosin and titin binding protein, from bovine skeletal muscle. Eur J Biochem 233:110–115

Obermann WMJ, Gautel M, Steiner F, Van der Ven P, Weber K and Fürst DO (1996). The structure of the sarcomeric M band: localization of defined domains of myomesin, M-protein and the 250 kD carboxy-terminal region of titin by immu-noelectron microscopy. J Cell Biol 134:1441–1453

Obermann WMJ, Gautel M, Weber K and Fürst DO (1997). Molecular structure of the sarcomeric M band: mapping of titin- and myosin-binding domains in myomesin and the identification of a potential regulatory phosphorylation site in myomesin. EMBO J 16:211–220

Obinata T, Reinach FC, Bader DM, Masaki T, Kitani S and Fischman DA (1984). Immunochemical analysis of C-protein isoform transitions during the develop-ment of chicken skeletal muscle. Developmental Biol 101:116–124

Offer G, Moos C and Starr R (1973). A new protein of the thick filaments. Extraction, purification, and characterization. J Mol Biol 74:653–676

Ohtsuka H, Yajami H, Maruyama K and Kimura S (1997a). The N-Terminal Z Re-peat 5 of connectin/Titin Binds to the C-Terminal Region of a-Actinin. Biochem Biophys Res Comm 235:1–3

Ohtsuka H, Yajima H, Maruyama K and Kimura S (1997b). Binding of the N-terminal 63 kDa portion of connectin/titin to alpha actinin. FEBS Lett 401:65–67

Okagaki T, Weber FE, Fischman DA, Vaughan KT, Mikawa T and Reinach FC (1993). The major myosin-binding domain of skeletal muscle MyBP-C (C-protein) resides in the COOH-terminal, immunoglobulin C2 motif. J Cell Biol 123:619–626.

Paterson HF, Self AJ, Garrett MD, Just I, Aktories K and Hall A (1990). Microinjec-tion of recombinant p21rho induces rapid changes in cell morphology. J Cell Biol 111:1001–1007.

Pearson RB and Kemp BE (1991). Protein kinase phosphorylation site sequences and consensus specifity motifs: tabulations. Meth Enzymol 200:62–81

Peckham M, Young P and Gautel M (1997). Constitutive and Variable Regions of Z-disk Titin/Connectin in Myofibril Formation: A Dominant-negative Screen. Cell Struct Funct 22:95–101

Pfuhl M, Winder SJ and Pastore A (1994). Nebulin, a helical actin binding protein. EMBO J 13:1782–1789

Pfuhl M, Winder SJ, Castiglione Morelli MA, Labeit S Pastore A (1996). Correlation between conformational and binding properties of nebulin repeats. J Mol Biol 257:367–384.

Politou A, Millevoi S, Gautel M, Kolmerer B and Pastore A (1997). SH3 in muscle: solution structure of the nebulin SH3. J Mol Biol (in press)

Pomiès P, Louis H and Beckerle M (1997). CRP1, a LIM domain protein implicated in muscle differentiation, interacts with a-actinin. J Cell Biol 139: 157–168

Puceat M, Hilaldandan R, Strulovici B, Brunton LL and Brown JH (1994). Differen-tial regulation of protein kinase C isoforms in isolated neonatal and adult rat cardiomyocytes. J Biol Chem 269:16938–16944

Ramocki MB, Johnson SE, White MA, Ashendel CL, Konieczny SF and Taparowsky EJ (1997). Signaling through mitogen-activated protein kinase and Rac/Rho does not duplicate the effects of activated Ras on skeletal myogenesis. Mol Cell Biol 17:3547–3555

Ridley AJ, Paterson HF, Johnston CL, Diekmann D and Hall A (1992). The small GTP-binding protein rac regulates growth factor-induced membrane ruffling. Cell 70:401–410

Rottbauer W, Gautel M, Zehelein J, Labeit S, Franz WM, Grünig E, Brown BD, Vollrath B, Mall G, Dietz R and Katus HA (1997). Novel splice donor site mutation in the cardiac myosin-binding protein-C gene in familial hypertrophic cardiomyopathy. Characterization of transcript and protein. J Clin Invest 100:475–482

Rowe RWD (1973). The ultrastructure of the Z discs from white, intermediate and red fibres of mammalian striated muscle. J Cell Biol 57:261–277

Schafer DA, Hug C and Cooper JA (1995). Inhibition of CapZ during myofibrillogenesis alters assembly of actin filaments. J Cell Biol 128:61–70

Schafer D, Jennings P and Cooper J (1996). Dynamics of capping protein and actin assembly in vitro: uncapping barbed ends by polyphosphoinositides. J Cell Biol 135:168–179

Schäfer BW and Heizmann C (1996). The S100 family of EF-hand calcium-binding proteins: functions and pathology. Trends Biochem Sci 21:134–140

Schaub MC, Hefti M, Harder BA and Eppenberger HM (1997). Various hypertrophic stimuli induce distinct phenotypes in cardiomyocytes. J Mol Med 75:901–920

Schlender KK and Bean LJ (1991). Phosphorylation of chicken cardiac C-protein by calcium/calmodulin-dependent protein kinase II. J Biol Chem 266:2811–2817

Schroeter JP, Bretaudiere JP, Sass RL and Goldstein MA (1996). Three-dimensional structure of the Z band in a normal mammalian skeletal muscle. J Cell Biol 133:571–583

Schultheiss T, Choi J, Lin ZX, DiLullo C, Cohen-Gould L, Fischman D and Holtzer H (1992). A sarcomeric a-actinin truncated at the carboxyl end induces the breakdown of stress fibers in PtK2 cells and the formation of nemaline-like bodies and breakdown of myofibrils in myotubes. Proc Natl Acad Sci USA 89:9282–9286

Sebestyén MG, Wolff JA and Greaser ML (1995). Characterization of a 5.4 kb cDNA fragment from the Z-line region or rabbit cardiac titin reveals phosphorylation sites for proline-directed kinases. J Cell Sci 108:3029–3037

Sebestyén MG, Fritz JD, Wolff JA and Greaser ML (1996). Primary structure of the kinase domain region of rabbit skeletal and cardiac titin. J Muscle Res Cell Motil 17:343–348

Shetty KT, Link WT and Pant HC (1993). cdc2-like kinase from rat spinal cord specifically phosphorylates KSPXK motifs in neurofilament proteins: isolation and characterization. Proc Natl Acad Sci USA 90:6844–6848

Sjöström M and Squire JM (1977). Fine structure of the A-band in cryosections: the structure of the A-band of human sskeletal fibers from ultrathin cryo-sections negatively stained. J Mol Biol 109:49–68

Sommerville LL and Wang K (1987). In vivo phosphorylation of titin and nebulin in frog skeletal muscle. Biochem Biophys Res Comm 147:986–992

Songyang Z, Blechner S, Hoagland N, Hoekstra MF, Piwnica-Worms H and Cantley LC (1994). Use of an oriented peptide library to determine the optimal substrates of protein kinases. Current Biology 4:973–982

Songyang Z, Lu KP, Kwon YT, Tsai LH, Filhol O, Cochet C, Brickey DA, Soderling TR, Bartleson C, Graves DJ, DeMaggio AJ, Hoekstra MF, Blenis J, Hunter T and Cantley LC (1996). A structural basis for substrate specificities of protein Ser/Thr kinases: Primary sequence preference of Casein Kinase I and II, Phosphorylase Kinase, Calmodulin-dependent Kinase II, CDK5 and Erk1. Mol Cell Biol 16:6486–6493.

Sorimachi H, Freiburg A, Kolmerer B, Ishiura S, Stier G, Gregorio CC, Labeit D, Linke WA, Suzuki K and Labeit S (1997). Tissue-specific expression and alpha-actinin binding properties of the Z-disc titin: implications for the nature of vertebrate Z-discs. J Mol Biol. 270:688–695

Soteriou A, Gamage M and Trinick J (1993). A survey of interactions made by the giant protein titin. J Cell Sci 104:119–123

Squire J (1981). The Structural Basis of Muscular Contraction. Plenum Press, New York and London. 364–375

Sutherland CJ, Esser KA, Elsom VL, Gordon ML and Hardeman EC (1993). Identification of a program of contractile protein gene expression initiated upon skeletal muscle differentiation. Dev Dynam 196:25–36

Terai M, Komiyama M and Shimada Y (1989). Myofibril assembly is linked with vinculin, a-actinin, and cell-substrate contacts in embryonic cardiac myocytes. Cell Mot Cytoskeleton 12:185–194

Thomas SM, Soriano P and Imamoto A. (1995). Specific and redundant roles of Src and Fyn in organizing the cytoskeleton. Nature 276:267–271

Thorburn J and Thorburn A (1994). The tyrosine kinase inhibitor, genistein, prevents alpha-adrenergic-induced cardiac muscle cell hypertrophy by inhibiting activation of the Ras-MAP kinase signaling pathway. Biochem Biophys Res Commun 202:1586–1591

Thorburn J, Xu S and Thorburn A (1997). MAP kinase- and Rho-dependent signals interact to regulate gene expression but not actin morphology in cardiac muscle cells. EMBO J 16:1888–1900

Tokuyasu KT and Maher PA (1987a). Immunocytochemical studies of cardiac myofibrillogenesis in early chick embryos.I. Presence of immunofluorescent titin spots in premyofibril stages. J Cell Biol 105:2781–2793

Tokuyasu KT and Maher PA (1987b). Immunocytochemical studies of cardiac myofibrillogenesis in early chick embryos.II. Generation of a-actinin dots within titin spots at the time of the first myofibril formation. J Cell Biol 105:2795–2801

Trinick J (1994). Titin and nebulin: protein rulers in muscle? Trends. Biochem Sci 19:405–409

Trinick J (1996). Titin as a scaffold and spring. Current Biology 6:258–260

Tsai L-H, Delalle I, Caviness VS, Chae T and Harlow E (1994). p35 is a neural-specific regulatory subunit of cyclin-dependent protein kinase 5. Nature 371:419–423

Turnacioglu KK, Mittal B, Sanger JM and Sanger JW (1996). Partial characterization of zeugmatin indicates that it is part of the Z-band region of titin. Cell Motil Cytoskel 34:108–121

Valle G, Faulkner G, De Antoni A, Pacchioni B, Pallavacini A, Pandolfo D, Tiso N, Toppo S, Trevisan and Lanfranchi G (1997). Telethonin, a novel sarcomeric protein of heart and skeletal muscle. FEBS Lett 415:163–168

Van der Loop FTL, van der Ven PFM, Fürst DO, Gautel M, van Eys GJJM and Ramaekers FCS (1996). Integration of titin into the sarcomeres of cultured differentiating human skeletal muscle cells. Eur J Biol 69:301–307

Van der Ven PFM, Schaart G, Croes HJE, Jap PHK, Ginsel LA and Ramaekers FCS (1993). Titin aggregates associated with intermediate filaments align along stress fiber-like structures during human skeletal muscle cell differentiation. J Cell Sci 106:749–759

Vaughan KT, Weber FE, Einheber S and Fischman DA (1993). Molecular cloning of chicken myosin-binding protein (MyBP) H (86-kDa protein) reveals extensive homology with MyBP-C (C-protein) with conserved immunoglobulin C2 and fibronectin type III motifs. J Biol Chem 268:3670–3676

Vibert P, Edelstein SM, Castellani L and Elliot BW (1993). Mini-titins in striated and smooth molluscan muscles: structure, location and immunological crossreactivity. J Muscle Res Cell Motil 14:598–607

Vigoreaux JO (1994). The muscle Z band: lessons in stress management. J Muscle Res Cell Motil 15:237–255

Vinkemeier U, Obermann W, Weber K and Fürst DO (1993). The globular head domain of titin extends into the center of the sarcomeric M band. J Cell Sci 106:319–330

Vojtek AB, Hollenberg SM and Cooper JA (1993). Mammalian Ras interacts directly with the serine/threonine kinase. Raf Cell 74:205–214

Vojtek AB and Cooper JA (1995). Rho family members: activators of MAP kinase cascades. Cell 82:527–529

Wang K, McClure J and Tu A (1979). Titin: Major myofibrillar components of striated muscle. Proc Natl Acad Sci USA 76:3698–3702

Wang SM, Greaser M, Schultz E, Bulinsky JC, Lin JJ-C and Lessard J (1988). Studies on cardiac myofibrillogenesis with antobodies against titin, tropomyosin and myosin. J Cell Biol 107:1075–1083

Wang K and Wright J (1988). Architecture of the sarcomere matrix of skeletal muscle: Immunoelectron microscopic evidence that suggests a set of parallel inextensive nebulin filaments anchored at the Z-line. J Cell Biol 107:2199–2212

Wang K, McCarter R, Wright J, Beverly J and Ramirez-Mitchell R (1993). Viscoelasticity of the sarcomere matrix of skeletal muscles. Biophys J 64:1161–1177

Wang K, Knipfer M, Huang Q, van Heerden A, Hsu L, Gutierrez G, Quia X and Stedman H (1996). Human skeletal muscle nebulin sequence encodes a blueprint for thin filament architecture: Sequence motifs and affinity profiles of tandem repeats and terminal SH3. J Biol Chem 271:4303–4314

Wang B, Golemis E and Kruh GD (1997). ArgBP2, a multiple Src Homology 3 Domain-containing, Arg/Abl-interacting protein, is phosphorylated in v-Abl-transformed cells and localized in stress fibers and cardyocyte Z-disks. J Biol Chem 17542–17560

Watanabe G, Saito Y, Madaule P, Ishizaki T, Fujisawa K, Morii N, Mukai H, Ono Y, Kakizuka A and Narumiya S (1996). Protein kinase N (PKN) and PKN-related protein rhophilin as targets of small GTPase Rho. Science 271:645–648

Watkins H, Conner D, Thierfelder. L, Jarcho JA, MacRae C, McKenna WJ, Maron BJ, Seidman JG and Seidman CE (1995). Mutations in the cardiac myosin-binding protein-C on chromosome 11 cause familial hypertrophic cardiomyopathy. Nature Genet 11:434–437

Weisberg A and Winegrad S (1996). Alteration of myosin cross bridges by phosphorylation of myosin-bindig protein C in cardiac muscle. Proc Natl Acad Sci USA 93:8999–9003

Weintraub H, Dwarki VJ, Verma I, Davis R, Hollenberg S, Snider L, Lassar A and Tapscott SJ (1991). Muscle-specific transcriptional activation by MyoD. Genes & Development 5:1377–1386

Whiting A, Wardale J and Trinick J (1989). Does titin regulate the length of thick filaments? J Mol Biol 205:263–268

Wu H and Parsons JT (1993). Cortactin, an 80/85-kilodalton pp60src substrate, is a filamentous actin-binding protein enriched in the cell cortex. J Cell Biol 120:1417–1426

Xia H, Winokur ST, Kuo W-L, Altherr M and Bredt D (1997). Actinin-associated LIM-protein: Identification of a domain interaction between PDZ and spctrin-like motifs. J Cell Bol 139:507–515

Yajima H, Ohtsuka H, Kawamura Y, Kume H, Murayama T, Abe H, Kimura S and Maruyama K (1996). A 11.5-kb 5'-terminal cDNA sequence of chicken breast muscle connectin/titin reveals its Z line binding region. Biochem Biophys Res Comm 223:160–164

Yamaguchi M, Izumimoto M, Robson RM and Stromer MH (1985). Fine structure of wide and narrow vertebrate muscle Z-lines. J Mol Biol 184:621–643

Yamamoto K and Moos C (1983). The C-proteins of rabbit red, white and cardiac muscles. J Biol Chem 258:8395–8401

Yasuda M, Koshida S, Sato N and Obinata T (1995). Complete primary structure of chicken cardiac C-protein (MyBP-C) its expression in developing striated muscles. J Mol Cell Cardiol 27:2275–2286

Yoon H and Boettiger D (1994). Expression of v-Src alters the expression of myogenic regulatory factor genes. Oncogene 9:801–807

Young P, Ferguson C, Bañuelos S and Gautel M (1998). Molecular structure of the sarcomeric Z-disk: two types of titin interactions lead to an asymmetrical sorting of α-actinin. EMBO J 17:1614–1624

The Elastic Filament System in Myogenesis

A. B. Fulton

Department of Biochemistry University of Iowa, Iowa City, IA 52242, USA

Contents

*"There is one major unsolved problem. How do
these fibrillar elements, which produce movement,
themselves get shunted to the right place in the
cell, at the right time, to produce just the right
movement? ...Here we know so little we can only
wait and see what will turn up."*

F.H.C. Crick, 1977

1.1 Introduction

A mechanical role for titin (connectin) is reasonably well established [1, 2, 3, 4]. A variety of observations suggest that titin also participates in the morphogenesis of the myofibril. The changing patterns of titin in developing muscle have now been analyzed in heart and skeletal muscle, studying both embryos and cultured cells. For the most part, the different systems offer similar results. Briefly, titin is present and periodic when the first periodic arrays of myosin and α-actinin appear. Titin may be mandatory for such periodic arrays, since it is one of the few proteins unique to and ubiquitous in striated muscle. Titin is a stronger candidate for this morphological role during development than is nebulin, since some heart muscles lack nebulin. Titin mRNA becomes highly localized during this process, with regions of the mRNA co-localized with protein domains. This may reflect a role for mRNA localization in myofibril assembly. Several new molecular biological probes re-inforce and extend the conclusions obtained with inhibitors of cytoskeletal components, that the myofibril may be acting as a tensegrity structure during myofibrillogenesis.

1.2 The Elastic Filament System in Embryonic Myogenesis

1.2.1 Cardiac Development

1.2.1.1 In Avians

The particular patterns of titin in early myogenesis are complex, but they offer important clues to the course of morphogenesis. Detailed, developmental studies of titin using immunofluorescence on chicken embryonic hearts reveal the following sequence (see Fig. 1 for a précis; summarized from [5, 6]. At the earliest stages that myofibril proteins can be detected, titin is diffuse, in random punctate arrays, and in linear punctate arrays that have irregular spacing (1–1.7 μm). At this time, α-actinin, muscle-specific

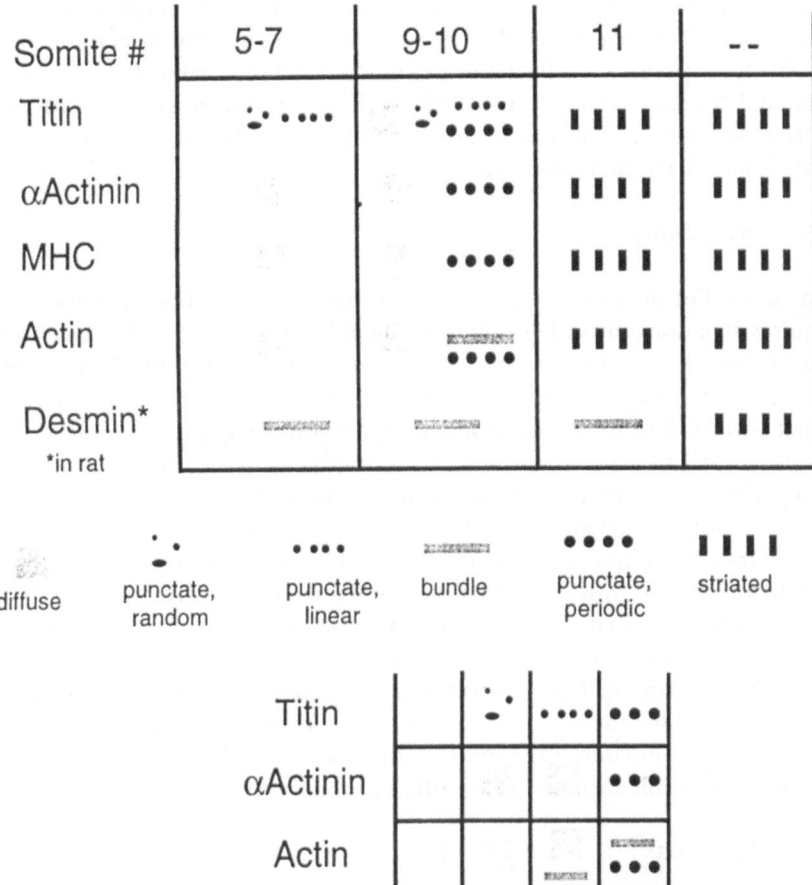

Fig. 1. The diagram indicates the various degrees of organization (diffuse, punctate, linear punctate array or sarcomeric, as shown in the key) seen for the myofibril proteins listed. Top panel shows stages in heart muscle development; bottom panel shows skeletal development

myosin heavy chain (myosin, hereafter) and F-actin are all diffuse. The diffuse titin is more prevalent earlier, and the linear arrays later; thus, this appears to be a developmental sequence. The linear arrays are not always associated with the diffuse, veil-like F-actin at the cell borders: some other structure must provide the linear substrate for these titin arrays.

At the 9–10 somite stage, a new pattern of titin appears, a periodic punctate array; the less mature patterns decrease in abundance. Periodic titin is associated with periodic α-actinin, and with both periodic and non-periodic actin. Some of the non-periodic actin is associated with periodic myosin. Thus periodic arrays abruptly appear that contain titin, α-actinin, and (probably) myosin; actin becomes periodic a few hours later. For titin, myosin, and α-actinin, the spacing of the earliest periodic arrays is about 1.6–1.7 μm. At the 11 somite stage, only two hours later, all four proteins appear in well spaced, striated myofibrils.

1.2.1.2 In Rodents

Similar studies on mouse embryos show that desmin is fibrillar early, but acquires striations much later than the myofibril proteins [7]. The relationships between titin, α-actinin, and actin are summarized in the lower panel of Fig. 1. Diffuse or randomly punctate titin has no α-actinin or actin associated with it. The linear, aperiodic arrays of titin may or may not have a bundle or veil of actin associated. The periodic array of titin is associated with periodic α-actinin, and actin that may or may not be periodic.

This is not consistent with a proposal (based on in vitro observations; [8] that discrete I-Z-I complexes form at a distance from the nascent myofibril and then associate with it. Instead, it suggests a co-operative registration of most myofibril proteins, followed by increased stability or interactions of the actin filaments to bring them into register. It can in any case occur swiftly in the embryo. These observations raise new questions as well. Why, for example, does myosin not associate with titin in the aperiodic linear arrays? In the punctate, randomly positioned arrays, is the titin collapsed into an inaccessible form? If so, how does titin unfurl soon after?

1.2.1.3 In Rabbits

The expression and intracellular distribution patterns of muscle-specific proteins during rabbit embryo development (7–13 dpc) are not identical to that seen in rodents [9]. Titin appears first in the embryonic rabbit heart. Upon differentiation (myocyte and myotube formation), dot-like titin reorganizes into a cross-striated pattern (in 9- to 30-somite embryos) via a short-lived filamentous intermediate. Tropomyosin expression followed titin expression in the heart anlagen and tropomyosin became striated later. Myosin and desmin were arranged into cross-striations after titin and tropomyosin, but were not fully organized in 13-dpc embryos. Since heart beats are detected in 3-somite embryos, the striated organization of titin, tropomyosin, myosin and desmin does not appear essential for contraction. Perhaps surprisingly, the sequence of expression of muscle-specific and intermediate filament proteins during cardiomyogenesis appears to be species-dependent.

1.2.2 Skeletal Muscle Development

1.2.2.1 In Avians

Avian embryos have richly provided for studies of cardiac development in the embryo and for studies of skeletal muscle *in vitro*. Studies of skeletal muscle development in the embryo using avian systems have recently been reviewed [10]; this review also summarizes cardiac development in chickens. The great ease of obtaining embryos, the abundance of skeletal muscle, and its inherent organization greatly facilitate such studies, although numerically they have not been exploited to the same extent as the heart. Additional insights have come from avian cells *in vitro*, which will be discussed below (section B).

1.2.2.2 In Rodents

The development of skeletal muscle during mouse embryogenesis [11] begins in the myotome. At gestation day (g.d.) 9, cells in the myotome domain of cervical somites begin to stain positively for myofibril proteins. Desmin is detected first, followed by titin, muscle-specific actin and myosin heavy chains, and finally by nebulin. At g.d. 9.5 fibrillar desmin is visible, but the other myogenic proteins are distributed diffusely. In some prefusion myoblasts (g.d. 11 and 12) small, immature myofibrils stain periodically for myosin, nebulin, and titin epitopes that reside near the Z line. Other titin epitopes, found at the A-I junction, still stain diffusely. After fusion, all titin epitopes in myotubes (g.d. 13 and 14) show the periodic myofibrillar pattern. The mostly longitudinal arrangement of desmin filaments seen earlier reorganizes (g.d. 17 and 18) to distinct Z line connected striations.

1.3 The Elastic Filament System in Myogenesis In Vitro

1.3.1 Cardiomyocyte Development

In cultured chick cardiac cells, myofibrillogenesis begins with linear structures that contain non-periodic α actin. These linear structures were classified as either central or peripheral; central structures contained titin developing a sarcomeric pattern while peripheral ones had a more uniform, weakly staining titin distribution. The peripheral, non-periodic linear structures with uniform actin and diffuse titin are perhaps the structural homolog of the cables observed in chicken skeletal muscles (see below). The central structures had sarcomeric patterns of titin and myosin at multiple sites. The appearance of sarcomeric titin patterns coincided with or preceded sarcomere periodicity of either alpha actin or muscle tropomyosin. The early patterns of titin distribution in myofibrillogenesis are consistent with a role for titin in filament alignment during sarcomere assembly [12, 13]. Several others have also emphasized the importance of titin for myofibrillogenesis in cultured cardiac cells [14, 15, 16, 17, 18].

Primary cultures of rat myocardial cells have been used to examine expression and organization of muscle-specific proteins during differentiation and dedifferentiation. In differentiating muscle cell cultures, the redistribution of desmin, actin and myosin followed the differentiation patterns observed *in vivo*, but always showed a delay when compared to titin. The slower rate of differentiation in vitro facilitated this discrimination. While titin was observed to be organized first during differentiation of the cell lines, during dedifferentiation in cultured rat myocard cells titin retained organization the longest. This is consistent with titin being an early player in myogenic differentiation, both *in vivo* and *in vitro* [19].

1.3.2 Skeletal Myoblast Development

A developmental study of skeletal muscles revealed that round post-mitotic myoblasts are the earliest ones to express titin [20]; some express titin before expressing myosin, zeugmatin (originally described as a Z line protein, but now believed to be a region of titin associated with the Z line [21], or α-actinin. The ability to detect titin before "zeugmatin" is consistent with other reports of cryptic epitopes of titin that are detected only after certain stages of development. All of these myofibril proteins are punctate quite early; they are co-localized, perinuclear, and stably associated with the cytoskeleton. When these young myoblasts become bipolar, the myofibril proteins extend in fibrillar or punctate linear arrays. Only somewhat later is actin detectably periodic. One advantage of cultured skeletal muscle, compared to cardiac cells, is that these myofibrils form completely *de novo;* in contrast, the myofibrils in cardiac cells in part reform from pre-existing components. A final clue that titin has a morphogenetic role is that titin is overproduced early in myotube development [22].

Early studies of myofibrillogenesis in the embryo noted the close conjunction between titin and myosin and the subsequent registration of desmin with the myofibril [9, 7, 11]. However, some in vitro studies did not reproduce this correlation [23].

Titin assembly and integration into sarcomeres of cultured human skeletal muscle cells has been studied at high resolution with antibodies to four titin epitopes [24]. In postmitotic mononuclear myoblasts, these epitopes of titin are found in a punctate pattern. During elongation and fusion of the cells, the titin spots associate with linear arrays and finally arrive at the Z-line, the A-I junction or the A-band. It appears that during this transition the large titin molecule unfolds, the N terminus migrating toward the Z-line and the C terminus towards the M-line. In mature myotubes all four titin epitopes attained their appropriate cross-striated patterns. While titin unfolds, other components of the Z-line and the A-band also move to specific positions in the nascent sarcomere, after appearing as dot-like aggregates during initial stages of muscle cell. In these cultured cells, the Z-line associated intermediate filament, desmin, only becomes cross-striated later.

1.4 Reconciling Developmental Patterns Observed in Embryos and In Vitro

It may be taken for granted that the pattern of development that occurs in the embryo is the authentic pattern, the one against which any studies in vitro must be compared. For the most part, the patterns of organization and rearrangement observed in the embryo are reproduced *in vitro*. When apparent discrepancies have been reported, three factors can be considered to reconcile the two systems. First, the embryonic process is much swifter and more synchronous than any *in vitro* system, accomplishing in a few hours what usually requires many hours or even days *in vitro*. This difference may actually provide a window into the mechanisms of development, because transient stages which might last for a few minutes in the embryo may persist for one or a few hours in culture, thus permitting their detection. An example of such transient intermediates being detected in vitro are the cells which contained cables of titin without linear myosin, or sarcomerically organized titin with linear, co-axial arrays of myosin [25].

The second source of apparent discrepancies is that titin, a huge molecule, can evoke a variety of antibodies, and those directed against different epitopes sometimes reveal different patterns. When one antibody reveals a periodic pattern and a second shows none, it is possible that other proteins are obscuring the second epitope or that it is not appropriately modified, for example, phosphorylated. When one antibody reveals a periodic pattern and a second shows a diffuse pattern, it is possible that two populations of titin are present in the cell, perhaps differing in age and therefore in organization.

A third source of apparent discrepancies may arise from a peculiarity of the cardiac myocyte. Cardiac cells already contain myofibril when isolated from the embryo; thus, it is difficult at times to distinguish between *de novo* assembly and the re-extension of a pre-existing fibrillar structure.

1.5 Messenger RNA Location During the Organization of the Elastic Filament System

The "embryonic determinants" of the nineteenth century have in many cases proven to be mRNAs that are distributed in organized patterns in the oocyte or egg. In recent years it has come to be appreciated that in somatic cells, the mRNA for some cytoskeletal proteins is also organized [26]. The mRNAs for titin and nebulin [27], vimentin [28], desmin and vinculin [29], and several other muscle proteins [30, 31, 32] are organized in specific, sometimes highly specific, patterns which change during development.

Of all of these organized mRNAs, the mRNA for titin is detected the earliest and shows the earliest pattern of organization [27]. It is first found in the mononucleated myoblasts in punctate patterns near the nucleus, in some cases before titin protein is detected. In the young myotube, when the titin protein is in slender strands that show no periodicity, the titin mRNA is dispersed in the cytoplasm, in the finely granular pattern that has been described for mRNA transport granules. When the titin protein becomes punctate and periodic in its linear arrays, the titin mRNA is seen colocalized as a linear but non-periodic array.

Somewhat later in development, the slender myofibrils that show a resolved doublet of titin staining at the AI junction begin to align laterally with each other and come into registration at the their Z lines. It is only at this time that the mRNA for titin is present as a periodic array [27]. The probe used for these studies encodes peptide sequences found in the A band; it is at the A band specifically that the probe hybridizes. Thus, for titin, not only are the mRNA and protein co-localized; at least one domain of the protein is co-localized with the specific segment of mRNA that encodes it.

One interpretation for the delay with which the titin mRNA acquires an organized pattern takes account of the kinetics with which these structures form. The transition from a linear, non-periodic array of titin to one with periodicities can occur in a few hours. However, the average lifetime of the titin peptide is approximately 70 hours [22]. Therefore, the titin present in periodic structures is largely "old" titin, that has had no connection to the mRNA for at least a day or more.

It is striking that so many myofibril mRNA acquire highly organized patterns, some with a precision of better than 0.5 μ in position. Work in progress has shown alterations of cell shape and motility with altered mRNA position. Thus it is highly likely that these mRNA patterns are required for the faithful assembly of at least some of these components.

We had earlier established that during muscle development in culture, titin synthesis is closely related temporally with its assembly. Newly-synthesized titin rapidly forms contacts with the cytoskeleton; labeling of nascent titin molecules with tritiated puromycin indicates that these contacts form during translation. Similar observations have been made for myosin assembly [33, 34].

Experiments employing a chemical cross-linking reagent have implicated the identity of the site of this initial contact as either pre-existing titin or muscle specific myosin heavy chain molecules. Once this contact is made, chemical cross-linking and detergent extraction have been used to demonstrate that two proteins, titin and myosin heavy chain (mhc), form a cytoskeletal structure which retains morphological features of intact myotubes. The association between titin and mhc is detectable at the earliest stages of myofibrillogenesis, before mhc becomes periodically registered [33]. It seems likely that at least one function of the highly precise organization of titin mRNA is to facilitate cotranslational assembly.

1.6 Probing the Development of the Elastic Filament System

1.6.1 Pharmacological Probes

Observations *in vitro* of cells regenerating myofibrils after prolonged drug treatment lead to a model that proposed that A bands and I-Z-I bundles formed separately and came together through the intermediation of a "stress fiber-like structure" [35, 36]. This proposal stimulated many additional studies in a variety of systems. The more recent observations, summarized above, use the natural synchrony of the heart or the earliest stages of *in vitro* development and show that titin appears before other myofibril proteins. Titin can organize into an aperiodic linear array even when other myofibril proteins appear diffuse, and is found in periodic patterns when other proteins are present in linear arrays.

The largely synchronous development *de novo* of myofibrils described earlier [37] allows the mechanism of sarcomere formation to be studied by short term treatment with drugs. Since both myofibrils and their precursor structures contain actin, the anti-microfilament drug, cytochalasin D, was of interest. In addition, various reports of microtubule involvement in the assembly process [23, 36, 38–40] focused interest on the microtubule depolymerizing drug, nocodazole, and the microtubule stabilizing drug, taxol. It was desirable to test previous reports of the effects of these drugs after eliminating, as far as possible, the possibility of secondary effects during long term incubations. In addition, in several nonmuscle systems, behavioral interactions between microfilaments and microtubules have been reported [41–47]. The interactions usually represent synergy; that is, when a particular process is affected by either class of drug alone, the use of both classes of drugs frequently leads to less inhibition than with either drug used separately.

For close analysis, it was necessary to quantitate the morphological effects of the drugs. For this reason, samples were scored for morphological effects and compared at several cell densities and ages of cultures. These quantitative measures of the interactions have allowed framing a model which appears to account for all of the observations, including the interactions between the two classes of drugs.

The common design for the experiments was to take synchronized cultures at the time that cables (a linear array of myosin and titin) or sarcomeres formed, fix samples to provide control values before treatment, and allow other samples to develop in the presence or absence of the drug for either 6 or 16 hours. All samples were then processed for immunofluorescence with an anti-myosin antibody specific for skeletal myosin heavy chain. Cells were scored by immunofluorescence as containing either diffusely distributed myosin or titin, cables or sarcomeres. Cells which contained sarcomeres also often still contained cables. Many cells that contained cables also contained diffuse myosin or titin; however, a cell was scored as positive for the most complex structure present in it. This scoring system is conservative, in that if the structure in question was present at all, the cell was scored as positive. Therefore, if the effect of a drug were to reduce the number of sarcomere chains in a cell without eliminating them, the cell would still be scored as containing sarcomeres. Thus, a reduction in the frequency of structures represents a complete absence of those structures for the cells so scored.

1.6.1.1 Formation of Cables

"Recruitment" is the rearrangement of diffuse muscle-specific proteins into cables, linear non-periodic arrays. Recruitment and the stability of these cables once formed was studied by treating muscle cultures engaged in recruitment with drugs for 6 hours (Fig. 2). From Fig. 2, it is clear that after a 6 hour period, half as many cells treated with nocodazole contain cables of

Titin in cables, 24 hours

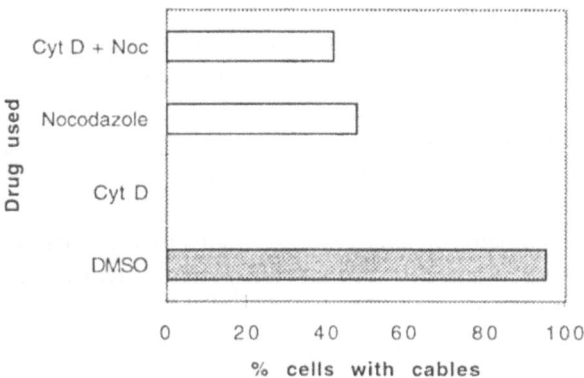

Myosin in cables, 24 hours

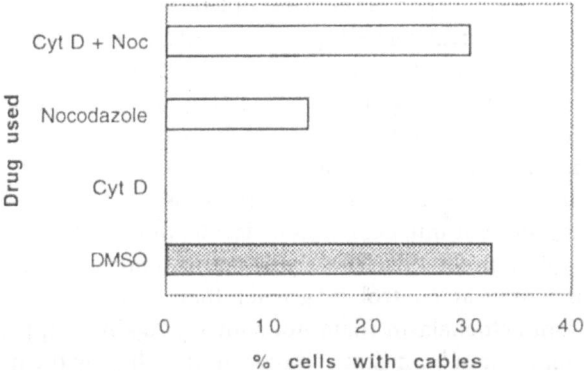

Fig. 2. Graphs show the ability of cables (linear, non-periodic arrays) to form in the presence of anticytoskeletal drugs. Cells begin exposure to drugs at 24 hours after fusion was initiated; drug treatment is for six hours. Cytochalasin is highly disruptive of the cables. More cells have organized titin at this time than have organized myosin. Only the cables containing both titin and myosin display the surprising synergy in which nocodazole protects from disruption by cytochalasin

titin; separate experiments indicate that this is a failure to recruit titin to form new cables. Cytochalasin causes a loss of pre-existing cables from most cells. However, nocodazole protects the existing cables against the effects of cytochalasin. Taxol, which has an opposite effect on microtubules from that of nocodazole, both allows recruitment of myosin into cables to proceed normally and protects against the effects of cytochalasin (data not shown).

Strikingly, the stress fibers in adjacent fibroblasts are not protected by either antimicrotubule drug; this suggests that the cables observed in skeletal muscle cultures are the structural homologs of the "premyofibril" seen in cardiomyocytes. No microtubules that could be detected by immunofluorescence remain after treatment with nocodazole.

Comparing the top and bottom panels of Fig. 2 reveals several other aspects of recruitment. At twenty four hours after fusion has been initiated, more cells have titin in cables than have myosin organized in cables. Myosin organization is more sensitive to nocodazole than titin organization is, consistent with the observations *in vivo* that titin can be organized without myosin present. When in the non-periodic array of the cable, both proteins are very sensitive to perturbations of thin filaments. Finally, the only cables of titin that are protected from the effects of cytochalasin by nocodazole appear to be those that also contain myosin.

1.6.1.2 Sarcomere Initiation

The muscle-specific myosin and titin present in cables undergoes a cooperative rearrangement into a periodic array of sarcomeres [25]. Initiation of sarcomere formation was studied by treating 72 hour cultures with the anticytoskeletal drugs for 6 hours (Fig. 3). Cells treated with nocodazole frequently lose the periodicity of the arrays of myosin, leaving within the cell myosin-containing cables that are often wavy. Treating cells with cytochalasin causes nearly total loss of sarcomeres. As in the case of myosin and titin recruitment into cables, nocodazole, though it disrupts sarcomeres by itself, confers protection from cytochalasin upon sarcomeres. Taxol both permits assembly at control or greater than control rates and confers protection from cytochalasin (data not shown). Again, it appears to be the sarcomere that contain both titin and myosin that display the drug interaction.

The sensitivity of sarcomeres to the various drug treatments changed during their development. Nocodazole and cytochalasin are most destructive to younger sarcomeres. Sarcomeres become largely resistant to the effects of nocodazole with the passage of time, and some other stability has been gained, since the protection against cytochalasin by both taxol and nocodazole is substantially greater in the older cells.

These results confirm and substantially extend earlier studies. The brief (6 h) treatments, permitted here by the synchronous cultures, reduce the possibility that disruption by antimicrotubule agents [48, 49] is a secondary effect of long (3 day) incubation with the drug. Others have shown [50] that rounding of cells does not prevent myofibril formation. These results show that cytochalasin not only prevents the generation of periodic myosin arrays (confirming [48, 49], but can disassemble them once formed. These results extend previous work by distinguishing between recruitment of myosin and titin into cables and initiation of periodic sarcomeres. However, the observations of interaction between the filament systems are both striking and novel.

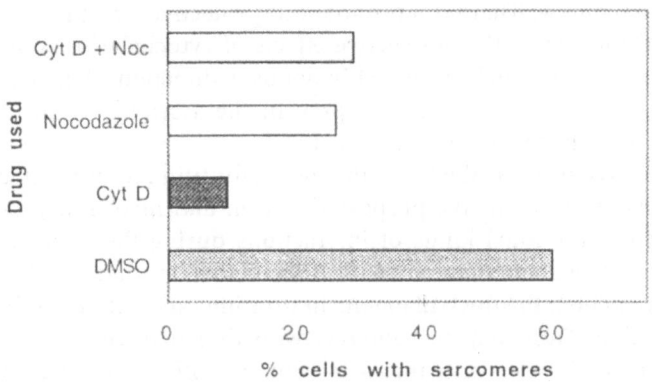

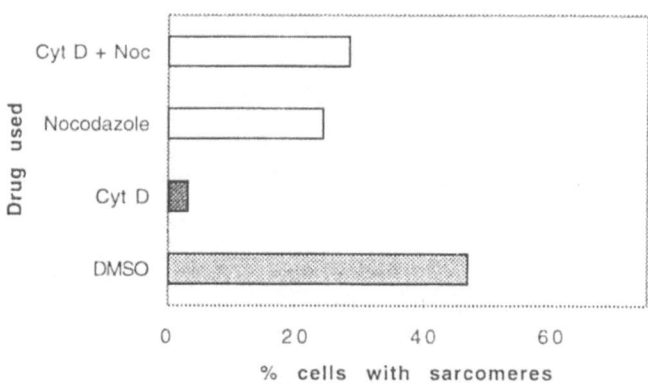

Fig. 3. Graphs show the ability of sarcomeres to form in the presence of anticytoskeletal drugs. Cells begin exposure to drugs at72 hours after fusion was initiated; drug treatment is for six hours. Cytochalasin is highly disruptive of these young sarcomeres (which will become resistant to drugs later in development). Somewhat more cells have sarcomeric titin at this time than have sarcomeric myosin. Only the sarcomeres containing both titin and myosin display the surprising synergy in which nocodazole protects from disruption by cytochalasin

Several models of sarcomere formation, such as the intrinsic geometry model [51] and the directed polymerization model [52], have been proposed. These models however provide no role for microtubules and no explanation of why both nocodazole and taxol, which act in opposite directions on microtubules, should counteract the effects of cytochalasin. Any model adequate to the data should account for the following features of the results described above. There is clearly an involvement of microtubules in both recruitment of proteins into cables and initiation of the periodicity of the sarcomere. There is a pronounced protective effect by both nocodazole and taxol against the destructive effects of cytochalasin, even though these two anti-microtubule drugs act in opposite directions. Finally, in some cases, the drugs used do not merely prevent the assembly of structures but actually cause the loss of existing structures.

We propose the following model for titin recruitment and sarcomere initiation (Fig. 4). We propose that titin and myosin engage in at least three (possibly four) kinds of interactions during the formation of sarcomeres. Titin filaments interact with thick filaments (perhaps through M line components); the thick filaments in turn interact with microfilaments within the cable. Titin may also interact with thin filaments through Z line components. Myosin also interacts with microfilaments that extend outside the cable. Microtubules interact with titin or thick filaments in two ways. The first is a direct interaction, either between titin or myosin and tubulin or contacts mediated by a microtubule-associated protein capable of interact-

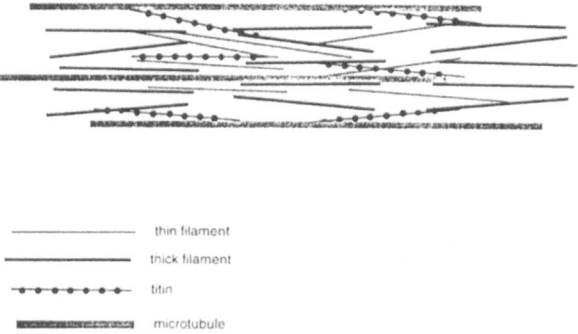

——————— thin filament

——————— thick filament

-•-•-•-•-•-•-•- titin

▬▬▬▬▬ microtubule

Fig. 4. Tensegrity model of myofibrillogenesis: The structural elements are indicated in the key. The model suggests the multiple and dynamic interactions between titin, thick filaments, thin filaments and microtubules that are proposed to account for the pharmacological sensitivities of the overall structure. Tension bearing elements (titin) and tension generating elements (myosin-microfilament complexes) interact with compression bearing elements (microtubules) in a dynamic and interconnected strucure during cable formation and the initiation of sarcomeres. Time point shown is at early stages of alignment of thick filaments, for convenience

ing with titin or myosin without the intermediation of microfilaments. Direct contact of this kind has been proposed [36], and the possibility that such contacts deliver myosin to myofibrillar structures has tentatively been suggested [38]. Secondly, microtubules interact with myosin through a microfilament-dependent connection. These indirect microtubule interactions with myosin comprise some of the interactions between thick and thin filaments outside the cable proposed previously.

These postulated interactions can be used to compose the following scenario for recruitment and initiation. Microtubules deliver titin and myosin into the cables by some direct interaction. Once within the titin-actin-myosin-containing cable, myosin is held in the cable by the actin-myosin interactions within the cable. The other interactions, such as microtubule-titin interactions and microtubule-microfilament-myosin interactions continue but do not normally predominate, so that the majority of titin myosin delivered to a cable remains there. Cytochalasin unbalances these interactions, so that the microfilament interactions with myosin and therefore titin within the cable no longer predominate. This unbalancing effect is consistent with the numerous observations that cytochalasin causes a disruption and disorganization of microfilament relationships without causing net depolymerization.

The combination of drugs, nocodazole with cytochalasin, leaves cables intact because without microtubules the actin-containing networks do not generate adequate tension to remove myosin from the cables. In this case, the persistent cables are wavy. Taxol is presumed to protect against cytochalasin because the direct contacts between microtubules and myosin or titin persist and counteract the unbalanced tensions generated by cytochalasin.

We further suggest that the direct microtubule contacts with myosin or titin are required to bring both thick and titin filaments into register during sarcomere initiation. Presumably later addition of other auxiliary proteins, such as M line proteins and components of the Z disk, accounts for the eventual resistance of sarcomeres to the drugs. The destructive effects of cytochalasin and the protection offered against cytochalasin by both taxol and nocodazole would be explained in the same way as for cables. One important point to emerge from the protection offered by these two drugs is that there appears to be little spontaneous disassembly of these structures that occurs with the active mediation of cytoplasmic filaments. This dynamic and interactive model of titin organization during myofibrillogenesis suggests that the myofibril may be a tensegrity structure (see below (F)).

1.6.2 Molecular Genetic Probes

Studies directed originally at zeugmatin, a myofibril protein, revealed that zeugmatin is part of the Z-Band region of titin and plays an role in the assembly of the Z-bands and myofibrils. Z1.1, a 1.8 kb cDNA from a chicken cardiac expression library and isolated with an anti-zeugmatin antibody, turned out to be 60% identical at the amino acid level to a segment of the Z-band region of human cardiac titin (connectin). This homology and immunological data suggested that zeugmatin is part of the N-terminal portion of chicken titin. Z1.1 product expressed in non-muscle cells colocalized with the alpha-actinin in focal adhesions and stress fiber dense bodies. A fusion construct (Z1.1GFP) linked the Z1.1 kb cDNA to the cDNA for green fluorescent protein (GFP). Although Z1.1 encodes only 362 of the amino acids in the Z-band region of titin, the Z1.1GFP fusion protein targets *in vivo* to the alpha-actinin rich Z-bands of contracting myofibrils. A dominant negative phenotype was observed in living cells expressing high levels of this Z1.1GFP fusion protein. Myofibrillogenesis was inhibited and preexisting myofibrils in these cells were disassembled. These data indicate that the Z-band region of titin (connectin) plays an important role in organizing and maintaining the structure of the myofibril [53].

There is clearly a role for thin filament interactions in myofibrillogenesis. This role might depend upon components that interact in a polar manner, such as the barbed-end actin-binding protein, CapZ. Actin-binding activity of CapZ was inhibited in cultured developing myotubes by two different methods. A monoclonal antibody that prevents the interaction of CapZ and actin disrupts the cables of actin filaments during the early stages of myofibril formation. Second, myotubes expressing a CapZ mutant that does not bind actin have a delay in the time that a striated pattern of actin or organization of α-actinin at Z-disks occurs. Notably, the developmental rhythm of titin and myosin forming into sarcomeres is not altered [54]. These *in vitro* results are consistent with developmental observations that suggested a primary role for titin. Presumably, lateral interactions with microfilaments, which do not depend upon capZ, still allow tension to be generated.

Expressing the myogenic transcription factor MyoD induces myogenesis in non-muscle cell cultures, including expression of muscle-specific proteins. MyoD-transfected cells showed expression of desmin, titin and dystrophin. Cell fusion and in some cases a striated titin pattern also occur. This opens the possibility of developing stable lines to test the effects on myofibrillogenesis of various mutations in myogenic proteins [55, 56].

1.6.3 Potential Sites of Interaction

A complete understanding of the development of the myofibril will entail identifying sites of molecular interactions that contribute to these structures. The M line is one of the two anchorage sites for titin. One of the titin immunoglobulin domains, m4, binds to a myomesin region in domains My4–My6. This binding is affected by phosphorylation. If myomesin is phosphorylated at this site, either by cAMP-dependent kinase or by similar or identical activities in muscle extracts, it no longer associates with titin. It is possible that this interaction controlled by phosphorylation plays a role in sarcomere formation and/or turnover [57]. It may also contribute to some of the developmentally cryptic epitopes of titin.

Other M line interactions may be controlled by phosphorylation. The 5.7 kb of titin sequences from the M-line include immunoglobulin-C2 repeats, separated by interdomain insertion sequences. An interdomain insertion contains four KSP repeats that resemble the multi-phosphorylation repeats of neurofilament subunits H and M. In vitro assays detected high levels of titin KSP phosphorylating kinases , but only in developing muscle. The kinase activity is depleted by antibodies against cdc2 kinase, suggesting the titin KSP kinase is structurally related to cdc2 kinase. Titin C-terminal phosphorylation by SP-specific kinases may be regulated during differentiation to control assembly of M-line proteins into the myofibril [58].

1.7 The Role of the Elastic Filament System in Myogenesis: Toward a Tensegrity Model of the Myofibril

Tensegrity structures are a class of structures first analyzed by Buckminster Fuller. They are made of interacting elements, some of which bear tension and others bear compression. In every day life, examples include geodesic domes, bicycle wheels, umbrellas and most radio towers. Tensegrity structures have the advantage of great strength for a given amount of material and the ability to transmit mechanical stresses and signals across the structure.

The cytoskeleton contains tension bearing components (microfilaments, intermediate filaments) and compression bearing components, and these are extensively cross-linked. It is not surprising then that tensegrity models of the cytoplasm have been proposed and explored. Several recent reviews [59–61] present an overview of these studies. A model of the mitotic spindle as a tensegrity structure has been proposed [62]. A theoretical treatment of tensegrity structures modeled the behavior of cytoskeletal structures more closely than did conventional gel models [63]; its emphasis on "prestress," the amount of tension in the system before mechanical perturbations may provide insights into myofibrillogenesis. The ability of a tensegrity structure in the cytoskeleton to act as a mechanotransducer has also been explored [64].

For titin to play a role in organizing the myofibril, titin must interact precisely with other myofibril components. Titin domain patterns appear to correlate with the architecture of the thick filament at its end. The spacing of six fibronectin-3 (fn3) domains at the A-I junction is regular and consistent with titin serving as a ruler along the thick filament, with each domain measuring about 4 nm in length. An alteration in the interaction between the titin fn3 domains and myosin towards the end of the thick filament may contribute to forming the crossbridge gap and therefore aid in terminating the thick filament [65]. Finely tuned differences in splicing that modify the M region of titin may account for M line differences in different muscles which are conserved across species [66], suggesting that even minor differences in structure here are relevant to the functionality of the architecture. This would be consistent with the fine tuning observed in other myofibril families [67].

One question as yet unresolved is whether the direct microtubule contacts needed for cables are the same as the direct contacts involved in initiating sarcomeres. The drug studies presented here offer no means of distinguishing between these alternatives. In addition, it is possible that the microtubule contacts are not to myosin or titin *per se* but to myosin or titin binding proteins. Finally, this model suggests that the periodicity of the sarcomeres is due to the extension of titin molecules with the load generated being initially borne by microtubules; later protein additions to the myofibril stabilize the structure and make it independent of microtubules.

This model differs in several respects from that proposed by Dlugosz et al. [68] for myofibril formation in cardiac myocytes. They proposed a "template" of cytoplasmic myosin onto which muscle myosin was deposited. They did not include a stage of linear, aperiodic myosin, nor did they envision any mechanical role for microtubules.

The principal implication of this model is that microtubules and microfilaments co-operate in generating both cables and sarcomeres. Moreover, these structures once formed do not represent a static, definitive structure, like those of a virus or a ribosome. Rather, for a considerable period of time, these structures continue to be dependent upon the appropriately balanced interactions between these two cytoskeletal systems. Myofibril development continues until the structures become resistant to drug treatment. The resemblance of this model to a tensegrity structure should be clear.

Similar drug interferences have also been observed in seven non-muscle cell types [41–47]. Although no comparably detailed model has been presented for these examples, Joshi et al. suggest that the behavior they observe is best accounted for by a tensegrity model. Tensegrity structures are, as stated before, ones in which tension elements connect compression elements; the interaction of these elements confers specific properties upon the structure as a whole. The drug interactions seen here during sarcomere initiation make the sarcomere a strong candidate for a tensegrity structure, since only a highly interactive model can account for the mutual interference of drugs that, when applied singly, disassemble the sarcomere. Only further studies of many cell types can show how wide spread tensegrity structures are in cellular organization.

For a tensegrity model to apply, tension elements need to be connected to compression bearing elements. Microtubules have long been known to be able to bear compression. A detailed study of the time course of myofibrillogenesis documented the appearance of longitudinally arrayed bundles of microtubules just before the non-periodic cables of myofibrillar proteins became periodic [39].

What might be the tension bearing elements in the myofibril? Relaxing isolated cardiac myocytes quickly recover their slack length. The force that underlies passive relengthening, or restoring force, may be contributed by titin. Skinned rat cardiac myocytes were shortened and allowed to relax to their original slack sarcomere length. The ability to relengthen was lost when cells were treated with trypsin, which degraded titin without significantly affecting other proteins. Immunoelectron microscopy confirmed that the segment of titin in the I band, which is considered elastic, was missing. Directly measured restoring force of myocytes was reduced when titin had been degraded with trypsin. Titin appears to be the strongest candidate for providing a continuing source of tension in the myofibril, that is independent of contractile state [69].

Another requirement for most tensegrity structures is anchorage into a substrate that can dissipate mechanical forces. It appears that in muscle this role may be played by integrins. Two prominent sites of anchorage are at the myotendinous junction (MTJ) and the costameres, which encircle the interior membrane of the cell to link Z-disks to the sarcolemma. The α V subunit is first found in a diffuse staining pattern, with some associated with nascent myofibrils. Later, it is found striated at the costamere and sometimes at the M-line after α-actinin and titin become striated, but before desmin. That the α V integrin subunit is at the costamere suggests a role in anchoring the myofibril laterally. A developmental role for integrins is further supported by the observation that antibodies to the β 1 subunit inhibit myofibril assembly [70]. Thus, we are gaining increasing confidence that the interactions needed for a tensegrity structure are active in the myofibril, and that this most ordered of biological structures begins its life in a dynamic interplay of balanced forces.

References

1. Squire JM (1997) Architecture and function in the muscle sarcomere. Curr Opin Struct Biol 7:247–257
2. Maruyama K, Kimura S, Ohasi K, Kuwano Y (1996) Connectin, an elastic protein in muscle. Identification of "titin" with connectin. J Biochem 89:701–719
3. Trinick J (1996) Titin as a scaffold and spring. Curr Biol 6:258–260
4. Wang K (1996) Titin/connectin and nebulin: giant protein rulers of muscle structure and function. Adv Biophys 33:123–134
5. Tokuyasu KT, Maher PA (1987) Immunocytochemical studies of cardiac myofibrillogenesis in early chick embryos. II. Generation of alpha-actinin dots within titin spots at the time of the first myofibril formation. J Cell Biol 105:2795–2801
6. Tokuyasu KT, Maher PA (1987) Immunocytochemical studies of cardiac myofibrillogenesis in early chick embryos. I. Presence of immunofluorescent titin spots in premyofibril stages. J Cell Biol 105:2781–2793
7. Schaart G, Viebahn C, Langmann W, Ramaekers F (1989) Desmin and titin expression in early postimplantation mouse embryos. Dev 107:585–596
8. Lin ZX, Eshleman J, Grund C, Fischman DA, Masaki T, Franke WW, et al. (1989) Differential response of myofibrillar and cytoskeletal proteins in cells treated with phorbol myristate acetate. J Cell Biol 108:1079–1091
9. van der Loop FT, Schaart G, Langmann W, Ramaekers FC, Viebahn C (1992) Expression and organization of muscle specific proteins during the early developmental stages of the rabbit heart. Anatomy & Embryology 185:439–450
10. Shimada Y, Komiyama M, Begum S, Maruyama K (1996) Development of connectin/titin and nebulin in striated muscles of chicken. Adv Biophys 33:223–233
11. Furst DO, Osborn M, Weber K (1989) Myogenesis in the mouse embryo: differential onset of expression of myogenic proteins and the involvement of titin in myofibril assembly. J Cell Biol 109:517–527
12. Wang SM, Greaser ML, Schultz E, Bulinski JC, Lin JJ, Lessard JL (1988) Studies on cardiac myofibrillogenesis with antibodies to titin, actin, tropomyosin, and myosin. J Cell Biol 107:1075–1083
13. Handel SE, Greaser ML, Schultz E, Wang SM, Bulinski JC, Lin JJ, et al. (1991) Chicken cardiac myofibrillogenesis studied with antibodies specific for titin and the muscle and nonmuscle isoforms of actin and tropomyosin. Cell & Tiss Res 263:419–430
14. Terai M, Komiyama M, Shimada Y (1989) Myofibril assembly is linked with vinculin, alpha-actinin, and cell-substrate contacts in embryonic cardiac myocytes in vitro. Cell Motil & Cytoskel 12:185–194
15. Schultheiss T, Lin ZX, Lu MH, Murray J, Fischman DA, Weber K, et al. (1990) Differential distribution of subsets of myofibrillar proteins in cardiac nonstriated and striated myofibrils. J Cell Biol 110:1159–1172

16. Komiyama M, Maruyama K, Shimada Y (1990) Assembly of connectin (titin) in relation to myosin and alpha-actinin in cultured cardiac myocytes. J Mus Res & Cell Motil 11:419–428

17. Rhee D, Sanger JM, Sanger JW (1994) The premyofibril: evidence for its role in myofibrillogenesis. Cell Motil Cytoskel 28:1–24

18. Komiyama M, Kouchi K, Maruyama K, Shimada Y (1993) Dynamics of actin and assembly of connectin (titin) during myofibrillogenesis in embryonic chick cardiac muscle cells in vitro. Dev Dyn 196:291–299

19. van der Loop FT, van Eys GJ, Schaart G, Ramaekers FC (1996) Titin expression as an early indication of heart and skeletal muscle differentiation in vitro. Developmental re-organisation in relation to cytoskeletal constituents. J Mus Res Cell Motil 17:23–36

20. Colley NJ, Tokuyasu KT, Singer SJ (1990) The early expression of myofibrillar proteins in round postmitotic myoblasts of embryonic skeletal muscle. J Cell Sci 95:11–22

21. Turnacioglu KK, Mittal B, Sanger JM, Sanger JW (1996) Partial characterization of zeugmatin indicates that it is part of the Z-band region of titin. Cell Motil Cytoskel 34:108–121

22. Isaacs WB, Kim IS, Struve A, Fulton AB (1989) Biosynthesis of titin in cultured skeletal muscle cells. J Cell Biol 109:2189–2195

23. Hill CS, Duran S, Lin ZX, Weber K, Holtzer H (1986) Titin and myosin, but not desmin, are linked during myofibrillogenesis in postmitotic mononucleated myoblasts. J Cell Biol 103:2185–2196

24. van der Loop FT, van der Ven PF, Furst DO, Gautel M, van Eys GJ, Ramaekers FC (1996) Integration of titin into the sarcomeres of cultured differentiating human skeletal muscle cells. Eur J Cell Biol 69:301–307

25. Fulton AB, Alftine C (1997) Organization of protein and mRNA for titin and other myofibril components during myofibrillogenesis in cultured chicken skeletal muscle. Cell Struc & Func 22:51–58

26. Fulton AB (1993) Spatial organization of the synthesis of cytoskeletal proteins. J Cell Biochem 52:148–152

27. Fulton AB, Alftine C (1997) Organization of protein and mRNA for titin and other myofibril components during myofibrillogenesis in cultured chicken skeletal muscle. Cell Struc & Funct

28. Cripe L, Morris E, Fulton AB (1993) Vimentin mRNA location changes during muscle development. Proc Natl Acad Sci USA 90:2724–2728

29. Morris EJ, Fulton AB (1994) Rearrangement of mRNAs for costamere proteins during costamere development in cultured skeletal muscle from chicken. J Cell Sci 107:377–386

30. Taneja KL, Singer RH (1990) Detection and localization of actin mRNA isoforms in chicken muscle cells by in situ hybridization using biotinated oligonucleotide probes. J Cell Biochem 44:241–252

31. Dix DJ, Eisenberg BR (1990) Myosin mRNA accumulation and myofibrillogenesis at the myotendinous junction of stretched muscle fibers. J Cell Biol 111:1885–1894

32. Russell B, Dix DJ (1992) Mechanisms for intracellular distribution of mRNA: in situ hybridization studies in muscle. [Review] [69 refs]. Am J Physiol 262:C1–C8

33. Isaacs WB, Kim IS, Struve A, Fulton AB (1992) Association of titin and myosin heavy chain in developing skeletal muscle. Proc Natl Acad Sci USA 89:7496–7500
34. Isaacs WB, Fulton AB (1987) Cotranslational assembly of myosin heavy chain in developing cultured skeletal muscle. Proc Natl Acad Sci USA 84:6174–6178
35. Lu MH, Dilullo C, Schultheiss T, Holtzer S, Murray JM, Choi J, et al. (1992) The vinculin/sarcomeric-alpha-actinin/alpha-actin nexus in cultured cardiac myocytes. J Cell Biol 117:1007–1022
36. Forry-Schaudies S, Murray JM, Toyama Y, Holtzer H (1986) Effects of colcemid and taxol on microtubules and intermediate filaments in chick embryo fibroblasts. Cell Motil Cytoskel 6:324–338
37. Denning G, Fulton AB (1986) A simple trypsin resistance assay for muscle and other cell fusion. J Histochem Cytochemi 34:959–962
38. Holtzer H, Forry-Schaudies S, Dlugosz A, Antin P, Dubyak G (1985) Interactions between IFs, microtubules, and myofibrils in fibrogenic and myogenic cells. Ann NY Acad Sci 455:106–125
39. Lin Z, Lu MH, Schultheiss T, Choi J, Holtzer S, Dilullo C, et al. (1994) Sequential appearance of muscle-specific proteins in myoblasts as a function of time after cell division: evidence for a conserved myoblast differentiation program in skeletal muscle. Cell Motil Cytoskel 29:1–19
40. Lowrey AA, Kaufman SJ (1989) Membrane-cytoskeleton associations during myogenesis deviate from traditional definitions. Exp Cell Res 183:1–23
41. De Petris S (1974) Inhibition and reversal of capping by cytochalasin B, vinblastine and colchicine. Nature 250:54–56
42. Cheung HT, Cantarow WD, Sundharadas G (1978) Colchicine and cytochalasin B (CB) effects on random movement, spreading and adhesion of mouse macrophages. Exp Cell Res 111:95–103
43. Solomon F, Magendantz M (1981) Cytochalasin separates microtubule disassembly from loss of asymmetric morphology. J Cell Biol 89:157–161
44. Gard DL, Cha BJ, King E (1997) The organization and animal-vegetal asymmetry of cytokeratin filaments in stage VI Xenopus oocytes is dependent upon F-actin and microtubules. Dev Biol 184:95–114
45. van Deurs B, von Bulow F, Vilhardt F, Holm PK, Sandvig K (1996) Destabilization of plasma membrane structure by prevention of actin polymerization. Microtubule-dependent tubulation of the plasma membrane. J Cell Sci 109:1655–1665
46. Ingber DE, Prusty D, Sun Z, Betensky H, Wang N (1995) Cell shape, cytoskeletal mechanics, and cell cycle control in angiogenesis. J Biomech 28:1471–1484
47. Roy SG (1993) Role of stress fibers in the association of intermediate filaments with microtubules in fibroblast cells. Cell Biol Int 17:645–652
48. Croop J, Holtzer H (1975) Response of myogenic and fibrogenic cells to cytochalasin B and to colcemid. I. Light microscope observations. J Cell Biol 65:271–285
49. Holtzer H, Croop J, Dienstman S, Ishikawa H, Somlyo AP (1975) Effects of cytochaslasin B and colcemide on myogenic cultures. Proc Natl Acad Sci USA 72:513–517
50. Puri EC, Chiquet M, Turner DC (1979) Fibronectin-independent myoblast fusion in suspension cultures. Biochem Biophys Res Commun 90:883–889
51. Fischman DA (1970) The synthesis and assembly of myofibrils in embryonic muscle. Curr Top Dev Biol 5:235–280

52. Shimada Y, Obinata T (1977) Polarity of actin filaments at the initial stage of myofibril assembly in myogenic cells in vitro. J Cell Biol 72:777–785

53. Turnacioglu KK, Mittal B, Dabiri GA, Sanger JM, Sanger JW (1997) Zeugmatin is part of the Z-band targeting region of titin. Cell Struct Funct 22:73–82

54. Schafer DA, Hug C, Cooper JA (1995) Inhibition of CapZ during myofibrillogenesis alters assembly of actin filaments. J Cell Biol 128:61–70

55. Roest PA, van der Tuijn AC, Ginjaar HB, Hoeben RC, Hoger-Vorst FB, Bakker E, et al. (1996) Application of in vitro Myo-differentiation of non-muscle cells to enhance gene expression and facilitate analysis of muscle proteins. Neuromuscul Disord 6:195–202

56. Choi J, Costa ML, Mermelstein CS, Chagas C, Holtzer S, Holtzer H (1990) MyoD converts primary dermal fibroblasts, chondroblasts, smooth muscle, and retinal pigmented epithelial cells into striated mononucleated myoblasts and multinucleated myotubes. Proc Natl Acad Sci USA 87:7988–7992

57. Obermann WM, Gautel M, Weber K, Furst DO (1997) Molecular structure of the sarcomeric M band: mapping of titin and myosin binding domains in myomesin and the identification of a potential regulatory phosphorylation site in myomesin. EMBO J 16:211–220

58. Gautel M, Leonard K, Labeit S (1993) Phosphorylation of KSP motifs in the C-terminal region of titin in differentiating myoblasts. EMBO J 12:3827–3834

59. Ingber DE (1993) Cellular tensegrity: defining new rules of biological design that govern the cytoskeleton. J Cell Sci 104:613–627

60. Ingber DE, Dike L, Hansen L, Karp S, Liley H, Maniotis A, et al.. (1994) Cellular tensegrity: exploring how mechanical changes in the cytoskeleton regulate cell growth, migration, and tissue pattern during morphogenesis. Int Rev Cytol 150:173–224

61. Ingber DE (1997) Tensegrity: the architectural basis of cellular mechanotransduction. Annu Rev Physiol 59:575–599

62. Pickett-Heaps JD, Forer A, Spurck T (1997) Traction fibre: toward a "tensegral" model of the spindle. Cell Motil Cytoskeleton 37:1–6

63. Stamenovic D, Fredberg JJ, Wang N, Butler JP, Ingber DE (1996) A microstructural approach to cytoskeletal mechanics based on tensegrity. J Theor Biol 181:125–136

64. Wang N, Butler JP, Ingber DE (1993) Mechanotransduction across the cell surface and through the cytoskeleton. Science 260:1124–1127

65. Bennett PM, Gautel M (1996) Titin domain patterns correlate with the axial disposition of myosin at the end of the thick filament. J Mol Biol 259:896–903

66. Kolmerer B, Olivieri N, Witt CC, Herrmann BG, Labeit S (1996) Genomic organization of M line titin and its tissue-specific expression in two distinct isoforms. J Mol Biol 256:556–563

67. Schiaffino S, Reggiani C (1996) Molecular diversity of myofibrillar proteins: gene regulation and functional significance. Physiol Rev 76:371–423

68. Dlugosz AA, Antin PB, Nachmias VT, Holtzer H (1984) The relationship between stress fiber-like structures and nascent myofibrils in cultured cardiac myocytes. J Cell Biol 99:2268–2278

69. Helmes M, Trombitas K, Granzier H (1996) Titin develops restoring force in rat cardiac myocytes. Circ Res 79:619–626

70. McDonald KA, Lakonishok M, Horwitz AF (1995) Alpha v and alpha 3 integrin subunits are associated with myofibrils during myofibrillogenesis. J Cell Sci 108:2573–2581

Structure and Assembly of the Sarcomeric M Band

D. O. Fürst[1], W. M. J. Obermann[2] and P. F. M. van der Ven[1]

[1] Department of Cell Biology, University of Potsdam, Germany
[2] Department of Cellular Biochemistry, Max-Planck-Institute
for Biochemistry, Martinsried, Germany

Contents

1 Introductory Remarks

The motile machinery of cross-striated muscle cells is organized in a remarkably ordered and stable arrangement of thin actin- and thick myosin-based filaments. This particular structure, highly spezialized for the performance of rapid contractions, has evolved to meet the demands of animals for intentional movement at a higher speed. Variations in filament structure and organization appear to reflect adaptations to specific needs of the respective animal and to the specialization of the particular muscle within the animal´s body (see e.g. Squire, 1981).

The striking regularity of thin and thick filaments within the sarcomere is not simply the result of the self-assembly properties of their major constituting proteins alone, but rather involves specific interactions with a cytoskeletal lattice (Small et al. 1992). The most obvious structures in this context are the M bands and Z-disks, which are involved in packing thick and thin filaments, respectively. The sole component that integrates both compartments is obviously the giant protein titin, also called connectin (reviewed e.g. by Fürst and Gautel, 1995; Labeit et al. 1997; Maruyama, 1994; Trinick, 1994; Wang, 1996). The goal of this chapter is to briefly summarize the structure and the components of the sarcomeric muscle M band (with emphasis on mammalian systems) and to review the latest achievements in the molecular characterization of M band constituents as well as our knowledge about the assembly of this structure.

2 Structural Analysis of the Sarcomeric M Band

Although the existence of a dark zone that traverses the sarcomere in its middle was already reported in the last century (Dobie, 1849), it took another 100 years until this structure received more attention with the advent of electron microscopy techniques. An electron-dense band about 75 nm wide was observed in the cross-bridge-free bare zone center (Hall et al. 1946; Draper and Hodge, 1949). Higher resolution cross sections revealed three to five bridge-like structures (called "M-bridges") at regular intervals, which seem to connect the thick filaments into their characteristic hexagonal lattice. In addition, these M-bridges were thought to be linked by "M-filaments" that run parallel to the thick filaments (Knappeis and Carlsen, 1968). Negatively contrasted ultrathin cryosections after all provided a lot more of structural detail. The currently used terminology is based on these results (Sjöström and Squire, 1977). Thus, in longitudinal orientation up to 17 electron dense stripes arranged in lateral symmetry can be resolved in the entire M-region. The sketch in Fig. 1 summarizes this nomenclature, where the most prominent lines are called M1 (the central line), M4 / M4´ (22 nm distant on either side of M1) and M6 / M6´ (at a distance of 44 nm from M1). Three-dimensional image reconstruction revealed an even more complex M band architecture. Luther and Squire (1978) identified Y-shaped structures that could cross-link adjacent M-filaments (see also Fig. 1b).

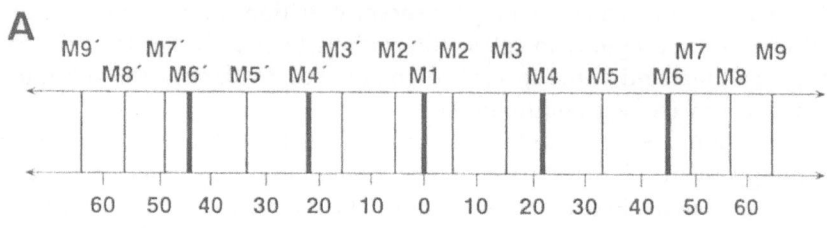

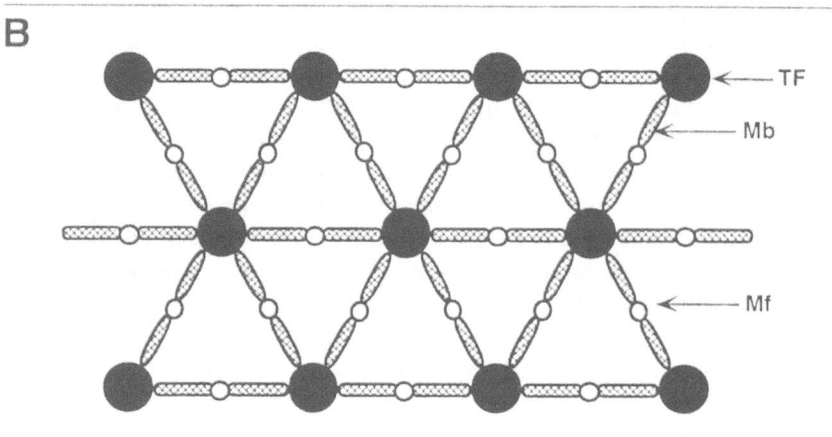

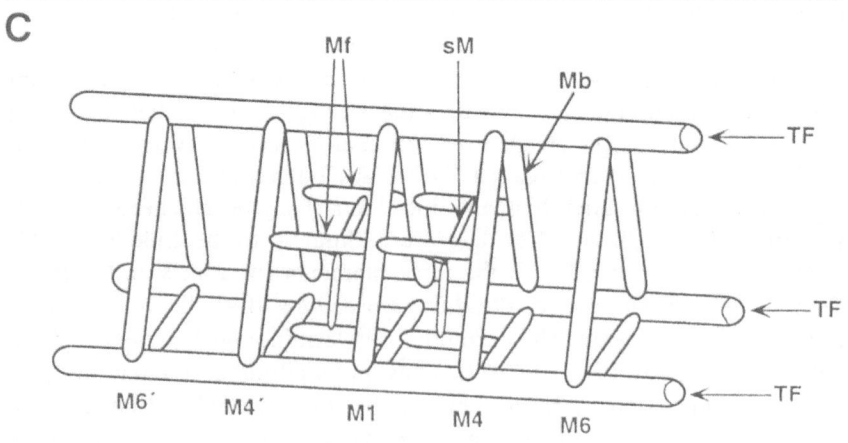

A systematic comparison of different muscles unravelled a surprising complexity of M band ultrastructure that seems to reflect mainly adaptations to distinct physiological fiber characteristics (Edman et al. 1987). As a general guideline one can conclude that slow type I fibers have M bands with four strong M-lines (M4, M4′, M6, M6′; M1 absent), while fast type II fibers show three M-lines (M1, M4, M4′; M6, M6′ absent). An M band with 5 M-lines seems to correlate with fibers of intermediate speed (Sjöström & Squire, 1977; Edman et al. 1988; Luther & Crowther, 1984; Luther et al. 1981; see also review by Squire et al. 1987). More recently it was found that the morphology of the M band in cardiac muscles can be correlated roughly with heartbeat frequency. Thus, in general cardiac M bands give a five line pattern, in which the M1-line is relatively much stronger than the other M-lines (Pask et al. 1994). These findings also reinforced the view that at least some of the constituting M band proteins could also have a physiological instead of solely a structural role.

3 Protein Components of the M Band

Until now, five proteins have been described to be localized specifically in the M band: the muscle isoform of creatine kinase, M-protein, myomesin, skelemin and titin. The major role of creatine kinase is most likely a physiological one in energy metabolism, hence it will be reviewed only very briefly below. Skelemin is thought to be involved in the linkage of myofibrils to intermediate filaments rather than serving a function in establishing the actual M band structure. Thus, we feel that it is justified to keep the discussion of this protein to a minimum. Instead we will focus on a description of M-protein and myomesin and discuss here briefly the older literature and the resulting knowledge about the basic biochemical properties of these

---

Fig. 1. Schematic presentation of the appearance of the M band of vertebrate stiated muscle sarcomeres in electron micrographs. Panel A gives approximate locations and the nomenclature of the striations revealed in negatively contrasted ultrathin longitudinal cryosections (vertical bars). The most prominent lines, also referred to as M-bridges, are printed in bold. The numbers below this scheme show the distances in nm from the sarcomere center, the M1-line (according to Sjöström and Squire, 1977). Panel B gives a schematic drawing of a cross-section at the level of the M4-lines. TF, thick filaments; Mb, M-bridge; Mf, M-filament (Knappeis and Carlsen, 1968; Luther and Squire, 1978; Luther et al. 1981; Luther and Crowther, 1984). Panel C is a three-dimensional model of the M band as deduced from electron microscopical data. M-bridges (Mb) which appear at the level of M1-, M4- and M6-lines are connected in longitudinal direction by M-filaments (Mf). M-filaments, in turn, are bridged by Y-shaped secondary M-bridges around the M3-line (according to Knappeis and Carlsen, 1968; Luther and Squire, 1978; Squire et al. 1987)

polypeptides. Since the most recent studies were based on details of the primary structures, these will be discussed in sections 5 and 6 of this chapter.

3.1 Creatine Kinase

The family of the creatine kinases (CK) consists in vertebrates of at least five isoenzymes that are expressed in a tissue-, developmental stage-, and intracellular compartment-specific manner (Eppenberger et al. 1967; Hossle et al. 1988; Wallimann et al. 1992). While two of these isoforms are restricted to the mitochondrial intermembrane space (Jacobs et al. 1964; Jacobus and Lehninger, 1973; Hossle et al. 1988), the other three isoforms are cytosolic enzymes (Eppenberger et al. 1967; Wallimann et al. 1986; Friedman and Perryman, 1991). All the distinct CK isoenzymes are expressed from different genes that show a high degree of conservation, indicating that the genes have descended from a single ancestral gene. The degree of identity between the different subunits ranges between 60 and 65%, while the identity level between identical subunits in different species is significantly higher (Mühlebach et al. 1994). The main function of creatine kinases is to catalyze the reversible exchange of high-energy phosphate between phosphocreatine (PCr) and ATP, and thus, the regulation of intracellular ATP concentrations. In this way, a temporary surplus of energy can be stored as PCr, preventing increases of local intracellular ATP concentrations that might lead to undesirable influences on fundamental processes (Meyer et al. 1984). On the other hand, from the pool of PCr, ATP can be regenerated from ADP in order to meet any raised energy requirements of the cell. Since usually energy is required at specific sites within the cell, one might expect that ATP is regenerated or stored in close vicinity to these sites (see e.g. reviews by Wyss et al. 1992; Wallimann, 1994).

In adult skeletal muscles, where energy requirements fluctuate enormously, both temporally and spatially, MM-CK is partially present in the cytosol as a soluble enyme. Part of the MM-CK pool is, however, localized specifically at i) I-bands (Wallimann et al. 1989; Wegmann et al. 1992), together with ATP-producing glycolytic complexes (Arnold and Pette, 1970), ii) the sarcolemma where it regenerates ATP for Na^+/K^+-ATPase (Sharov et al. 1977; Saks et al. 1977; Grosse et al. 1980), iii) the sarcoplasmatic reticulum (Baskin and Deamer, 1970; Kahn et al. 1972) where it provides energy for the ATP-dependent Ca^{2+} pumps (Levitzki et al. 1978; Rossi et al. 1990) and iv) the M band, where it rephosphorylates ADP produced by the Mg^{2+}-ATPase during contraction. The MM-CK in the latter location (~5 to 10% of the total cellular amount depending on the muscle type; Turner et al. 1973; Wallimann et al. 1977) correlates well with the appearance of an electron dense M band (Wallimann et al. 1977). Detailed analysis at higher resolution using specific antibodies allowed to correlate MM-CK to M4, M4′-bridges (Strehler et al. 1983; Wallimann et al. 1983). Conversely, myofibrils loose both electron density in the M1, M4, M4′-bridges and M band-bound MM-CK upon extraction with solutions of low ionic strength (Wallimann et al. 1977; Strehler et al. 1983).

Interestingly, only the muscle-specific homodimeric MM-CK isoform has the potential to bind to the M band, while both the brain-specific isoform BB-CK and heterodimers of both enzymes (i.e. MB-CK) cannot bind (Perriard et al. 1978; Wallimann et al. 1983). In an extensive study this M band targeting of MM-CK could be ascribed to an amino-terminal epitope of the protein (Stolz, 1997). Since the three-dimensional structure of the homologous mitochondrial isoenzyme was resolved recently (Fritz-Wolf et al. 1996), the extension of the dimeric enzyme and the orientation of specific sites can now be modelled rather nicely. Thus the dimer is 9.2 x 4.2 nm wide, with the M band binding epitopes reaching out at both ends of the longitudinal axis. While this would be sufficient to account for the dimensions of half an M4, M4′-bridge in a resting sarcomere, it cannot be directly correlated with longer M-bridges as a result of the increased distance of thick filaments during contraction. At present, the only way out of this conflict is the postulation of reversible and dynamic rearrangements. An indication that this could indeed be the case is the identification of a putative secondary M band binding site at the back of the MM-CK dimer (Stolz, 1997). These findings strengthen the concept that the M band does not only have a structural role but also imply a more dynamic and physiological nature of this compartment.

The introduction of an MM-CK null mutation into the germline of mice, recently allowed for more detailed studies on the function of the enzyme. Unexpectedly, these mice did not show any obvious phenotypic abnormalities. Also the levels of free ATP, PCr and P_i were found to be normal in their resting muscles, and upon electrostimulation PCr was hydrolyzed and the concentration of P_i was raised. It remained unclear which enzyme took over the role of MM-CK (Van Deursen et al. 1993; Van Deursen et al. 1994) until it was found out that in double mutants (MM-CK/ScCKmit) PCr was no longer used to regenerate hydrolyzed ATP (Steeghs et al. 1997). This observation supported the idea that in MM-CK null mutants PCr was redirected through mitochondrial CK (Van Deursen et al. 1993; Van Deursen et al. 1994). Other adaptations of muscles to the lack of MM-CK included the increase of mitochondrial capacity and glycogen content. These adaptations could, however, not completely compensate for the absence of MM-CK, as evident from the finding that – obviously due to the lack of buffering capacity – knock-out mice lack burst activity. Additional experiments, in which mice were generated that have MM-CK activities between 16 and 50%, showed that the level of the reduction correlated with the decrease in MM-CK activity (Van Deursen et al. 1994).

3.2 M-Protein

M-protein was identified in the first systematic biochemical searches for M band-specific proteins. Masaki et al. (1968) described it as the antigen that was responsible for M band crossreactivity of their α-actinin antiserum (Masaki et al. 1967). Interestingly, the authors also reported that "this substance markedly accelerated the lateral association of myosin aggregates" (Masaki et al. 1968). The subsequent purification of the respective protein yielded a 165 kDa polypeptide. Immunocytochemistry using an antiserum against the purified protein confirmed its M band localization (Masaki & Takaiti, 1974).

A revised isolation procedure revealed that the M-protein preparation described above was still contaminated with glycogen debranching enzyme (Trinick and Lowey, 1977). In the same study, M-protein was unequivocally shown to be an intrinsic constituent of the M band by immunoelectron microscopy. This observation was further specified in a series of subsequent studies (Strehler et al. 1980; Eppenberger et al. 1981; Strehler et al. 1983). Consistently, the M6/M6′ lines and the region between them were decorated. The lack of higher resolution data that would allow to describe the exact arrangement of M-protein within the M band can be attributed to two circumstances. First, polyclonal antibodies were used, which most likely reacted with a larger number of epitopes that cover the entire polypeptide. As a result, spatial resolution in the electron microscope was lost. Second, it became evident that these antisera also crossreacted with myomesin, another major structural protein of the M band, which in turn was discovered later via the use of monoclonal antibodies (Grove et al. 1984; see below).

Biochemical parameters of M-protein were most thoroughly studied by Trinick and Lowey (1977). Apparently M-protein behaves as a monomer in solution and has a sedimentation coefficient of 5.1 S. Circular dichroism spectra revealed no appreciable amount of α-helix and no enzymatic activity could be ascribed to M-protein.

At the same time, several attempts were undertaken to investigate the predicted interaction of M-protein with other M band constituents. The results of these studies, however, were not conclusive, since partially conflicting results were obtained. On the one hand, analytical ultracentrifugation revealed a 1 : 1 association of M-protein with MM-CK (Mani and Kai, 1978a) and, in addition, binding of M-protein to the S2 portion of myosin (Mani and Kai, 1978b; Herasymowych et al. 1980). Although the latter result was subsequently confirmed by fluorescence spectroscopy by the same authors (Mani and Kai, 1981), other laboratories could not confirm these findings and concluded "that such interactions are either weak or absent *in vitro*" (Woodhead and Lowey, 1983; Bähler et al, 1985). Doubt was cast on these results because they were obtained with buffers of unphysiologically high salt concentrations. When similar experiments were performed under more native conditions, essentially negative results were reported (Woodhead and Lowey, 1983). Furthermore, affinity chromatography revealed a weak interaction of M-protein with the LMM-fragment of myosin, but once more no binding with MM-CK (Woodhead and Lowey, 1983). An immunoelectron microscopic investigation clearly demonstrated a strong association of M-protein with native isolated thick filaments. If, however, M-protein was extracted by increasing the ionic strength, it was no longer possible to obtain re-binding of M-protein. This was interpreted as an indication for a weak interaction between M-protein and myosin (Bähler et al. 1985). The surprising finding that M-protein is a component of the conspicuous globular head structure of isolated titin molecules (Nave et al. 1989; Vinkemeier et al. 1993) reinforced the interest in efforts to elucidate the contribution of M-protein to the formation of the M band structure (see below).

3.3 Myomesin

Ironically it were the problems with efforts to specifically localize M-protein at high resolution that led to the discovery of another M band constituent. This protein turned out to be more widely distributed and of greater importance than M-protein itself. In the course of raising mAbs against M-protein, Grove et al. (1984) also obtained antibodies that reacted with a distinct and immunologically unrelated 185 kDa polypeptide which they called "myomesin". M-protein and myomesin were shown to elute in two distinct peaks on ion exchange columns. Since, however, myomesin was not purified to homogeneity at that time, the protein could not yet be characterized (Grove et al. 1984). Subsequent studies that focussed on the investigation of embryonic expression patterns clearly established that myomesin is expressed in all cross-striated muscle fibers from the earliest stages of myofibrillogenesis onwards (see below). Thus it was speculated that myomesin might be important for the assembly and structural maintenance of thick filaments (Bähler et al. 1985).

It was not until recently that a procedure for the purification of myome-sin from bovine skeletal muscle was described, which allowed a biochemi-cal-biophysical characterization of the protein (Obermann et al. 1995). Cir-cular dichroism spectra showed a high degree of β-structure and analytical ultracentrifugation yielded an s_{20} value of 5.1 S as sedimentation coefficient. In electron micrographs the protein appeared as a short, flexible and seg-mented rod with a contour length of ~50 nm and a diameter of ~4 nm. These data closely resemble the values reported for M-protein (see above). The availability of purified protein now also allowed a reevaluation of bind-ing assays. Indeed, in a solid phase overlay assay myomesin was shown to bind to the LMM portion of myosin and this binding could be ascribed to the amino-terminal 240 residues of myomesin (Obermann et al. 1995).

Similar to M-protein, myomesin was also found to interact strongly with titin in its globular M band end (Nave et al. 1989). These polypeptides also seem to be involved in the formation of titin dimers and higher oligomers. One can conclude, therefore, that the major role of myomesin is a participa-tion in the linkage of titin molecules to the thick filament. The recent mo-lecular characterization of myomesin elucidated at least some aspects of its interactions with titin and myosin (see section 6.1.1).

3.4 Titin

The giant protein titin seems to be the major component responsible for the integration of the thick and thin filament systems of the sarcomere, since individual titin molecules span an entire half sarcomere. This was most firmly established by a panel of monoclonal antibodies that recognized dif-ferent epitopes along the titin polypeptide and that in immunoelectron mi-croscopy decorated sarcomeres at distinct positions along the whole half sarcomere (Fürst et al. 1988; Fürst et al. 1989a). In biochemical experiments it was subsequently shown that under non-denaturing extraction conditions a fraction of myomesin and M-protein remained firmly bound to the con-spicuous globular head of titin (Nave et al. 1989). Sequencing of titin cDNA portions put the amino-terminus towards the Z-disc (Labeit and Kolmerer, 1995; Yajima et al. 1996; Gautel et al. 1996), while carboxy-termini were shown to reside in the M band (Gautel et al. 1993; Obermann et al. 1996).

Several, more general aspects of the biology of titin will be covered in other chapters of this book in great detail (and have been covered by a num-ber of review article in the past), therefore we find it justified to leave it at this short statement at this point. An appretiation of the cDNA portion of titin that encodes its M band domains, the exact localization and its interac-tions will be reviewed in sections 4, 5 and 6, respectively.

3.5 Skelemin

Skelemin was initially identified as a pair of closely related high molecular mass proteins of 220 and 200 kDa, respectively. It was shown to copurify with the intermediate filament proteins desmin and vimentin, and was detected in cardiac, skeletal and smooth muscle (Price, 1984). Immunofluorescence microscopy located skelemin to the periphery of M bands in skeletal muscle, which suggested a role in the linkage of the sarcomeres to the intermediate filament cytoskeleton (Price, 1985). Although some aspects of the recently published cDNA sequence (Price and Gomer, 1993; see also below) may imply such an association, no firm proof has been published so far. Unfortunately, it was not possible to establish a protocol for the purification of native skelemin that would allow for the characterization of its biochemical properties.

4 Cloning of M Band Protein cDNAs and Genes

Recent years have been characterized by enormous progress in the elucidation of the primary structures of a great number of structural and regulatory proteins. On the one hand this allowed to identify unexpectedly complex protein families. On the other hand, a better understanding of the protein-protein interactions that lead to the formation of supramolecular structures was enabled. Both points also apply to the proteins that comprise the sarcomeric cytoskeleton. As to the first point, it turned out that these proteins form an intracellular branch of the immunoglobulin superfamily. Thus, the elucidation of the primary structure of the *Caenorhabditis elegans* unc-22 gene, which encodes an invertebrate titin homologue called twitchin (Benian et al. 1989), broke the dogma that immunoglobulin motifs were limited to cell surface and extracellular proteins (Williams and Barclay, 1988). Later it was shown that a common, modular building plan is shared by all these proteins, because they consist mainly of two kinds of almost globular ~100 amino acid residue repeat domains: (a) fibronectin type III domains (Fn-like; Potts and Campbell, 1994), and (b) immunoglobulin cII motifs (Ig-like; Williams and Barcley, 1988; Holness and Simmons, 1994). Both repeats share important features with their extracellular counterparts, as has been demonstrated in several structural studies at least for the latter motifs. The polypeptide chain of both domains folds into a beta-barrel with seven strands that enclose an invariant tryptophane (Holden et al. 1992; Politou et al. 1994). The modular building principle is also reflected by the shape of purified and negatively stained or rotary shadowed molecules in electron micrographs, in which they all appear as segmented rods (Hartzell et al.

1985; Woodhead and Lowey, 1982; Nave et al. 1989; Fürst et al. 1992; Vinke-meier et al. 1993; Obermann et al. 1995).

Sequence conservation between all the members of this protein family is mostly limited to relatively few residues that specify the particular three-dimensional fold of the barrels, while a higher degree of variability is al-lowed for residues localized on the external surfaces (Williams et al. 1989). This provides a versatile modular protein organization in which seemingly minor amino acid exchanges (or modifications) on the module surface can result in quite distinct ligand binding properties. At the same time their modular organization allows to dissect these proteins into smaller, better manageable units in order to approach structure – function correlation at different levels.

4.1 M-Protein cDNA Sequences and Gene Structure

The complete cDNA sequence encoding M-protein has been cloned from chicken pectoralis muscle (Noguchi et al. 1992), human skeletal muscle (Vinkemeier et al. 1993) and murine skeletal muscle (Steiner et al. 1998). Concurrently, a protein of molecular mass ~165 kDa was predicted, whose expression is limited to cardiac and fast skeletal muscle fibers. Almost 95 % of the sequence consist of either Fn-like repeats (motif 1) or Ig-like domains (motif 2). These are arranged in the order 1 – 1 – 2 – 2 – 2 – 2 – 1 – 1 – 1 – 1 – 1 (see Fig. 2). The amino-terminal ~110 amino acids, however, do not show any significant homology to other sequences in the databanks. Inter-estingly, the degree of inter-species conservation is significantly higher within the repeat domains as compared to the unique amino-terminal re-gion (Steiner et al. 1997).

The first ~80 residues are extremely basic, and secondary structure pre-dictions would allow for longer stretches of α-helix in the first 150 residues. A closer inspection of the primary sequence indicated two potential target sites for cAMP-dependent protein kinase A (PKA): the serine at position 39 (located in the sequence motif KKRAS) and the serine at position 76 (in the motif KRVS) (Vinkemeier et al, 1993). Since both these sites are well con-served between all the species studied, a potential participation in some regulatory mechanism seems plausible.

Using the cloned cDNA, the expression of M-protein was reinvestigated at the mRNA level both in chicken (Noguchi et al. 1992) and mouse (Steiner et al. 1997). In general, both studies agree with previous data at the protein level (Grove et al. 1985; Grove et al. 1987; Carlsson et al. 1990; see also above). Thus, the expression patterns in cardiac muscle differ strongly from those in skeletal muscles: while the level of cardiac M-protein expression

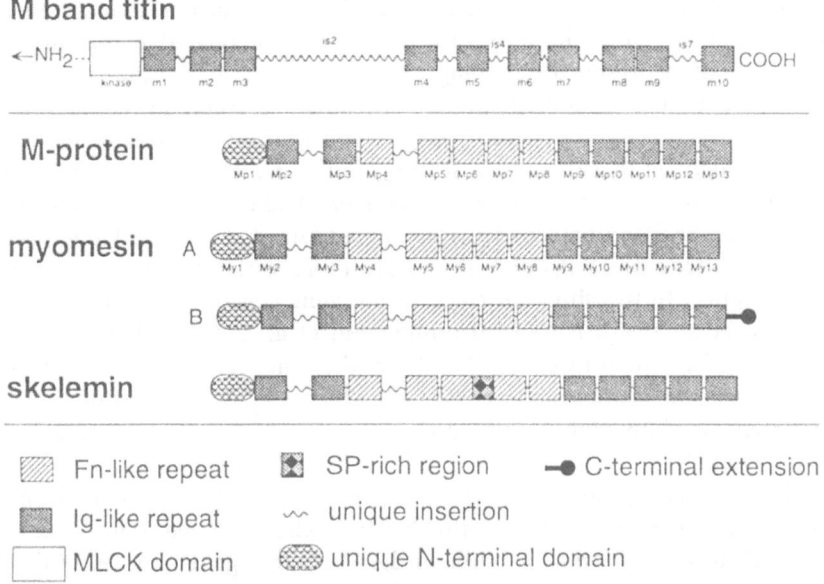

Fig. 2. Domain organization of structural proteins of the sarcomeric M band. The diagrams show the composition of the major, presently known cytoskeletal proteins that constitute the M band. All of them are intracellular members of the immunoglobulin superfamily. M-protein, myomesin (A: mammalian form; B: longer splice variant found in avian muscle) and skelemin form a subgroup of this family that shares a common domain structure. Compiled from: titin – Gautel et al. 1993; M-protein – Noguchi et al. 1992 and Vinkemeier et al. 1993; myomesin – Vinkemeier et al. 1993 and Bantle et al. 1996; skelemin – Price and Gomer, 1993

increased steadily and strongly from the fetus to the adult animal, the expression in skeletal muscles appeared biphasic. Initial M-protein expression was observed in all primary generation myotubes, before it was down-regulated around birth. Re-expression in the adult animal was limited to fast skeletal and cardiac muscle fibers, while all slow skeletal muscle fibers remained negative (Noguchi et al. 1992; Steiner et al. 1997). These data indicated complex events that regulate the expression of M-protein. Since these results were identical to those obtained at the protein level (Grove et al. 1985; Grove et al. 1987; Carlsson et al. 1990), gene activity must be regulated at the transcriptional level.

To allow for a thorough analysis of the factors that could specify such subtle expression differences, the gene encoding murine M-protein was recently investigated in detail (Steiner et al. 1998). First, the complete exon – intron organization of this gene was determined. The M-protein gene spans 75 kbp and is divided into 37 exons and 36 introns (Fig. 3). Several aspects of intron positioning that largely can be related to the modular structure of M-protein are strikingly similar to a number of the extracellular members of the immunoglobulin superfamily. Essentially all the repeat domains are encoded by two exons, except for two Ig-like domains which are encoded by three exons, a feature that is novel for this domain type (Steiner et al. 1998). Interestingly, all domain-flanking introns are in phase I, i. e. the corresponding codon is interrupted after the first nucleotide, while internal intron positions and phases are more variable. This corresponds, for instance, to the situation faced in the genes encoding human perlecan (Cohen et al. 1993), N-CAM (Cunningham et al. 1987) or lymphocyte T4 protein (Littman and Gettner, 1987). On the other hand, a novel feature of the murine M-protein gene is that the short sequence insertions located between neighbouring Ig-like dimains are encoded by separate exons (Steiner et al. 1998).

In contrast, the murine M-protein gene is strikingly different from the portion of the human titin gene that has been characterized so far (Kolmerer et al. 1996; see section 4.3). While in the M-protein gene only 7% of the sequence are coding, 67% are coding in the case of titin. Also, no correlation of the exon – intron organization to the domain structure as described here was found (Fig. 3).

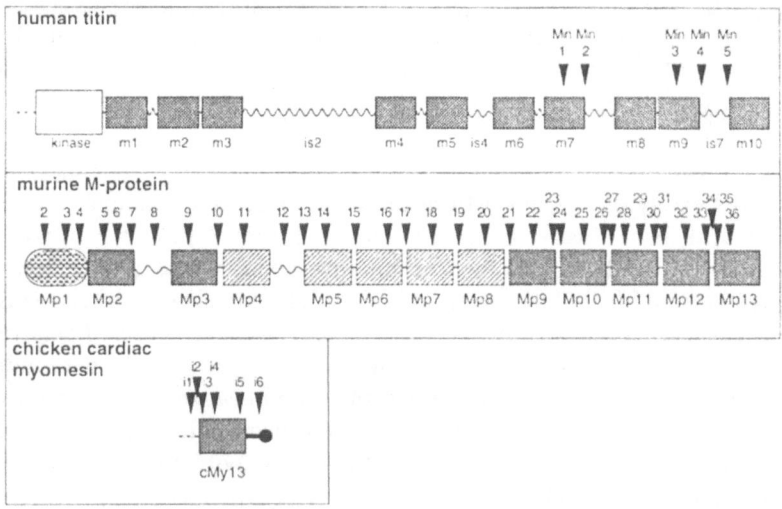

Fig. 3. Partial structure of the genes coding for human titin, murine M-protein and chicken cardiac myomesin. Molecular models correspond to drawings in Fig. 2. Documented intron positions are indicated by arrowheads. Compiled from: titin – Kolmerer et al. 1996; M-protein – Steiner et al. 1998; chicken cardiac muscle myomesin – Bantle et al. 1996

The M-protein gene organization can be considered a typical example for a mosaic protein which is thought to have developed by endoduplication events and exon shuffling (Patthy et al. 1991). In such a case, strict conservation of intron phases at domain borders is mandatory in order to prevent a shift of the reading frame and a premature stop. Since both, gene size and organization, differ so strongly between M-protein and titin, it is obvious that both genes developed independently and at different time scales. Clearly, more information about the gene structure of other members of the intracellular branch of the immunoglobulin superfamily is needed, before further conclusions about their evolution can be drawn.

The complex transcriptional regulation of M-protein gene activity that was proposed on the basis of the expression studies described above also justified a systematic search for regulatory elements in the putative promoter region (Steiner et al. 1998). Sequencing of ~1 kbp upstream of the transcriptional start point yielded, in addition to canonical TATA- and CAAT-boxes, a wealth of general and muscle-specific transcription factor binding motifs. In a reporter gene analysis three regions that harbor a MEF2-site (Fickett, 1996) and several E-boxes (binding sites for transcription factors of the MyoD family; Lassar et al. 1989) turned out to determine the muscle-specific transcription of M-protein (Steiner et al. 1997). The particular expression pattern of M-protein, however, requires the interaction with additional factors. Since the biphasic expression of herculin/myf-6 in the mouse embryo (Bober et al. 1991; Hinterberger et al. 1991) equals the appearance of M-protein, this protein is a plausible candidate for being an important regulator of M-protein expression.

Two more sites of potential interest were revealed in the sequence analysis but their significance has not yet been shown *in vivo*: an AP-1 box (Lee et al. 1987) and a TPA-responsive element (Okuda et al. 1990) could be of importance in light of the dramatic effects of phorbol esters on muscle cells in vitro (Croop et al. 1982; Lin et al. 1987). It is feasible that the interference of these substances with the transcriptional activation of certain sarcomeric proteins is a prime cause for the observed morphological changes.

4.2 Myomesin cDNA Sequences

Complete cDNAs coding for myomesin have been cloned from human skeletal muscle (Vinkemeier et al. 1993) as well as from chicken cardiac muscle (Bantle et al. 1996), while a partial cDNA sequence is available from rat heart myomesin (Bantle et al. 1996). A comparison of these sequences with each other and with M-protein (see above) revealed a number of interesting features. First, it became evident that myomesin and M-protein share the same core domain organization of Ig-like and Fn-like repeats (see Fig. 2). The extremely high degree of relatedness between myomesin and M-protein is also reflected by the fact that the percentage of identical amino acid residues within the same species is above 50. Even the length of all the inter-domain linkers is conserved between myomesin and M-protein. Major differences, therefore, are limited to both terminal molecular regions. The amino-terminus of myomesin shows no relation to M-protein or to any other known sequence. A conspicuous repeated hexapeptide motif (KQSTAS) is found in variable copy numbers in both mammalian sequences but not in the avian sequence. If indeed the conserved serine and threonine residues are targets of a yet unidentified protein kinase, as was proposed (Vinkemeier et al. 1993), this would indicate species-specific differences in the regulation of certain myomesin activities and/or binding properties (see section 6.1.2).

A second, species-specific difference was identified in the carboxy-terminal region of myomesin. While all mammalian cDNA sequences and the sequence from chicken skeletal muscle terminate at almost identical positions shortly after the last Ig-domain, a larger chicken cardiac-specific transcript gives rise to a polypeptide that is 78 amino acid residues longer (Bantle et al. 1996). This avian muscle-specific length heterogeneity seems to be related to differential splicing in the 3′ region of the single chicken myomesin mRNA, giving rise to tissue-specific transcripts of different length (Fig. 3; Bantle et al. 1996). A definitive answer will have to await detailed analysis of the structure of a mammalian myomesin gene. Interestingly, a preliminary observation from *in situ* hybridization on eight day old chicken embryos indicated in addition differences in temporal expression patterns of the observed isoforms. While a cardiac sequence-specific probe was found to be expressed exclusively in cardiac muscle tissue, a skeletal muscle sequence-specific probe reacted mainly with skeletal muscle and to a lesser extent with cardiac muscle (Bantle et al. 1996).

4.3 Titin M Band cDNA Portion and Its Partial Gene Structure

Since an appreciation of the complete primary structure of titin and its implications for sarcomere structure and physiology are the focus of other chapters of this volume, we will restrict ourselves here on the information that is necessary for an understanding of M band architecture.

The carboxy-terminal end of titin was shown to reside in the M band (Gautel et al. 1993). A serine/threonine protein kinase domain that is very similar to the kinase domains in invertebrate titin analogues (Benian et al. 1989; Ayme-Southgate et al. 1991; Heierhorst et al. 1994) and in smooth muscle myosin light chain kinase (Olson et al. 1990; Pearson et al. 1991) denotes a clear transition from an ordered pattern of Ig-like and Fn-like repeats in the outer A band regions (Labeit et al. 1992) to a unique and complex arrangement of Ig-like domains (see Fig. 2). A total of 10 Ig-motifs was identified, which together with the kinase domain and several unique insertions form a polypeptide of ~250 kDa. Interestingly, the sequences of these M band Ig-like domains are somewhat divergent from those of other A band regions (Gautel et al. 1993). Again, conservation is restricted to those residues which are important for the correct folding of the characteristic β-barrels (Holden et al. 1992). Another distinctive feature of this molecular region of titin is the presence of up to ~500 residues long interdomain insertions. Of particular interest is the insertion located between Ig-domains m5 and m6 (is4) because it harbors four repeats of the tetra-peptide VKSP. This short motif was found to serve as a serine phosphorylation site in neurofilament-H and -M proteins (Lee et al. 1988) and – in the case of titin – turned out to be phosphorylated in developing but not in differentiated myocytes. The responsible kinase is at least structurally related to cdc2 kinase. This result was taken indicative for a possible role of M band titin phosphorylation in the regulation of sarcomere assembly (Gautel et al. 1993).

A second region of particular interest is the insertion between the two most carboxy-terminally situated Ig-like domains (is7). This insertion was revealed to represent one out of two specific binding sites for the muscle-specific calpain isoform p94 (Sorimachi et al. 1995). It was speculated that the presence of two binding sites on the titin polypeptide indicated a dual role for the titin – p94 interaction: the prime binding site (located in the I band portion) might be involved in the turnover of titin; the second site (is7) could sequester and stabilize p94, since its soluble form is usually degraded rapidly (Sorimachi et al. 1995). The finding that is7 is differentially spliced in some muscles, makes the situation even more complex (Kolmerer et al. 1996). Alltogether, this means that the M band portion of titin could elegantly provide means to control titin's own stability in a fiber-type specific manner.

The recently characterized 3´ end of the human titin gene (Kolmerer et al. 1996) showed that the M band portion is encoded by only six exons (Mex1 to Mex6) which are interrupted by five introns, indicating an unusually compact gene strucure: 67% of the sequences were found to be coding, while only 33% are intronic. The largest exon is over 5 kbp in size and makes up more than half of the M band titin-coding portion. Of the exons only Mex5, encoding the insertion is7, was found to be differentially spliced (Fig. 3; see above). This exon-skipping occurs in a muscle type-specific manner but does not depend on the developmental stage in mouse embryos. Both splice variants usually coexist, however, at variable ratios. As a rule of thumb, Mex5 is mostly skipped in fast twitch muscles, while both forms are about equimolar in slow twitch fibers. In cardiac muscle, there seems to be a trend to a correlation of Mex5 inclusion with lower heart beat frequency (Kolmerer et al. 1996). Considering that is7 (encoded by Mex5) binds to p94 (see above) an involvement of this sequence in titin stability and/or turnover presently is the most plausible explanation (Sorimachi et al. 1995). Interesting in this context is also the finding that the Mex5 flanking intron Min4 has remained strongly conserved during evolution (Kolmerer et al. 1996). The observed lack of larger sequence variations in this region of titin further supports the idea of a direct involvement in M band structure formation and maintenance. It is, however, unlikely that differential splicing of titin is directly involved in establishing a morphologically distinct M band appearance.

4.4 Skelemin cDNA Sequence

A complete skelemin cDNA sequence was reported from mouse skeletal muscle (Price and Gomer, 1993). Analysis of this sequence showed that skelemin belongs to the same group of intracellular immunoglobulin superfamily proteins as M-protein and myomesin, because it is built from Ig-like and Fn-like repeats in an identical order. The only specificities unique to skelemin seem to be two desmin intermediate filament-like motifs that were predicted to specify desmin binding and a serine/proline-rich region of 107 amino acid residues length of unknown function (Price and Gomer, 1993).

Several serious points, however, require clarification before the sequence published can be definitely ascribed to the protein skelemin that was investigated in previous papers from the same laboratory. First, skelemin was originally described as a protein doublet with apparent molecular masses of 220 and 200 kDa and an isoelectric point near 5.0 (Price, 1984; Price, 1987). The cDNA sequence now predicts a single polypeptide with a calculated mass of 185 kDa and a pI of 5.8 (Price and Gomer, 1993). It is presently not clear whether this enormous difference can be accounted for only by posttranslational modifications.

Second, the original skelemin antibody was reported to detect a protein not only in skeletal and cardiac muscle samples, but also in smooth muscle, myoepithelial cells (Price, 1987) and certain glial cells within the brain (see Introduction to Price and Gomer, 1993). In Northern blots, skelemin mRNAs were detected exclusively in heart and skeletal muscle tissues, but not in brain and liver. Unfortunately, smooth muscle and myoepithelial cells were not included in this investigation.

Third, the so-called "intermediate filament-like regions" in the alignments shown also allow for a different interpretation: the first motif, consisting of the 46 amino acids amino-terminal to the first Ig-like domain, shows identical residues in only 13 positions (16 if conservative substitutions are included). In the desmin α-helix (as typical for any "classical" α-helical coiled coil) positions a and d in the heptamer repeats are occupied by hydrophobic residues in nine out of twelve cases (see e.g. Geisler and Weber, 1982). In contrast, in the aligned skelemin sequence only four out of the twelve possible residues are hydrophobic. The second IF-like motif is a short linker located between the last two carboxy-terminal Ig-like domains and would allow for maximally two heptade repeats. Without any further experimental evidence it is not clear whether one or both regions could indeed assign for the observed tight association of skelemin with desmin (Price, 1984).

Fourth, it turned out that the translated murine cDNA sequence is almost totally colinear with the sequences of myomesin and M-protein. The highest identity level was revealed with human myomesin; the value of 90% (Bantle et al. 1996) is in the order of the species drift between mouse and man. Thus, the skelemin sequence published (Price and Gomer, 1993) is at least one of a particular subclass of the immunoglobulin superfamily which includes M-protein and myomesin. There remains the possibility that the conspicuous serine/proline-rich region specifies the distinctive feature which is responsible for the observed entirely different binding characteristics of skelemin. The enormous degree of similarity between skelemin and myomesin could also indicate that both proteins are splice variants from one and the same gene and the absence or presence of the serine/proline-rich region decides about the subsequent targeting of the resulting protein within the myofibril. An answer to this question will have to await the analysis of the gene(s) as well as protein-protein-binding studies.

5 A New, Molecular Model of the M Band Begins to Emerge

The structural M band model that is most favoured presently is based on excellent high resolution electron microscopy in combination with image analysis (see 2). Although several proteins have been described as prime molecular constituents of the M band, their corresponcence to specific structural entities has remained impossible. Only for MM-CK a correlation to M4/M4′-bridges was suggested (see section 3.1). A first attempt towards an understanding of the molecular layout of titin, myomesin and M-protein was published recently (Obermann et al. 1996). In this study, a panel of sequence-assigned antibodies directed against these proteins was used in immunoelectron microscopy on extracted muscle fiber preparations. The epitope positions of the antibodies used were chosen in order to give positional information of a representative fraction of molecular regions of the respective proteins. Modelling was also facilitated, since a wealth of information is available about the structure and the dimensions of the domains that largely constitute titin, myomesin and M-protein (see section 4). This method allowed to trace single Ig-like or Fn-like domains (or even shorter peptide sequenes) at the sarcomere level in the nanometer resolution range. As a consequence, it was possible to "model" titin, myomesin and M-protein molecules into the M band and to propose the structure shown in Fig. 4. Let us first discuss the data for titin: The MLCK-like kinase domain was positioned to the edge of the M band near the M6/M6′ line (Obermann et al. 1996). The correctness of the measured positioning of the kinase domain is confirmed by the following circumstantial evidence. Along the cross-bridge zone of the A band the titin Ig-like and Fn-like domains are arranged in an eleven-domain super-repeat pattern (Labeit et al. 1992). Immediately adjacent in carboxy-terminal direction the last super-repeat is followed by seven domains in a particular pattern that is conserved between vertebrate titin and *Caenorhabditis elegans* twitchin (Benian et al. 1989). Since the dimensions of individual domains are well known (see section 4), these seven domains can be predicted to span a distance of approximately 28 nm. This, in turn, is in excellent agreement with the measured value of 29 nm for the distance of the kinase domain to the last myosin cross-bridge (Obermann et al. 1996). These results make the kinase domain even more enigmatic than it was already, since it can be stated now unequivocally that it is far away from the first row of myosin heads. Furthermore, a possible involvement in the phosphorylation of myomesin or M-protein seems also ruled out (see below). These data imply, therefore, that either the titin kinase domain interacts with a yet unidentified – most likely soluble – substrate, or otherwise that interactions and locations are too dynamic to draw a static model.

Fig. 4. Arrangement of titin, myomesin and M-protein in the sarcomeric M band as proposed by immunoelectron microscopical results. Panel A gives a scale bar (in nm) and the positions of the most prominent M band striations (see also Fig. 1A). Panel B shows the layout of titin and myomesin molecules. Panel C presents the location of M-protein around the M1-line. In order not to overburdon the figures, only two molecules of each kind are shown. Panel D is a compilation of the domain types used in panels A to C. Letters in the molecular models indicate the decoration positions of the individual antibodies used. Adopted from Obermann et al. 1996

The location of three more epitopes makes an extended nature of the entire region carboxy-terminal to the kinase-domain most plausible. This would mean that titin molecules cross the sarcomere center and extend ~60 nm into the neighbouring sarcomere half, resulting in a total antiparallel overlap of almost 120 nm. Interestingly, secondary structure analyses

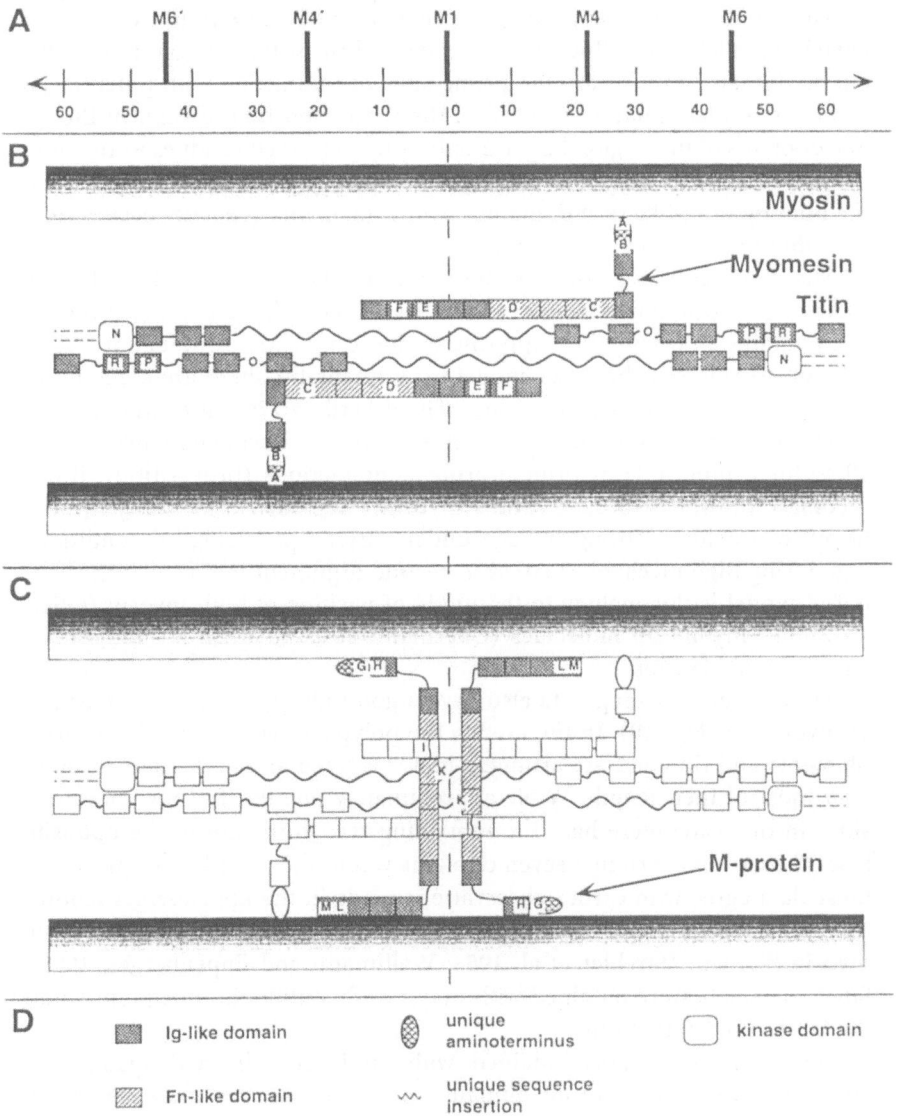

predicted for is2, the 490-residue-long insertion located between Ig-like domains m3 and m4, long stretches of potential for α-helix and coiled-coil formation (Gautel et al. 1993; Obermann et al. 1996). Therefore, substantial intermolecular titin-titin interactions within the region of is2 can be expected. These might explain the formation of titin dimers and higher oligomers adhering strongly at their M band ends as revealed in electron micrographs (Nave et al. 1989).

The model would also predict an orientation of titin´s Ig-like domain β-barrels parallel to the thick filament axis. This, in turn, suggests that the lateral domain surfaces are the prime sites of interaction with potential ligands. A similar situation is faced in the C-zone portion of titin. In the Ig-like domains of this region larger numbers of conserved residues were found in the predicted β-strands (Freiburg and Gautel, 1996). Thus it seems that the binding of Ig-like titin domains to thick filaments generally occurs side-on rather than to inter-strand loops.

The data described would, in theory, also allow for a so-called "U-turn model". This would require is2 to loop back, giving the remaining seven Ig-like domains an orientation opposite to the rest of titin. Although it will be difficult to rule out the latter model experimentally, there are some arguments that favour the former model. The described 120 nm overlap of titin molecules approaching the M band from opposite sarcomere halves would allow for a multitude of protein-protein interactions (titin – titin, titin – myomesin, titin – M-protein, titin – myosin, etc.). Such a versatility is more likely to provide a strong linkage under varying physiological conditions and during high mechanical stress. A second argument in line with the proposed model is the analogy to the mode of packing of both myosin (Offer, 1987) and myomesin molecules (Obermann et al. 1996) in an antiparallel and staggered fashion.

The immunolabelling data also gave a good idea about the organization of myomesin (Fig. 4B). In the model, the polypeptides run parallel to thick filaments and titin and are arranged in a staggered and antiparallel fashion over most of their length. While approximately one third of myomesin resides in one sarcomere half, the remaining two thirds are in the opposite half. The carboxy-terminal seven domains would then overlap. In principle, these data agree with earlier observations made in the Eppenberger laboratory which were suggestive for a localization of myomesin parallel to the myosin filament (Strehler et al. 1983; Wallimann and Eppenberger, 1985). Since their data were obtained with polyspecific antibodies this point could not be clarified at that time.

Since the decorations obtained with antibodies directed against the amino-terminal three to four domains of myomesin practically cannot be discerned from one another, this region is expected to bend towards the surface of the thick filaments. This orientation perpendicular to the thick filament axis is supportedby protein biochemical data that provided evidence that this part of myomesin is involved in the binding to the rod portion of myosin (Obermann et al. 1995; Obermann et al. 1997). The described organization would therefore make myomesin an ideal protein to crosslink the sarcomere cytoskeleton (represented in the M band essentially in addition by titin) with the thick filaments of the contractile apparatus. Since both titin and myomesin are found in all vertebrate sarkomeres, this kind of structural organization could indeed be envisioned to provide a general scaffold. In the ultrastructural model of the M band that was discussed above (Luther and Squire, 1978; see section 2) this "universal" complex of myomesin and titin could correspond to the so-called M-filaments that run parallel to thick filaments between M4 and M4′ bridges.

A major component that is most likely involved in differential M band appearance is obviously M-protein, since it is found only in cardiac and fast skeletal muscle cells (see section 3.2). Immunolabeling data using polyclonal antibodies have already suggested that M-protein could constitute the M1-lines (Carlsson et al. 1990). Recent work using several sequence-assigned antibodies agrees with this proposed localization at least for the central part of M-protein (Fig. 4C). Both amino- and carboxy-terminal domains would then be organized along neighbouring thick filaments (Obermann et al. 1996). In excellent agreement with this structural model, a myosin binding site was identified in the amino-terminal region of M-protein (see section 6.1.2; Obermann et al. 1998). In summary, these data imply that M-protein could build an additional link between thick filaments and constitute the M1-lines of cardiac and fast skeletal muscle fibers to meet stronger strain. In accordance with this suggestion, it was exactly these fibers that were described to have prominent M1-lines in electron micrographs (Pask et al. 1994).

6 Regulation of M Band Assembly

6.1 Protein-Protein Interactions Between M Band Proteins

The cDNA cloning and sequencing of M band proteins has revealed their domain structure and thus made it feasible to dissect these polypeptides into smaller, better manageable units, in order to describe and characterize various binding sites by biochemical experiments. At the same time, the emerging new, molecular model of the M band put several constraints on the intermolecular interactions that are possible, required the predictions in the model were indeed correct. It was therefore an exciting venture to analyze the implied mutual binding sites between titin, myomesin, M-protein and myosin. In summary, one can say now that the biochemical data that have been obtained so far are in excellent agreement with the proposed model and hence can be seen as further evidence for its correctness.

6.1.1 Mapping of Titin- and Myosin-Binding Sites in Myomesin

In the case of myomesin it was only recently possible to map both titin- and myosin-binding sites and to identify phosphorylation as the means to control one of these interactions *in vitro* (Obermann et al. 1997). Using recombinant fragments of myomesin, the titin-binding region could be confined to a construct that consists of three Fn-like domains (My4 to My6; Fig. 6A). All the fragments smaller than this were negative under the conditions of the solid phase binding assay employed. On the other side, domain m4 was established as the sole titin domain which shows interaction with myomesin (Obermann et al. 1997). This result shows parallels to the nature of the binding of C-protein to titin, where also a minimum of three C-protein domains are required for binding (Freiburg and Gautel, 1996). Obviously the strong binding between titin and its associated proteins that is revealed *in vivo*, is based on a larger number of weak cooperative interactions.

Particularly interesting was the characterization of a site within this titin-binding region, which is subject to phosphorylation. A single serine residue

```
                             *
human     AALDPAEKARLKSRPSAPWTGQIIVTEEEPSEGIV
mouse     .................H.................T..VI
bovine    ...................................
chicken   ......DR...R.H................A..V.
```

Fig. 5. Alignment of the sequence insertion located between domains My4 and My5 of myomesin and the corresponding sequence of skelemin. The first line gives in single letter code the complete translated sequence of the 35 amino acids long insertion deduced from a human cDNA (Vinkemeier et al. 1993). Aligned are the sequences of murine skelemin (Price and Gomer, 1993), bovine myomesin (Obermann et al, 1995) and chicken myomesin (Bantle et al. 1996) from the same region. Identical residues are shown as dots, amino acid exchanges relative to the human sequence are given in full. The bovine sequence was obtained by direct sequencing of a proteolytic fragment of the purified protein and stops three residues before the end of this insertion. The serine residue that is phosphorylated by PKA *in vitro* is printed in bold and marked by an asterisk. Note the extreme degree of cross-species conservation of this region which only allows for conservative substitutions

located in the ~35 amino acids long linker between domains My4 and My5 (Figs. 2 and 3) was revealed as the only residue that is phosphorylated by cAMP-dependent protein kinase A (PKA) *in vitro*. This linker should be readily accessible in solution, at least better than the densely packed Ig- and Fn-like domains. Additional significance to this sequence is added if the degree of inter-species conservation is considered. The alignment in Fig. 5

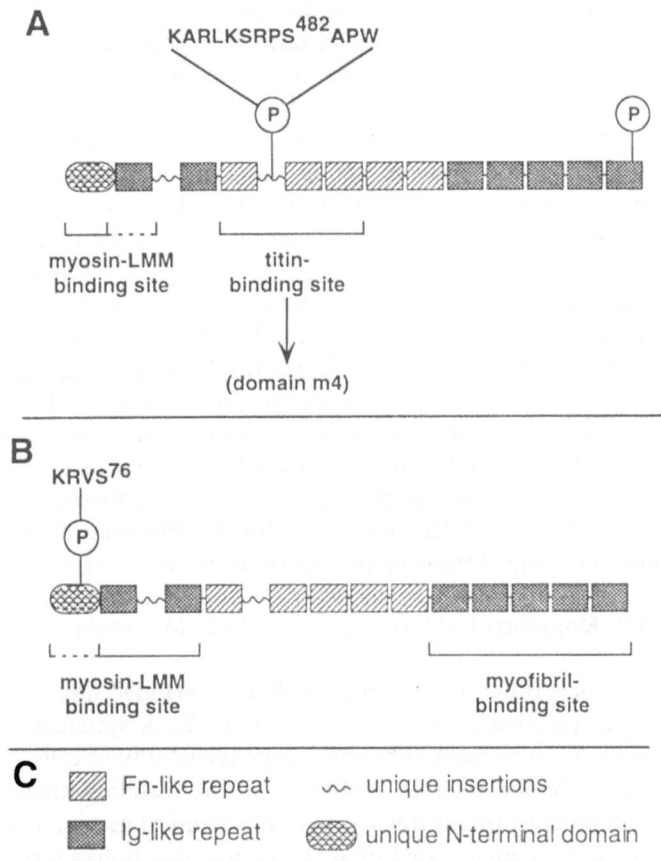

Fig. 6. Summary of documented binding sites in myomesin (A) and M-protein (B). Molecular models correspond to drawings in Fig. 2. Brackets indicate domains in which binding sites were located. Amino acid sequences of potentially regulatory PKA phosphorylation motifs are shown. Panel D gives the domain types used in panels A and B. Compiled from: myomesin – Obermann et al. 1995, Obermann et al. 1997; Auerbach et al. 1997; M-protein – Obermann et al. 1998

reveals that this region is extremely highly conserved from chicken to man and only very few conservative exchanges are allowed. The entire loop therfore has to be envisioned as potentially important in the context of the observed phosphorylation. It is presently not clear whether PKA itself or a closely related enzyme is active *in vivo*. If it was PKA, then the described sequence forms a novel target site. Previous work has established that two basic residues separated by a single aliphatic residue from a serine or threonine form the preferred target site for PKA (Pearson and Kemp, 1991). It was therefore not surprising that the K_m and v_{max} values are not among the best PKA substrates, they are, however, within the range of several other substrates. Such comparisons also have to consider the fact that usually soluble proteins are measured, for whom low K_m values are important if phosphorylation was to occur in the cytosol. In contrast, a high local concentration and stable anchorage in the M band might as well allow for the higher K_m values measured (Obermann et al. 1997).

Functionality is added to this observation by the finding that phosphorylation of this serine residue almost completely inhibits the binding of myomesin to titin. Thus it seems that a complex three-dimensional rearrangement occurs that is capable of controlling titin – myomesin binding (Obermann et al. 1997). The latter result has far reaching implications for the processes of sarcomere formation and turnover. In both cases it is necessary to control the assembly/disassembly of a highly complex macromolecular structure. A tight control of certain protein-protein bindings in a time- and space-specific manner is mandatory if these processes are to occur in an ordered way. The described phosphorylation of myomesin gives a first hint which signal transduction mechanisms could be involved.

6.1.2 Mapping of a Myosin-Binding Site in M-Protein

In the past different investigators have concluded different binding properties for purified M-protein (see section 3.2). A systematic approach to this question was facilitated by the recent availability of cDNA data (Noguchi et al. 1992; Vinkemeier et al. 1993) and the resulting chance to use defined recombinant protein fragments in biochemical assays. Thus it was possible to identify domains Mp2 to Mp3 (the first two Ig-like repeats) of human M-protein as a site binding to the central region of the LMM portion of myosin (Obermann et al. 1998; Fig. 6B). Since two "classical" target sequences for cAMP dependent protein kinase A (PKA) were found in the adjacent amino-terminal domain Mp1 by sequence inspection, this region was investigated in more detail. Indeed, one of the predicted sites as well as the purified protein could be phosphorylated *in vitro* at the serine residue in the motif KRVS

(Fig. 6B). Most importantly, this phosphorylation seems to regulate the interaction of M-protein with myosin: phosphorylated M-protein does no longer bind to myosin. In a more *in vivo* approach, some overlapping recombinant constructs were transiently overexpressed in cultured cells. Again, the construct Mp2 to Mp3 was the smallest fragment of this region that showed binding to myofibrillar structures (Obermann et al. 1998). These data imply that Mp2 to Mp3 are a constitutive myosin-binding site which is modulated by PKA phosphorylation in the amino-terminal domain Mp1. This has, like in the case of myomesin (see section 6.2.1), far reaching implications for the processes of sarcomere formation and turnover. Specific phosphorylation events are implied to regulate protein-protein bindings which are important for both the assembly and the disassembly of the sarcomere. It seems that a single signal transduction pathway could be involved in the regulation of two distinct protein interactions (myomesin – titin, M-protein – myosin) thus facilitating control.

The transient expression of defined regions of M-protein in cultured cells also defined a second portion which is capable of binding to myofibrils. The construct comprising domains Mp9 to Mp13 bound to the sarcomeric M band (Fig. 6B). This might be either a second myosin-binding site which could not be identified by the binding assays used, or otherwise, a titin-binding site (Obermann et al. 1998). The second possibility will be difficult to assess biochemically, because the titin fragment that is located in the region in question (is2) cannot be expressed and purified (M. Gautel, personal communication).

6.2 M Band Ultrastructure During Sarcomerogenesis

Postnatal development of, for instance, rat cardiomyocytes is accompanied by gradual advances in the maturity of developing myofibrils. Although specific M-band components can be detected in newborn rats, their cardiomyocytes show myofibrils with a light central band in the center of the A-bands. It is not until 5 days after birth that premature M-bands can be discerned as thin cross bridges (Anversa et al. 1981). The thick M-bridges that link myosin filaments in adult rat cardiomyocytes are still absent from the immature M-bands of 11 days old rats. Obviously, the neonatal mammalian heart functions normally long before the morphology of "mature" M-bands is fully developed.

In contrast, skeletal muscle fibers of neonatal rats do already contain myofibrils with their typical appearance as described in the adult animal. At this developmental stage, however, all the M-bands show five lines. During further maturation and the development of different fibers types, M-bands change their morphology into the one of type I and type IIB fibers. While type IIA fibers retain their typical five-line pattern, type I and type IIB fibers M-bands are composed of four and three lines, respectively (Carlsson and Thornell 1987). Thus, although the details of this developmental pattern in skeletal muscle fibers clearly differ from the one in cardiomyocytes, both mammalian skeletal and cardiac muscle M-bands are postnatally remodelled during functional maturation and/or differentiation of muscle.

6.3 Expression of M-Band Proteins During Muscle Development *in vivo*

Comparative analyses of protein expression patterns in whole animals have been impeded by various reasons, one of the most important ones being the non-synchronous development of distinct muscles. The main requirement for such studies is, therfore, a profound knowledge of how to reconstruct an embryos´ anatomy from serial sections or to use microdissection techniques to allow for comparability of the results obtained. This may be the reason why such investigations have been relatively sparse.

A first systematic analysis of the expression of M band proteins in chicken skeletal muscles included myomesin and M-protein. Thus, both myomesin and M-protein were found to accumulate in all myotubes at early stages (7 days in ovo; Grove et al. 1987). While myomesin is then continuously expressed during development in all fibers, M-protein could no longer be detected from 10 days in ovo until the development of secondary myotubes. Re-expression of M-protein subsequently occurs only in fibers deriving from such primary myotubes that upon innervation develop into fast twitch fibers. It remains, however, permanently absent from slow twitch fibers (Grove et al. 1987; Grove and Thornell, 1988).

In avian cardiac muscle, both myomesin and M-protein were detected in embryonic heart extracts from day 7 on. In the earliest stages, two isoforms of myomesin were found in the cardiomyocytes, both of which are larger than the skeletal muscle isoform (Grove et al. 1985).

The recent discovery of differential splicing that occurs in the 3´ end region of the mRNA encoding myomesin and gives rise to three transcripts of different length most likely explains this finding (Bantle et al. 1996).

Most recently, spatial and temporal expression patterns of myomesin during myofibril assembly were also studied in whole mount chicken embryo hearts using confocal laser scanning microscopy (Auerbach et al. 1997). In 9-somite embryos, i.e. when the heart starts beating, myomesin was already detected in cardiomyocytes and it was found almost exclusively in a sarcomeric pattern. This indicates that M-bands are assembled at this developmental stage and the expression of myomesin coincides with the localization of other sarcomeric proteins in a striated pattern.

The investigation of the expression patterns of myomesin, M-protein and MM-CK during development of the rat yielded similar results. Thus, myomesin is the first of these three proteins to be expressed (Carlsson et al. 1990). Myofibrils stained with myomesin antibodies were distinguished in 14 days old embryos, while M-protein was hardly detected and MM-CK was not found at all. During further embryonic development and in neonatal rats, all three proteins were detectable in the M band in all muscle fibers (Carlsson et al. 1990). When muscle fibers differentiate into type I and type II fibers, M-protein expression was exclusively revealed in the latter fibers, in contrast to myomesin and MM-CK both of which are still . Since upon denervation at birth M-protein remains to be synthesized, it was concluded that the innervation of muscle fibers by slow twitch motor units suppresses M-protein expression (Carlsson et al. 1990). Consistent with the results from other investigations, it was found that in the presence of M-protein myofibrils of adult fast skeletal muscles exhibit a central M1-line (Carlsson et al. 1990; Grove et al. 1985; Obermann et al. 1996).

Surprisingly few data are available on the embryonic expression of titin. Reports agree that titin is among the first muscle specific proteins which are expressed both in cardiac and skeletal myoblasts of chicken and mouse embryos. A careful correlation of *in situ* hybridization and immunofluorescence data demonstrated that titin protein could be detected approximately six hours after the first appearance of titin mRNA (Kolmerer et al. 1996). The expression of titin is preceded by desmin and occurs simultaneously with α-actinin. With a short delay these proteins are followed by myomesin and the sarcomeric isoforms of actin and myosin (Fürst et al. 1989). A temporal correlation and hence, most likely, an important role of titin in sarcomeric Z-disc assembly was suggested by data on the early development of chicken embryo heart (Tokuyasu and Maher, 1987a; Tokuyasu and Maher, 1987b). Comparable studies for the M band end of titin and the demonstration of a possible correlation of the specific titin localization with M band assembly together with myomesin and M-protein is, however, still lacking. So far, only data from studies in cultured cells are available (see below).

6.4 Expression of M Band Proteins in Cultured Cells

In principle, the *in vivo* data summarized above are in good agreement with studies on the expression of myofibrillar M band proteins in cultures of embryonic chicken cardiomyocytes (Schultheiss et al. 1990), embryonic

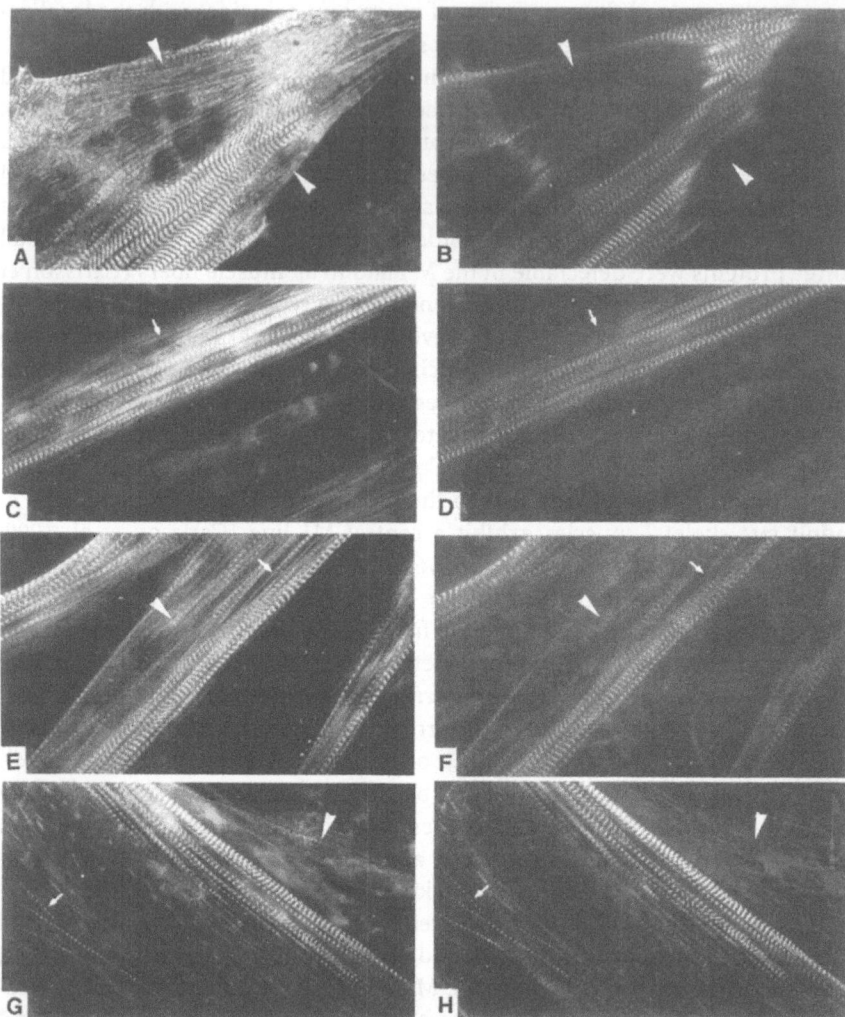

Fig. 7. Cultured human skeletal muscle cells double stained with antibodies directed against different regions of the titin molecule, myomesin and M-protein. An I band epitope-specific titin antibody reacts both with mature and developing non-striated (arrowheads) myofibrils (A), whereas anti-M band epitope titin antibodies do not stain non-striated myofibrils (B). Similarly, myomesin (D) and M-protein (F) can be detected only in cross-striated myofibrils, while Z-disc titin antibodies also detect non-striated myofibrils (arrows in C and E). Note that all myofibrils are already positive for titin, while M-protein appears at a slightly later developmental stage (arrowheads in E and F). Some cells containing myofibrils (as indicated by titin staining; G) are devoid of M-protein (arrowhead; H). Arrows in G and H mark a single M band stained by both antibodies. Magnification: 510 x

chicken myoblasts (Lin et al. 1994) and human satellite cells (Van der Ven and Fürst, 1997). All these studies agree that myomesin expression is delayed relative to the expression of sarcomeric α-actinin, titin and the sarcomeric isoforms of myosin heavy chain (MyHC) and actin. Interestingly, myomesin was only detected in M bands and never free in the cytoplasm (see Fig. 7). At the same time, carboxy-terminal epitopes of titin are not accessible for specific antibodies during the assembly of early non-striated myofibrils (Fig. 7; van der Ven and Fürst, 1997). This could indicate that M band protein binding sites are "hidden" and might explain why myomesin and M-protein are never found associated with non-striated myofibrils.

M-protein levels were consistently found to be lower than those of myomesin in early cultures of embryonic chicken myoblasts (Grove et al. 1985). Obviously, M-protein is accumulated at later developmental stages, but eventually all differentiated chicken myotubes contain both M-band proteins. In contrast, cultures of human satellite cells revealed fully developed myotubes in which myomesin is organized in a cross-striated pattern, but which at the same time are devoid of M-protein (Van der Ven and Fürst, 1997). These results confirmed earlier observations that cultured satellite cells from adult skeletal muscle are predestined to form a certain fiber type without being innervated (Barjot et al. 1995; Düsterhöft and Pette, 1993; Feldman and Stockdale, 1991).

The presence of myomesin in all muscle fibers and the integration of myomesin into sarcomeres at the time-point when the first striated myofibrils are perceptible (Schultheiss et al. 1990; Lin et al. 1994; Rhee et al. 1994; Van der Ven and Fürst, 1997), points to an important role of myomesin in the assembly of the M band and thus the connection of titin with the thick filaments and the stabilization of the thick filament lattice. M-protein on the other hand, is not only expressed later both *in vivo* and *in vitro*, but is also absent from a subset of myotubes during all stages of sarcomere assembly, indicating its less important general role in myofibrillogenesis.

6.5 Targeting of M-Band Proteins to their Specific Sarcomere Position

Regions of a certain protein that are responsible for its specific subcellular localization are often identified by the overexpression of recombinant fragments in cultured cells. This approach was also used to search for the domains of myomesin and M-protein which are necessary for their specific targeting to sarcomeric M bands in diffentiating muscle cells. Thus M-protein domains Mp2 and Mp3 were defined as a region sufficient for M band binding (Obermann et al. 1998). This observation is in line with *in vitro* binding data which indicated that the first two Ig-like domains of M-protein are the minimal requirement for binding to myosin (see section 6.1.2).

In a similar experiment, using cultured chicken cardiomyocytes, binding of My2, the first immunoglobulin-like domain of myomesin, to the M band was described (Auerbach et al. 1997; Fig. 6B). Biochemical data, on the other hand had indicated that at least the unique amino-terminal head domain (My1) of human myomesin should be involved in myosin binding (Obermann et al. 1997). This might be explained by species-specific differences in the respective amino-terminal domains, like for instance the missing conspicuous KQSTAS repeats in chicken myomesin. Furthermore, the entirely different assay conditions used presently make a final assessment difficult.

7 Conclusions

M bands and Z discs seem to be crucially involved in establishing the strikingly ordered array of thick and thin filaments, respectively. This requires control mechanisms that allow for the sequential assembly of contractile and structural proteins into the sarcomere. In addition, tightly controlled local disassembly processes have to provide a basis for the turnover and remodelling that has been observed (like, for instance, the exchange of developmental stage-specific sarcomere protein isoforms in a working muscle). Previous work has focussed on the characterization of a plethora of contractile and cytoskeletal proteins. More recent results give good reasons to hope that it may be possible to explain the various ultrastructural findings at a molecular level in the future. Likewise, they give a first hint at which signal transduction mechanisms could be involved in the control of the assembly of the supramolecular structure that makes us move.

Acknowledgements

The authors are particularly grateful to Prof. Klaus Weber (Max-Planck-Institute for Biophysical Chemistry, Göttingen) for his continuous support and interest in the topic "sarcomere cytoskeleton" which gave an invaluable impact to the field. Dr. Mathias Gautel (EMBL, Heidelberg) is thanked for an exciting collaboration and fruitful discussions. All the former and present lab mates are thanked for their help in making this project successful. This work was supported by grants from the Deutsche Forschungsgemeinschaft.

References

1. Anversa, P., G. Olivetti, P.-G. Bracchi, and A. V. Loud (1981) Postnatal development of the M-band in rat cardiac myofibrils. Circ Res 48:561–568
2. Arnold, H., and D. Pette (1970) Binding of aldolase and triosephosphate dehydrogenase to F-actin and modification of catalytic properties of aldolase. Eur J Biochem 15:360–366
3. Auerbach, D., B. Rothen-Rutishauser, S. Bantle, M. Leu, E. Ehler, D. Helfman, and J. C. Perriard (1997) Molecular mechanisms of myofibril assembly in heart. Cell Struct Funct 22:139–146
4. Ayme-Southgate, A., J. Vigureaux, G. Benian, and M. L. Pardue (1991) Drosophila has a twitchin/titin-related gene that appears to encode projectin. Proc Natl Acad Sci USA 88:7973–7977
5. Bähler, M., T. Wallimann, and H. M. Eppenberger (1985) Myofibrillar M-band proteins represent constituents of native thick filaments, frayed filaments and bare zone assemblages. J Muscle Res Cell Mot 6:783–800
6. Bantle, S., S. Keller, I. Haussmann, D. Auerbach, E. Perriard, S. Mühlebach, and J.-C. Perriard (1996) Tissue-specific isoforms of chicken myomesin are generated by alternative splicing. J Biol Chem 271:19042–19052
7. Barjot, C., M.-L. Cotten, C. Goblet, R. G. Whalen, and F. Bacou (1995) Expression of myosin heavy chain and of myogenic regulatory factor genes in fast or slow rabbit muscle satellite cell cultures. J Musc Res Cell Mot 16:619–628
8. Baskin, R. J., and D. N. Deamer (1970) A membrane-bound creatine phosphokinase in fragmented sarcoplasmic reticulum. J Biol Chem 259:14979–14984
9. Benian, G. M., J. E. Kiff, N. Neckelmann, D. G. Moerman, and R. H. Waterston (1989) Sequence of an unusually large protein implicated in the regulation of myosin activity in C. elegans. Nature 342:45–50
10. Bober, E., G. E. Lyons, T. Braun, G. Cossu, M. Buckingham, and H.-H. Arnold (1991) The muscle regulatory gene, Myf-6, has a biphasic pattern of expression during early mouse development. J Cell Biol 113:1255–1265
11. Carlsson, E., B. K. Grove, T. Wallimann, H. M. Eppenberger, and L.-E. Thornell (1990) Myofibrillar M-band proteins in rat skeletal muscles during development. Histochem 95:27–35
12. Cohen, I. R., S. Grässel, A. D. Murdoch, and R. V. Iozzo (1993) Structural characterization of the complete human perlecan gene and its promoter. Proc. Natl Acad Sci USA 90:10404–10408
13. Croop, J., G. Dubyak, Y. Toyama, A. Dlugosz, A. Scarpa, and H. Holtzer (1982) Effects of 12-O-tetradecanoyl-phorbol-13-acetate on myofibril integrity and Ca^{2+} content in developing myotubes. Dev Biol 89:460–474
14. Cunningham, B. A., J. J. Hemperly, B. A. Murray, E. A. Prediger, R. Brackenbury, and G. M. Edelman (1987) Neural cell adhesion molecule: structure, immunoglobulin-like domains, cell surface modulation, and alternative RNA splicing. Science 236:799–806
15. Dobie, W. M. (1849) Observations on the minute structure and the mode of contraction of voluntary muscle fibers. Ann Mag Nat Hist 3:109

16. Draper, M. H., and A. J. Hodge (1946) Studies on muscle with the electron microscope: I. The ultrastructure of of toad striated muscle. Austr J Exptl Biol Med Sci 27:465–503

17. Düsterhöft, S. and D. Pette (1993) Satellite cells from slow rat muscle express slow myosin under appropriate culture conditions. Differentiation 53:25–33

18. Edman, A.-C., J. M. Squire, and M. Sjöström (1988) Fine structure of the A-band in cryo-sections. J Ultrastruct Mol Struct Res 100:1–12

19. Eppenberger, H. M., D. M. Dawson, and N. O. Kaplan (1967) The comparative enzymology of creatine kinase isoenzymes. J Biol Chem 242:204–209

20. Eppenberger, H. M., J.-C. Perriard, U. B. Rosenberg, and E. E. Strehler (1981) The Mr 165,000 M-protein myomesin: a specific protein of croo-striated muscle cells. J Cell Biol 89:185–193

21. Feldman, J. L., and F. E. Stockdale (1991) Skeletal muscle satellite cell diversity: Satellite cells form fibers of different types in cell culture. Dev Biol 143:320–334

22. Fickett, J. W. (1996) Quantitative discrimination of MEF2 sites. Mol Cell Biol 16:437–441

23. Freiburg, A., and M. Gautel (1996) A molecular map of the interactions of titin and myosin-binding protein C: implications for sarcomeric assembly in familial hypertrophic cardiomyopathy. Eur J Biochem 235:317–323

24. Friedman, D. L., and M. B. Perryman (1991) Compartmentalization of multiple forms of creatine kinase in the distal nephron of the rat kidney. J Biol Chem 266:22404–22410

25. Fritz-Wolf, K., T. Schnyder, T. Wallimann, and W. Kabsch (1996) Structure of mitochondrial creatine kinase. Nature 381:341–345

26. Fürst, D. O., and M. Gautel (1995) The anatomy of a molecular giant: How the sarcomere cytoskeleton is assembled from immunoglobulin superfamily molecules. J Mol Cell Cardiol 27:951–960

27. Fürst, D. O., R. Nave, M. Osborn, and K. Weber (1989a) Repetitive titin epitopes with a 42 nm spacing coincide in relative position with known A band striations also identified by major myosin-associated proteins; an immunoelectron microscopical study on myofibrils. J Cell Biol 94:119–125

28. Fürst, D. O., M. Osborn, R. Nave, and K. Weber (1988) The organization of titin filaments in the half-sarcomere revealed by monoclonal antibodies in immunoelectron microscopy; a map of ten non-repetitive epitopes starting at the Z line extends close to the M line. J Cell Biol 106:1563–1572

29. Fürst, D. O., M. Osborn, and K. Weber (1989b) Myogenesis in the mouse embryo: differential onset of expression of myogenic proteins and the involvement of titin in myofibril assembly. J Cell Biol 109:517–527

30. Fürst, D. O., U. Vinkemeier, and K. Weber (1992) Mammalian skeletal muscle C-protein: purification from bovine muscle, binding to titin and the characterization of a full length human cDNA. J Cell Sci 102:769–778

31. Gautel, M., D. Goulding, B. Bullard, K. Weber, and D. O. Fürst (1996) The central Z-disk region of titin is assembled from a novel repeat in variable copy numbers. J Cell Sci 109:2747–2754

32. Gautel, M., K. Leonard, and S. Labeit (1993) Phosphorylation of KSP motifs in the C-terminal region of titin in differentiating myoblasts. EMBO J 12:3827–3834

33. Geisler, N., and K. Weber (1982) The amino acid sequence of chicken muscle desmin provides a common structural model for intermediate filament proteins. EMBO J 1:1649–1656

34. Grosse, R., E. Spitzer, V. V. Kupriyanov, V. A. Saks, and K. R. H. Repke (1980) Coordinate interplay between (Na^+/K^+)-ATPase and CK optimizes (Na^+/K^+)-antiport across the membrane of vesicles formed from the plasma membrane of cardiac muscle cells. Biochim Biophys A 603:142–156

35. Grove, B. K., L. Cerny, J.-C. Perriard, and H. M. Eppenberger (1985) Myomesin and M-protein: Expression of two M-band proteins in pectoral muscle and heart during development. J Cell Biol 101:1413–1421

36. Grove, B. K., B. Holmbom, and L.-E. Thornell (1987) Myomesin and M protein: differential expression in embryonic fibers during pectoral muscle development. Differentiation 34:106–114

37. Grove, B. K., V. Kurer, C. Lehner, T. C. Doetschmann, J.-C. Perriard, and H. M. Eppenberger (1984) A new 185,000-dalton skeletal muscle protein detected by monoclonal antibodies. J Cell Biol 98:518–524

38. Grove, B. K., and L.-E. Thornell (1988) Noncoordinate expression of M band proteins in slow and fast embryonic chick muscles. Muscle & Nerve 11:645–653

39. Hall, C. E., M. A. Jakus, and F. O. Schmitt (1946) An investigation of cross-striations and myosin filaments in muscle. Biol Bull 90:32–50

40. Hartzell, H. C., and W. S. Sale (1985) Structure of C-protein purified from cardiac muscle. J Cell Biol 100:208–215

41. Heierhorst, J., W. C. Probst, F. S. Vilim, A. Buku, and K. R. Weiss (1994) Autophosphorylation of molluscan twitchin and interaction of its kinase domain with calcium/calmodulin. J Biol Chem 269:21086–21093

42. Herasymowych, O. S., R. S. Mani, C. M. Kay, R. D. Bradley, and D. G. Scraba (1980) Ultrastructure studies on the binding of creatine kinase and the 165,000 molecular weight component of the M-band of muscle. J Mol Biol 136:193–198

43. Hinterberger, T. J., D. A. Sassoon, S. J. Rhodes, and S. F. Konieczny (1991) Expression of the muscle regulatory factor MRF4 during somite and skeletal myofiber development. Dev Biol 147:144–156

44. Holden, H. M., M. Ito, D. J. Hartshorne, and I. Rayment (1992) X-ray structure determination of telokin, the C-terminal domain of myosin light chain kinase, at 2.8 Å resolution. J Mol Biol 227:840–851

45. Holness, C. L., and D. L. Simmons (1994) Structural motifs for recognition and adhesion in memebers of the immunoglobulin superfamily. J Cell Sci 107:2065–2070

46. Hossle, J. P., J. Schlegel, G. Wegmann, M. Wyss, P. Böhlen, H. M. Eppenberger, T. Wallimann, and J.-C. Perriard (1988) Distinct, tissue-specific mitochondrial creatine kinases from chicken brain and striated muscle with a conserved CK framework. Biochem Biophys Res Comm 151:408–416

47. Jacobs, H., H. W. Heldt, and M. Klingenberg (1964) High activity of creatine kinase in mitochondria from muscle and brain and evidence for a separate mitochondrial isoenzyme of creatine kinase. Biochem Biophys Res Comm 16:516–521

48. Jacobus, W. E., and A. L. Lehninger (1973) Creatine kinase of rat heart, mitochondrial coupling of creatine phosphorylation to electron transport. J Biol Chem 248:4803–4810

49. Kahn, M. A., P. G. Holt, J. O. Knight, J. M. Papadimi, and B. A. Kakulas (1972) Creatine kinase, a histochemical study by gelatin film-lead precipitation technique. Histochem 32:49

50. Knappeis, G. G., and F. Carlsen (1968) The ultrastructure of the M line in skeletal muscle. J Cell Biol 38:202–211

51. Kolmerer, B., N. Olivieri, C. C. Witt, B. Herrmann, and S. Labeit (1996) Genomic organization of M-line titin and its tissue-specific expression in two distinct isoforms. J Mol Biol 256:556–563

52. Labeit, S., M. Gautel, A. Lakey, and J. Trinick (1992) Towards a molecular understanding of titin. EMBO J 11:1711–1716

53. Labeit, S., and B. Kolmerer (1995) Titins: giant proteins in charge of muscle ultrastructure and elasticity. Science 270:293–296

54. Labeit, S., B. Kolmerer, and W. A. Linke (1997) The giant protein titin: emerging roles in physiology and pathophysiology. Circ Res 80:290–294

55. Lassar, A. B., J. N. Buskin, D. Lockshon, R. L. Davis, S. Apone, S. D. Hauschka, and H. Weintraub (1989) MyoD is a sequence-specific DNA binding protein requiring a region of myc homology to bind to the muscle creatine kinase enhancer. Cell 58:823–831

56. Lee, V. M.-Y., L. Otvos, M. J. Carden, M. Hollosi, B. Dietzschold, and R. A. Lazzarini (1988) Identification of the major multiphosphorylation site in mammalian neurofilaments. Proc Natl Acad Sci USA 85:1998–2002

57. Lee, W., P. Mitchell, and R. Tjian (1987) Purified transcription factor AP-1 interacts with TPA-inducible enhancer elements. Cell 49:741–752

58. Lin, Z., J. R. Eshelman, S. Forry-Schaudies, S. Duran, J. L. Lessard, and H. Holtzer (1987) Sequential disassembly of myofibrils induced by myristate acetate in cultured myotubes. J Cell Biol 105:1365–1376

59. Lin, Z., M.-H. Lu, T. Schultheiss, J. Choi, S. Holtzer, C. DiLullo, D. A. Fischman, and H. Holtzer (1994) Sequential appearance of muscle-specific proteins in myoblasts as a function of time after cell division: evidence for a conserved myoblast differentiation program in skeletal muscle. Cell Mot Cytoskel 29:1–19

60. Littman, D. R., and S. N. Gettner (1987) Unusual intron in the immunoglobulin domain of the newly isolated CD4 (L3T4) gene. Nature 325:453–455

61. Luther, P. K., and R. A. Crowther (1984) Three-dimensional reconstruction from tilted sections of fish muscle M-band. Nature 307:566–568

62. Luther, P. K., P. M. G. Munro, and J. M. Squire (1981) Three-dimensional structure of the vertebrate muscle A-band. III. M-region structure and myosin filament symmetry. J Mol Biol 151:703–730

63. Luther, P. K., and J. M. Squire (1978) Three -dimensional structure of the vertebrate muscle M-region. J Mol Biol 125:313–324

64. Mani, R. S., and C. M. Kay (1978a) Interaction studies of the 165,000 dalton protein component of the M-line with the S2 subfragment of myosin. Biochim Biophys A 536:134–141

65. Mani, R. S., and C. M. Kay (1978b) Isolation and characterization of the 165 000 dalton protein component of the M-line of rabbit skeletal muscle and its interaction with creatine kinase. Biochim Biophys A 533:248–256

66. Mani, R. S., and C. M. Kay (1981) Fluorescence studies on the interaction of muscle M-line proteins, creatine kinase and the 165,000 dalton component, with each other and with myosin and myosin subfragments. Int J Biochem 13:1197–1200

67.   Maruyama, K. (1994) Connectin, an elastic protein of striated muscle. Biophys Chem 50:73–85

68.   Masaki, T., M. Endo, and S. Ebashi (1967) Localization of 6S component of a-actinin at Z-band. J Biochem 62:630–632

69.   Masaki, T., and O. Takaiti (1974) M-protein. J Biochem 75:367–380

70.   Masaki, T., O. Takaiti, and S. Ebashi (1968) "M-substance", a new protein constituting the M-line of myofibrils. J Biochem 64:909–910

71.   Meyer, R. A., H. Lee-Sweeney, and M. J. Kushmerick (1984) A simple analysis of the "phosphocreatine shuttle". Am J Physiol 250:C365–C377

72.   Mühlebach, S. M., T. Wirz, U. Brändle, and J.-C. Perriard (1996) Evolution of the creatine kinases. J Biol Chem 271:11920–11929

73.   Nave, R., D. O. Fürst, and K. Weber (1989) Visualization of the polarity of isolated titin molecules; a single globular head on a long thin rod as the M-band anchoring domain? J Cell Biol 109:2177–2188

74.   Noguchi, J., M. Yanagisawa, M. Imamura, Y. Kasuya, T. Sakurai, T. Tanaka, and T. Masaki (1992) Complete primary structure and tissue expression of chicken pectoralis M-protein. J Biol Chem 267:20302–20310

75.   Obermann, W., U. Plessmann, K. Weber, and D. O. Fürst (1995) Purification and biochemical characterization of myomesin, a myosin and titin binding protein, from bovine skeletal muscle. Eur J Biochem 233:110–115

76.   Obermann, W. M. J., M. Gautel, F. Steiner, P. F. M. van der Ven, K. Weber, and D. O. Fürst (1996) The structure of the sarcomeric M band: localization of defined domains of myomesin, M-protein and the 250 kDa carboxyterminal region of titin by immunoelectron microscopy. J Cell Biol 134:1441–1453

77.   Obermann, W. M. J., M. Gautel, K. Weber, and D. O. Fürst (1997) Molecular structure of the sarcomeric M band: mapping of titin- and myosin-binding domains of myomesin and the identification of a potential regulatory phosphorylation site in myomesin. EMBO J 16:211–220

78.   Obermann, W. M. J., P. F. M. van der Ven, F. Steiner, K. Weber, and D. O. Fürst (1998) Mapping of a myosin binding domain and a regulatory phosphorylation site in M-protein, a structural protein of the sarcomeric M band. Mol Biol Cell 9:829–840

79.   Offer, G. (1987) Myosin filaments. In Fibrous protein structure. J. M. Squire and P. J. Vibert, editors. Academic Press pp 307–356

80.   Okuda, A., M. Imagawa, M. Sakai, and M. Muramatsu (1990) Functional cooperativity between two TPA responsive elements in undifferentiated F9 embryonic stem cells. EMBO J 9:1131–1135

81.   Olson, N. J., R. B. Pearson, D. S. Needleman, M. Y. Hurwitz, B. E. Kemp, and A. R. Means (1990) Regulatory and structural motifs of chicken gizzard myosin light chain kinase. Proc Natl Acad Sci USA 87:2284–2288

82.   Pask, H. T., K. L. Jones, P. K. Luther, and J. M. Squire (1994) M-band structure, M-bridge interactions and contraction speed in vertebrate cardiac muscles. J Musc Res Cell Mot 15:633–645

83.   Patthy, L. (1991) Modular exchange principles in proteins. Curr Op Struct Biol 1:351–361

84.   Pearson, R. B., and B. E. Kemp (1991) Protein kinase phosphorylation site sequences and consensus specifity motifs: tabulations. Meth Enzymol 201:62–81

85.   Perriard, J.-C., M. Caravatti, E. R. Perriard, and H. M. Eppenberger (1978) Quantitation of CK isoenzyme transition in differentiating chicken embry-

onic breast muscle and myogenic cell cultures by immunoadsorption. Arch
Biochem Biophys 191:90–100

86. Politou, A., M. Gautel, M. Pfuhl, S. Labeit, and A. Pastore (1994) Immunoglobulin-type domains of titin: same fold, different stability? Biochem
33:4730–4737

87. Potts, J. R., and I. D. Campbell (1994) Fibronectin structure and assembly.
Curr Biol 6:648–655

88. Price, M. (1984) Molecular analysis of intermediate filament cytoskeleton – a
putative load-bearing structure. Am J Physiol 246:H566–H572

89. Price, M. G. (1987) Skelemins: cytoskeletal proteins located at the periphery
of M-discs in mammalian striated muscle. J Cell Biol 104:1325–1336

90. Price, M. G., and R. H. Gomer (1993) Skelemin, a cytoskeletal M-disc periphery protein, contains motifs of adhesion/recognition and intermediate filament proteins. J Biol Chem 268:21800–21810

91. Rhee, D., J. M. Sanger, and J. W. Sanger (1994) The premyofibril: evidence for
its role in myofibrillogenesis. Cell Mot Cytoskel 28:1–24

92. Saks, V. A., N. V. Lipina, V. G. Sharov, V. N. Smirnow, E. I. Chazov, and R.
Grosse (1977) The localization of MM-isoenzyme of creatine kinase on the
surface membrane of myocardial cellsand its functional coupling to ouabain-
inhibited Na^+/K^+ ATPase. Biochim Biophys A 465:550–558

93. Schultheiss, T., Z. Lin, M.-H. Lu, J. M. Murray, D. A. Fischman, K. Weber, T.
Masaki, M. Imamura, and H. Holtzer (1990) Differential distribution of sub-
sets of myofibrillar proteins in cardiac nonstriated and striated myofibrils. J
Cell Biol 110:1159–1172

94. Sharov, V. G., V. A. Saks, V. N. Smirnow, and E. I. Chazov (1977) An electron
microscopical histochemical investigation of the localization of creatine
kinase in heart cells. Biochim Biophys A 468:495–501

95. Sjöström, M., and J. M. Squire (1977) Fine structure of the A-band in cryo-
sections: the structure of the A-band of human skeletal muscle fibers from
ultra-thin cryo-sections, negatively-stained. J Mol Biol 109:49–68

96. Small, J. V., D. O. Fürst, and L.-E. Thornell (1992) The cytoskeletal lattice of
muscle cells. Eur J Biochem 208:559–572

97. Sorimachi, H., K. Kinbara, S. Kimura, M. Takahashi, S. Ishiura, N. Sasagawa,
N. Sorimachi, H. Shimada, K. Tagawa, K. Maruyama, and K. Suzuki (1995)
Muscle-specific calpain, p94, responsible for limb girdle muscular dystrophy
type 2A, associates with connectin through IS2, a p94-specific sequence. J Biol
Chem 270:31158–31162

98. Squire, J. M. (1981) The structural basis of muscular contraction. Plenum
Press, New York

99. Squire, J. M., P. K. Luther, and J. Trinick (1987) Muscle myofibril architecture.
In Fibrous Protein Structure. J. M. Squire and P. J. Vibert, editors. Academic
Press, London pp 423–450

100. Steeghs, K., A. Benders, F. Oerlemans, A. De Haan, A. Heerschap, W. Ruitenbeek, C. Jost, J. Van Deursen, B. Perryman, D. Pette, M. Brückwilder, J.
Koudijs, P. Jap, J. Veerkamp, and B. Wieringa (1997) Altered Ca^{2+} responses
in muscles with combined mitochondrial and cytosolic creatine kinase defi-
ciencies. Cell 89:1–20

101. Steiner, F., K. Weber, and D. O. Fürst (1998) Structure and expression of the
gene encoding murine M-protein, a sarcomere-specific member of the immunoglobulin superfamily. Genomics 49:83–95

102. Stolz, M. (1997) Chicken cytosolic muscle-type creatine kinase: ientification of the M-band binding domain and analysis of particular enzyme properties by site-directed mutagenesis. Swiss Federal Institute of Technology, Zürich

103. Strehler, E. E., E. Carlsson, H. M. Eppenberger, and L.-E. Thornell (1983) Ultrastructural localization of M-band proteins in chicken breast muscle as revealed by combined immunocytochemistry an ultramicrotomy. J Mol Biol 166:141–158

104. Strehler, E. E., G. Pelloni, C. W. Heizmann, and H. M. Eppenberger (1980) Biochemical and ultrastructural aspects of M_r 165,000 M-protein in cross-striated chicken muscle. J Cell Biol 86:775–783

105. Tokuyasu, K. T., and P. A. Maher (1987a) Immunocytochemical studies of cardiac myofibrillogenesis in early chick embryos. I. Presence of immuno-fluorescent titin spots in premyofibril stages. J Cell Biol 105:2781–22793

106. Tokuyasu, K. T., and P. A. Maher (1987b) Immunocytochemical studies of cardiac myofibrillogenesis in early chick embryos. II. Generation of a-actinin dots within titin spots at the time of the first myofibril formation. J Cell Biol 105:2795–2801

107. Trinick, J. (1994) Titin and nebulin: protein rulers in muscle? Trends Biochem Sci 19:405–409

108. Trinick, J., and S. Lowey (1977) M-protein from chicken pectoralis muscle: isolation and characterization. J Mol Biol 113:343–368

109. Turner, D. C., T. Wallimann, and H. M. Eppenberger (1973) A protein that binds specifically to the M-line of skeletal muscle is identified as the muscle form of creatin kinase. Proc Natl Acad Sci USA 70:702–705

110. Van der Ven, P. F. M., and D. O. Fürst (1997) Assembly of titin, myomesin and M-protein into the sarcomeric M band in differentiating human skeletal muscle cells in vitro. Cell Struct Funct 22:163–171

111. Van Deursen, J., A. Heerschap, F. Oerlemans, W. Ruitenbeek, P. Jap, H. Ter Laak, and B. Wieringa (1993) Skeletal muscles of mice deficient in muscle creatine kinase lack burst activity. Cell 74:621–631

112. Van Deursen, J., W. Ruitenbeek, A. Heerschap, P. Jap, H. Ter Laak, and B. Wieringa (1994) Creatine kinase in skeletal muscle energy metabolism: a study of mouse mutants with graded reduction in M-CK expression. Proc Natl Acad Sci USA 91:9091–9095

113. Vinkemeier, U., W. Obermann, K. Weber, and D. O. Fürst (1993) The globu-lar head domain of titin extends into the center of the sarcomeric M band. J Cell Sci 106:319–330

114. Wallimann, T. (1994) Dissecting the role of creatine kinase. Curr Biol 1:42–46

115. Wallimann, T., T. C. Doetschman, and H. M. Eppenberger (1983) Novel staining pattern of skeletal muscle M-lines upon incubation with antibodies against MM-creatine kinase. J Cell Biol 96:1772–1779

116. Wallimann, T., and H. M. Eppenberger (1985) Localization and function of M-line-bound creatine kinase. Cell Musc Motil 6:239–285

117. Wallimann, T., T. Schnyder, J. Schlegel, M. Wyss, G. Wegmann, A. M. Rossi, W. Hemmer, H. M. Eppenberger, and A. F. G. Quest (1989) Subcellular com-partmentalization of creatine kinase isoenzymes, regulation of CK and oc-tameric structure of mitochondrial CK: important aspects of the phosphoryl-creatine circuit. Progr Clin Biol Res 315:159–176

118. Wallimann, T., D. C. Turner, and H.-M. Eppenberger (1977) Localization of creatine kinase isoenzymes in myofibrils. J Cell Biol 75:297–317

119. Wallimann, T., G. Wegmann, H. Moser, R. Huber, and H. M. Eppenberger (1986) High content of creatine kinase in chicken retina: compartmentalized localization of creatine kinase isoenzymes in photoreceptor cells. Proc Natl Acad Sci USA 83:3816–3819
120. Wallimann, T., M. Wyss, D. Brdiczka, K. Nicolay, and H. M. Eppenberger (1992) Intrazellular compartmentalization, structure and function of creatine kinase isoenzymes in tissues with high and fluctuating energy demands: the "phosphocreatine circuit" for cellular energy homeostasis. Biochem J 281:21–40
121. Wang, K. (1996) Titin/connectin and nebulin: giant protein rulersof muscle structure and function. Adv Biophys 33:123–134
122. Wegmann, G., E. Zanolla, H. M. Eppenberger, and T. Wallimann (1992) In situ compartmentalization of creatine kinase in intact sarcomeric muscle: the acto-myosin overlap zone as a molecular sieve. J Musc Res Cell Motil 13:420–435
123. Williams, A. F., and A. N. Barclay (1988) The immunoglobulin superfamily – domains for cell surface recognition. Ann Rev Immunol 6:381–405
124. Williams, A. F., S. J. Davis, and A. N. Barclay (1989) Structural diversity in domains of the immunoglobulin superfamily. CSH Symp Quant Biol 54:637–647
125. Woodhead, J. L., and S. Lowey (1982) Size and shape of skeletal muscle M-protein. J Mol Biol 157:149–154
126. Woodhead, J. L., and S. Lowey (1983) An in vitro study of the interactions of skeletal muscle M-protein and creatine kinase with myosin and its subfragments. J Mol Biol 168:831–846
127. Wyss, M., J. Smeitink, R. A. Wevers, and T. Wallimann (1992) Mitochondrial creatine kinase: a key enzyme of aerobic energy metabolism. Biochim Biophys A 1102:119–166
128. Yajima, H., H. Ohtsuka, Y. Kawamura, H. Kume, T. Murayama, H. Abe, S. Kimura, and K. Maruyama (1996) A 11.5 kb 5´-terminal cDNA sequence of chicken breast muscle connectin/titin reveals its Z line binding region. Biochem Biophys Res Comm 223:160–164

The C-Protein (Myosin Binding Protein C) Family: Regulators of Contraction and Sarcomere Formation?

P. M. Bennett[1], D. O. Fürst[2] and M. Gautel[3]

[1]The Randall Institute, King´s College London, 26–29 Drury Lane, London WC2B 5RL, UK
[2]Department of Cell Biology, University of Potsdam, Lennéstr. 7a, 14471 Potsdam, Germany
[3]European Molecular Biology Laboratory, Postfach 102209, 69012 Heidelberg, Germany

Contents

1 Introduction

In the early 1970's, while investigating the purity of conventional preparations of muscle myosin, using the – then – new technique of SDS gel electrophoresis, Starr and Offer (1971) detected several proteins of chain weight
intermediate between that of the myosin heavy chain and light chains. These
they labelled alphabetically. Some were expected contaminants of a myosin
preparation, like actin, but several were then unknown. The major one of
these was band C. C-protein or myosin binding protein C (MyBP-C) as it has
become known (to avoid confusion with molecules such as C-reactive protein) has since been shown to be an essential component of the vertebrate
striated muscle thick filament. It binds strongly to myosin and is found in
specific regions in each half of the A-band (Pepe and Drucker, 1975; Craig
and Offer 1976; Dennis et al. 1984; Bennett et al. 1986) where stripes of extra
mass 43 nm apart have previously been observed in longitudinal sections of
muscle (Huxley 1967) and in A segments (Hanson et al. 1971; Fig. 1).

C-protein is one of the immunoglobulin family of proteins, in common
with a number of other myosin binding proteins such as myomesin, M-
protein and titin. It seems to have an essential role in filament formation
during myofibrillogenesis. Over the last 15 years its sequence, expression
during development and differentiation, and the relationship of isoform to
fibre type have been investigated. Since the discovery and sequence analysis
of titin the relationship to this giant protein, as well as to myosin in the thick

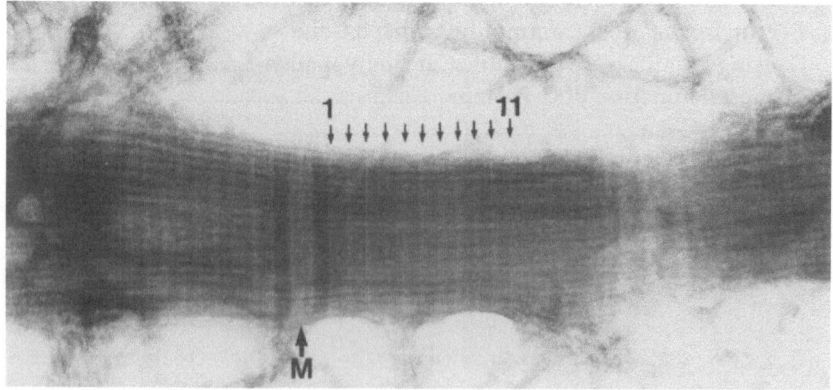

Fig. 1. Part of an A-segment from frog thigh muscle negatively stained with uranyl
acetate. The eleven stripes of extra material spaced 43 nm apart are marked, as is the
M-line. Magnification 55 000 x. Micrograph taken from Hanson et al. (1971)

filament is an interesting question. In mature muscle C-protein modulates contractile activity particularly in the heart. Its involvement in some of the cardiac familial myopathies has given an understanding of its function an increased relevance and urgency.

2 Occurrence of C-Proteins

C-protein has been found in all vertebrate striated muscles where it has been looked for. This includes mammals (rabbit: Offer, 1972; Offer et al. 1973; beef: Young & Davey, 1981; Callaway & Bechtel, 1981; Fürst et al. 1992; human: Fürst et al. 1992; rat: Jeacocke & England, 1980; mouse: Gautel et al. 1998; pig: Murakami et al. 1976), birds (chick: Pepe and Drucker, 1975; Reinach et al. 1982; Kasahara et al. 1994), amphibians (frog: Harzell and Titus 1982; axolotl: Ward et al. 1995a). It is present in both cardiac (Jeacocke & England 1980) and skeletal (Offer et al. 1973) muscles. It has not been found in vertebrate smooth muscle or in non-muscle cells. Neither has it been observed in non-vertebrate striated muscle such as scallop striated adductor or insect flight muscle. It is, therefore, strongly associated with the highly ordered A-bands of vertebrate striated muscle with their filaments of well defined length (1.6 μm) and diameter and their characteristic substructure. In this respect it may be that C-protein and titin go hand in hand in their involvement in control of thick filament assembly.

2.1 C-Protein Isoforms and their Distribution

In mature muscle, three clearly different isoforms of C-protein have been described. There is one form in cardiac muscle, mainly one form associated with fast (white) muscle such as rabbit psoas and chicken pectoralis, and one with slow (red) muscles such as rabbit soleus and chicken ALD (Jeacocke and England, 1980; Calloway and Bechtel, 1981; Young and Davey, 1981; Reinach et al. 1982; 1983; Yamamoto & Moos, 1983). The evidence for this until recently has been from biochemical and immunological observations and is now supported by genetic data (see section 3).

While these three isoforms explain many of the observations there is increasing evidence that there are more isoforms or variants. One of these is an 'embryonic' form (Gautel et al. 1998; Gilbert et al. 1996 and see section 2.2). There is also evidence for a number of proteins which have immunological similarities to slow C-protein: In chicken, four such variants have been distinguished by their differences in electrophoretic mobility (Takeno-Ohmara et al. 1989) and in the rat McCormick et al. (1994) have found two C-proteins of size intermediate between fast and slow. The rabbit protein, originally called X-protein, which was found to be very similar, if not identical, to C-protein extracted from slow red muscle may also be such a variant (Starr & Offer, 1983; Yamamoto & Moos, 1983). The differences observed may be due to splice variants such as seen in the two cardiac C-protein sequences determined by Yasuda et al. (1995) where one has an additional short insert. Alternatively, phosphorylation or posttranslational modifications may account for the additional isoforms.

In addition to these proteins which closely resemble the original C-protein, there is another smaller (55 kDa), but related protein. Called H-protein (MyBP-H), it is one of the original impurities found in myosin preparations by Starr and Offer (1971). It is related in that it has sequence homologies to C-protein, binds to myosin and is located in the C-protein region of the thick filament (Starr & Offer, 1983; Yamamoto, 1984). In chicken muscle a protein with similar properties and with apparent molecular weight of 86kDa has been characterised by Bähler et al. (1985a, b). cDNA cloning has identified it as chicken H-protein (Vaughan et al. 1993a, b). The different gel mobilities probably arise from a Pro-Ala-rich region at the amino-terminal end (see section 3).

While only cardiac C-protein has been demonstrated in cardiac muscle, (Bähler et al. 1985c; Obinata et al. 1985) the distribution of fast and slow C-protein isoforms and H-protein in adult skeletal muscle fibres is very variable. It is possible for one, two or all three proteins to be detected by immunofluorescence in any one fibre, although different muscles have different patterns of distribution (Reinach et al. 1982; 1983; Starr et al. 1985, Bähler et al. 1985b). The simplest situation is found in slow red muscles (type 1) where only slow C-protein and little or no H-protein has been detected in rabbit and mouse soleus (Starr et al. 1985; McCormick et al. 1994) and chicken ALD (Reinach et al. 1982; Bähler et al. 1985b). However, it should be noted that the cDNA for fast muscle H-protein hybridised to an mRNA in chicken ALD suggesting that another isoform of H-protein may be expressed here (Vaughan et al. 1993a). In fast (type 2) fibres the observations point to a difference, or continuum, between the white, glycolytic fibres and the red oxidative fibres. In the white fibres, e.g. most of those in chicken pectoralis and rabbit psoas, fast C-protein and frequently H-protein is present (Reinach et al. 1982; Bähler et al. 1985b; Starr et al. 1985), whereas in the faster redder fibres there is more slow-type C-protein (possibly a specific isoform or variant) and sometimes H-protein (Starr et al. 1985; McCormick et al. 1994, Takeno-Ohmara et al. 1989). This observation has led McCormick et al. (1994) to suggest that the expression of C-protein depends more on the metabolism of the fibre than its speed.

An interesting question is, at what stage is C-protein incorporated into the thick filament or sarcomere during myofibrillogenesis? Comparing the temporal expression of myosin (Sweeney et al. 1984) to that of C-protein, Bähler et al. (1985c) have suggested that C-protein is first expressed at a very similar time to myosin in chicken cardiac muscle and skeletal somatic myotomes. The same conclusion has now been reached directly by Gautel et al. (1998) in embryonic mouse (Fig. 2A). *In vitro* studies of cardiac myocytes have shown by immunofluorescence that C-protein is already incorporated into sarcomeres by the time they are formed (Schultheiss et al. 1990). That C-protein is necessary for sarcomere formation has been demonstrated by the elegant studies of Gilbert et al. (1996) on deletion mutants of C-protein in chicken myoblasts. They showed that the carboxy-terminal part of the C-protein molecule is essential for the correct formation of thick filaments and their incorporation into sarcomeres.

2.2 Developmental Expression Patterns

The expression of C-proteins during development is shown in Fig. 2 for a number of different muscles in chicken and mouse. The simplest situation is in cardiac muscle as described in mouse and chicken. Here the only isoform to be expressed is the cardiac one and it is present from the earliest times. It has been detected from day 2 in the chicken embryo (Kawashima et al. 1986) and day 8 in the mouse (Gautel et al. 1998; Fougerousse et al. 1998). A different conclusion was reached in the axolotl. At the initial heartbeat stage a slow isoform is expressed but is not retained in the adult heart where a specific cardiac form is found (Ward et al. 1996).

The situation is more complicated in the skeletal muscles. There is always a transitional expression of C-protein isoforms before the final distribution is established. The first work carried out by Fischman and his colleagues in the chicken is collated in Fig. 2B. In the developing embryonic muscle, the cardiac isoform (or an isoform sharing a cardiac epitope) is expressed from about day 3 in the somitic myotomes and continues into specific muscle formation until day 15 in the pectoralis and ALD (Obinata et al. 1985; Bähler et al. 1985c; Kawashima et al. 1986). The slow C-protein isoform kicks in at day 12 (Obinata et al. 1984; 1985). In the slow ALD this is the final situation. However, the fast, pectoralis muscle undergoes further change in that the fast C-protein isoform begins to be expressed at about day 17de. The slow isoform is slowly lost and has essentially disappeared by 14 days post hatch.

Unlike the chicken, mammalian (mouse) skeletal muscle does not exhibit transient expression of the cardiac isoform. However, using an antibody that

A

gestation day	8	8.5	9	9.5	10	10.5	11	11.5	12	13	14	15	16	17	18	nb	tissue
Theiler stage	12	13	14	15	16	17	18	19	20	21	22	23	24	25	26		
somite count	-7	-14	-20	-29	-34	-39	-44	>45									
desmin																	skeletal muscles
sarcomeric α-actinin																	
titin																	
sarcomeric myosin																	
sarcomeric actin																	
pan C-protein																	
slow C-protein																	
fast C-protein																	
desmin																	cardiac muscles
sarcomeric α-actinin																	
titin																	
sarcomeric myosin																	
sarcomeric actin																	
pan C-protein																	
cardiac C-protein																	

B

	0	5	10	15	20	1	7	14	adult	tissue
		in ovo				after birth				
cardiac C-protein										skeletal muscles
slow C-protein										
fast C-protein										
cardiac C-protein										cardiac m.

reacts with many if not all C-proteins (pan-C), in human and mouse muscles, it was shown that there is, expressed very early in skeletal muscle development, an as yet uncharacterised form of C-protein (Gautel et al. 1998). This isoform is possibly specific to embryonic muscle and may be equivalent to the cardiac isoform expressed early in developing chicken muscle. Subsequently, 14 days post gestation, the slow form is expressed in the somites. It is not until several days later that the specific distribution of fast and slow C-proteins characteristic of mature muscle begins to be established.

It is not clear what determines the isoform distribution of C-protein in the mature muscle. On the one hand, there is clearly some influence of the myosin type (McCormick et al. 1996), but where there is usually only one myosin heavy chain isoform in any one mature fibre there may be more than one C-protein isoform. Such expression may be correlated with the expression of other proteins which occur in more than one isoform within a fibre, such as myosin light chains or troponin (Dhoot et al. 1985). No direct correlation of this kind has been identified but it has been shown that when the myosin type changes during hypertrophy there is a concommitant change in the C-protein isoforms which are expressed (McCormick et al. 1994).

3 Cloning of C-Protein cDNAs and Genes

The existence of C-proteins with slightly different gel mobilities, depending on the muscle source, as well as differential immunological and ultrastructural localization data have suggested the existence of at least three distinct isoforms (Reinach et al. 1982, 1983; Yamamoto & Moos, 1983; Bennett et al. 1986). Both immunoblotting data and two-dimensional peptide maps essentially ruled out the possibility of extensive post-translational modifications accounting for these differences (Yamamoto & Moos, 1983). Clarity came with the availability of sequence data. Up to now, cDNA sequences for a slow skeletal muscle (Fürst et al. 1992; Weber et al. 1993), a fast skeletal muscle (Einheber & Fischman, 1989; Weber et al. 1993; Okagaki et al. 1993) and a

---

Fig. 2. Temporal expression patterns of C-protein and other myofibrillar proteins during mouse and chicken myogenesis. Expression as monitored by immunofluorescence microscopy is shown as a darkly shaded bar; lightly shaded bars reflect redistribution of certain proteins during myotube formation or fibre maturation. A) Data for mouse taken from Gautel et al. (1998) The upper half of the Figure shows expression patterns in body muscles of somitic origin; the lower half shows expression patterns in cardiac muscle. Gestation day and developmental stages of embryos according to Theiler (1972) are indicated at the top. B) Data for chicken C-protein taken from Obinata et al. (1984;1985), Bähler et al. (1985c) and Kawashima et al. (1986)

cardiac muscle (Kasahara et al. 1994; Gautel et al. 1995; Yasuda et al. 1995; Ward et al. 1995) isoform have been firmly established. One of the most exciting initial results revealed by these data was the fact that C-proteins are primarily composed of domains approximately 100 amino acids in length. These can be divided in two groups exhibiting significant homology to either fibronectin type III (Fn-like) or immunoglobulin cII (Ig-like) domains. The surprise arose, because until then the dogma existed that this kind of sequence motives was confined to cell surface and extracellular matrix proteins (Williams & Barclay, 1988). Now it is clear that Ig- and Fn-like domains are versatile sequence motifs found in a great number of invertebrate and vertebrate muscle proteins (see, for instance, the contributions about titin, the M-band proteins, myomesin and M-protein, and twitchin in this volume).

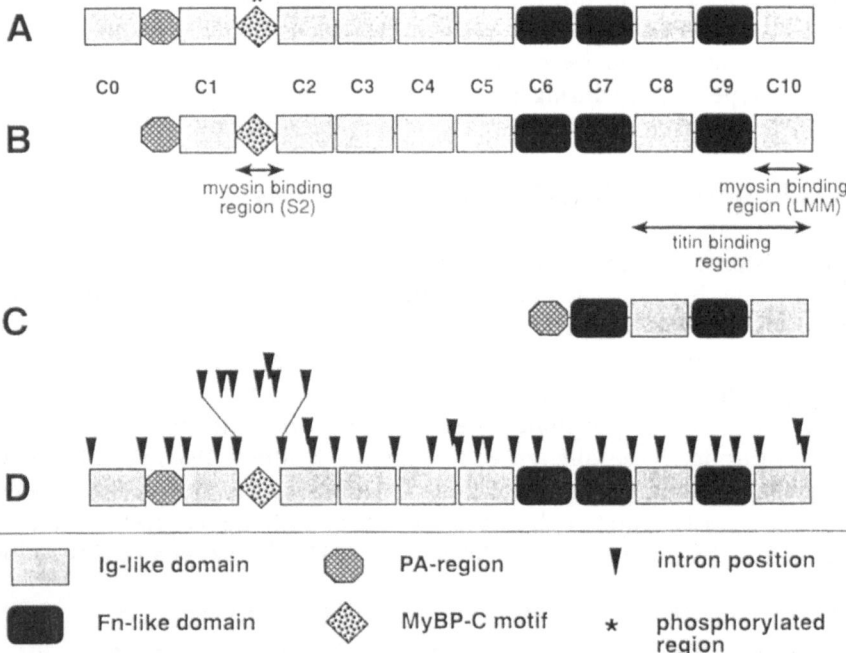

Fig. 3. Schematic representation of the structures of C-protein isoforms and of the closely related H-protein as deduced from translated cDNA sequences and the gene structure of the cardiac isoform. Panels A and B give the structures of cardiac and skeletal muscle C-protein isoforms, and panel C the structure of H-protein, respectively. Domain designations are shown between diagrams A and B. Panel D illustrates intron positions (arrowheads) in relation to the domain structure of human cardiac C-protein (Carrier et al. 1997) For ease of visability of the intron positions within the MyBP-C motif this region is enlarged. Domain symbols are explained in the box below the molecular models

Figure 3 shows a schematic representation of the domain organization of the three isoforms that have been characterized so far. Thus, slow and fast isoforms are composed of 10 Ig- and Fn-like domains which have been designated C1 to C10. C1 is flanked amino-terminally by a proline/alanine-rich unique sequence of approximately 50 amino acids (called the 'PA-region') and carboxy-terminally by another unique region approximately 100 residues in length (called the 'MyBP-C-motif'; Fig. 4). The cDNAs encoding human slow and fast C-protein isoforms predict proteins of molecular masses 126.5 and 128 kDa, respectively (Fürst et al. 1992; Weber et al. 1993). Deviations of these calculated masses from the values observed in SDS-PAGE are supposed to be caused by the PA-region (Weber et al. 1993). Cardiac C-protein is predicted to have a mass of 141 kDa, thus significantly bigger than the other isoforms (Kasahara et al. 1994; Yasuda et al. 1995). The reason is an additional Ig-like domain preceeding the PA-region, called C0 (Fig. 3A).

Comparisons of translated cDNA sequences from different species revealed that inter-species variability is lower within each individual isoform (identity levels between chicken and man are in the range of 70%) as compared to isoform variability within an individual species (identity levels around 50%). There are, however, distinct regions which exhibit a significantly higher degree of conservation, like, for instance, the carboxy-terminally situated domains, which were shown to bear binding sites for myosin heavy chain (Okagaki et al. 1993; Alyonycheva et al. 1997) and titin (Freiburg & Gautel., 1996). On the other hand, the PA-region shows the

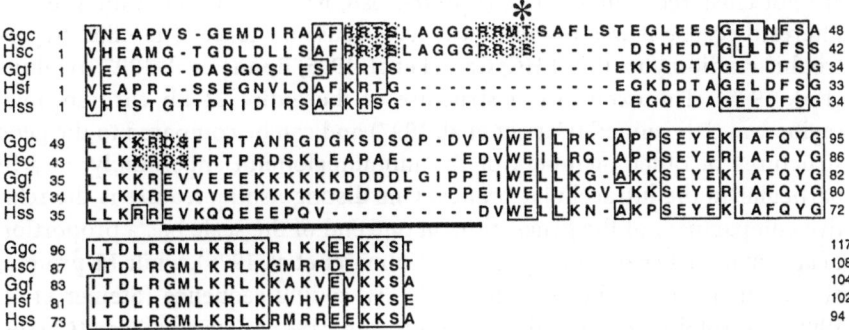

Fig. 4. Multiple sequence alignment of the MyBP-C motif of cardiac, slow skeletal and fast skeletal muscle isoforms from human and chicken. Residues conserved between all isoforms are highlighted as boxes. The key regulatory site is further emphasized by an asterisk. The region within the MyBP-C motif that shows the greatest variability between isoforms and species is underlined. Note that phosphorylation sites (shaded) are confined to the cardiac isoform

to reflect differences in regulatory interactions with other ligands. The MyBP-C motif shows regions of high conservation, interspersed with iso-form-specific insertions involved in signal transduction by phosphorylation (see below). For cardiac C-protein this was shown to be the case, since a PKA phosphorylation site involved in adrenergic regulation of contraction specific for this isoform was mapped to the MyBP-C motif (Gautel et al. 1995).

Sequencing also revealed another surprise. H-protein (also called 86 kDa protein in chicken), another myosin-binding protein located in the same thick filament region as C-protein (Starr & Offer, 1983; Bähler et al. 1985a, b), is composed of four Ig- and Fn-like domains subsequent to a unique amino-terminus (Vaughan et al. 1993a, b; Fig. 3). The order of these do-mains is identical to the arrangement at the carboxy-terminus of C-protein and the degree of similarity is, with 50% identical residues and another 17% conservative substitutions, comparatively high. Thus, H-protein can actually be envisioned as a low-molecular mass isoform of C-protein, or, at least it belongs to the same subgroup of intracellular immunoglobulin superfamily molecules. However, the N-terminal region, involved in interactions with the same ligand in all C-protein isoforms via the conserved MyBP-C motif, is distinctly different. Although H-protein shares a common myofibrillar sorting signal with C-protein, localised in the last four domains, its specific myofibrillar interactions have to be different.

Chromosomal mapping (slow C-protein: Chr. 12, fast C-protein: Chr. 19; cardiac C-protein: Chr. 11p11.2, H-protein 1q32.1) showed that – unlike, for instance, the myosin heavy chain family – C-protein and H-protein genes are not clustered. The cardiac C-protein isoform has attracted the greatest interest because it was found that defects cause a subset of familial hyper-trophic cardiomyopathy (FHC; Gautel et al. 1995; Bonne et al. 1995; Watkins et al. 1995). The exon-intron structure of the entire 21 kb chromosomal region was determined (Carrier et al. 1997) and can be compared to the gene structures of other members of the immunoglobulin superfamily, and in particular to the intracellular branch. One distinguishing feature is the rela-tive compactness of the genes: titin, at one end of the scale has a proportion of approximately 67% coding region (Kolmerer et al. 1996), while M-protein, at the other end of the scale has only 7% coding sequence (Steiner et al. 1998); C-protein takes a middle position with 21% coding sequence (Carrier et al. 1997). This, in turn, reflects fundamental differences in gene organiza-tion between titin and the other muscle Ig superfamily proteins. In the case of titin, almost no correlation of intron positions with domain structure was observed (Kolmerer et al. 1996). In contrast, Fn-like domains in cardiac C-protein are generally encoded by two exons with internal intron positions occurring only at two distinct positions. The introns that flank Fn-like domains strictly appear between the first and second base of the respective codon (i. e. phase I). This kind of organization is also found in the gene for murine M-protein (Steiner et al. 1998) and in the rat fibronectin gene (Odermatt et al. 1985; Schwarzbauer et al. 1987). The gene structure of Ig-like domains is not so strictly conserved. In cardiac C-protein, they are encoded by one to three exons with boundaries almost exclusively in phase I and often not correlated exactly with domain borders (Carrier et al. 1997). In contrast, M-protein Ig-like domains are almost exclusively encoded by two exons with intron positions well correlated with domain borders and in phase I.

In the future, a comparative analysis of promoter regions might give a better insight in mechanisms of muscle fiber type-specific gene regulation.

4 Biochemical and Biophysical Characterization of C-Protein

A summary of the biochemical characteristics of the C-protein molecule is shown in Table 1. The molecule has a molecular mass of 130 to 150 kDa and therefore consists of a single chain. It is a rod-like structure of a length of ~40 nm, consistent with it being formed from ten 4 nm Ig-like domains as indicated by the sequencing data. Electron microscopy shows that it is frequently bent into an L- or V-shape (Fig. 5).

C-protein is a polar protein; while the C-terminus mediates the association with the thick filament, the N-terminal region forms yet uncharacterized interactions and contains isoform-specific regions involved in signal transduction. Phosphorylation occurs at isoform-specific sites in a C-protein specific domain in the N-terminal region of the protein (Gautel et al. 1995). This 100 residue domain, the MyBP-C motif, is highly conserved between all isoforms of C-protein and between avians and mammals (Fig. 4). This suggests that it interacts with a common and equally conserved binding partner which, however, has not been identified to date. The extent of sequence identity between various C-protein isoforms and their cross-species conservation (Fig. 4) suggests further that the function of this module is specified by the conserved regions, and that the phosphorylation motifs found in the cardiac isoform represent an additional feature. Phosphorylation of the cardiac isoform can occur by cyclic-AMP dependent protein kinase (cAPK) or an unknown Ca^{2+}/Calmodulin dependent protein kinase (Hartzell and Titus, 1982; Hartzell and Glass, 1984; Schlender and Bean, 1991; Gautel et al. 1995). An isoform-specifc loop insertion, LAGGGRRIS, contains the major cardiac phosphorylation site for both enzymes (Gautel et al. 1995).

Table 1. Characteristics of C-protein molecule (for details see text)

Subunit structure	single chain	Offer et al. (1973)
Molecular weight	130–150 kd	Offer et al. (1973)
Extinction coefficient Crude C-protein	$E_{280}^{1\ mg/ml} = 1.09$	Starr & Offer (1982)
$S_{20,\ w}$	4.65 S	Offer et al. (1973)
Secondary structure	50% β-sheet (CD) immunoglobulin β-barrels	Offer et al. (1973) Einheber & Fischman (1990) Fürst et al. (1992)
Molecular shape ultracentrifuge electron microcopy	rod 3–4 nm in diameter axial ratio (10:1) 36 nm long 25 nm; 35 nm 44 nm (2 x 22) 32 nm 50 nm	Offer et al. (1973) Bennett et al. (1985) Harzell & Sale (1985) Swan & Fischman (1986) Fürst et al. (1992)
No of molecules/thick filament	40 36; 50	Offer et al. (1973) Morimoto & Harrington[a] (1973; 1974)

[a] assuming three myosin mols/14.3 nm.

The C-protein molecule has a length of about 40 nm (Offer et al. 1973; Bennett et al. 1985; Hartzell and Sale, 1985; Swan & Fischman, 1986; Fürst et al. 1992). This length is sufficient to bridge between thick and thin filaments, and indeed, in-vitro interaction with F-actin has been reported (Moos et al. 1978). On the other hand, it was reported that C-protein phosphorylation can directly affect the conformation of the myosin heads on isolated, cardiac thick filaments (Weisberg and Winegrad, 1996, 1998). This suggests rather that the phosphorylated MyBP-C motif interacts with a ligand on the thick filament, most plausibly myosin. Indeed, Starr and Offer (Starr and Offer,

---

Fig. 5. Electron micrographs of C-protein molecules rotary shadowed with platinum. (a) Rabbit slow C-protein plus myosin. Specimens prepared by the spraying technique. Many of the molecules are bent. Arrows indicate myosin molecules from which it can be seen that the C-protein rods are thicker than the myosin tail. Micrograph taken from Bennett et al. (1985). (b) Bovine slow muscle C-protein prepared by the spinning method which tends to pull the molecules straight. (Fürst et al. 1992) Magnification in (a) and (b) 100 000 x

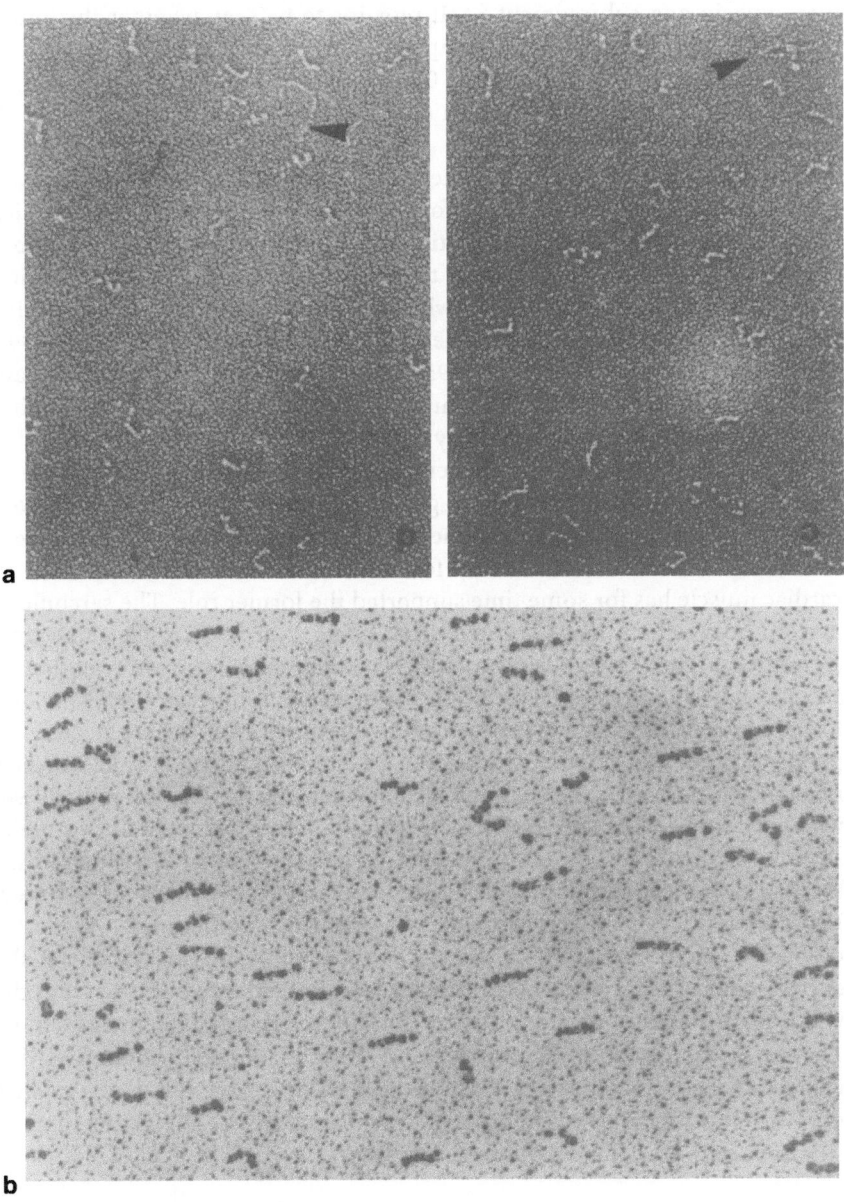

1978) described a weak interaction of full-length C-protein with heavy meromyosin and subfragment S2. It remains to be seen whether this interaction can be substantiated and mapped onto a defined region of C-protein different from the myosin-binding C-terminus. Recent results demonstrate that myosin S2 is indeed the ligand of the MyBP-C motif (M. Gruen and M. Gautel, unpublished results).

The C-terminal region of C-protein interacts strongly with myosin LMM (Okagaki et al. 1993) with some isoform-specific differences in affinity which may be relevant for the in-vivo sorting (Alyonycheva et al. 1997). Additionally, a weaker interaction between the three C-terminal domains and titin exists (Freiburg and Gautel, 1996) which may be responsible for the specific targeting of C-protein to the C-zone of the thick filament (Fürst et al. 1992; Labeit et al. 1992; Soteriou et al. 1993; Gilbert et al. 1996; Freiburg and Gautel, 1996). Whether, and how the three thick filament proteins myosin, titin and C-protein cooperate in a ternary complex is unclear.

One of the questions to be answered in developmental studies is whether C-protein is present at the early stages of filamentogenesis and is therefore capable of having a major role in the precise modelling of the thick filament or whether it binds to an already formed thick filament. The evidence in cardiac muscle has for some time supported the former role. The sarcomeric myosin heavy chain and C-protein both being expressed at the same time (Schultheiss et al. 1990). However the evidence for a similar situation in skeletal muscle is recent. Two pieces of evidence implicate C-protein in a formative role. One is suggested by Gautel et al. (1998) who detected the presence of an embryonic C-protein at the same time that titin and muscle myosin are expressed. This supports the evidence from Gilbert et al. (1996) who have transfected chick myoblasts with different fragments of C-protein, showing that cells that express the C-terminal fragments of C-protein which bind to myosin incorporate them into fibrils as soon as they are formed. However the length of the expressed fragment corresponding to the last 4 domains (C7–C10; for domain nomenclature see Fig. 3) is necessary for the correct formation of sarcomeric structure. Expressed fragments without the C-terminal domain are not sufficient for correct thick filament formation. That is, without this domain expressed, thick filament formation is disrupted and striated fibrils are not formed. Taken together this evidence points to the absolute requirement of C-protein to be present for the precise structure of the thick filament to be achieved.

5 Localization of C-Protein at the Electron Microscope Level

In electron micrographs of isolated A segments and of cryosections of muscle, 11 stripes of extra material ~43 nm apart were identified in each half of the A band (Hanson et al. 1971; Craig, 1977; Wilson and Irish, 1980; Sjöstrom & Squire, 1977; Fig. 1). These were designated 1 to 11, starting at the edge of the bare zone. In the first sections of rabbit psoas and chicken pectoralis muscle, immunolabelled with antibody to a crude C-protein preparation, the outer 9 of these stripes were decorated, i.e. stripes 3 to 11 (Pepe & Drucker, 1975; Craig and Offer 1976). Work since then, using better defined antibodies, has revealed a more complicated picture which correlates with the immunofluorescence data described above (Dennis et al. 1984; Bähler et al. 1985b; Bennett et al. 1986; section 2). The general conclusion is that the 9 stripe positions, 3 to 11, are occupied by a form of C-protein or H-protein. Whether there are other, as yet undescribed, components in these stripes is not known but it is unlikely that there are any major extra proteins to be

Table 2. Pattern of labelling on the accessory protein stripes in rabbit and chicken muscles. Data taken from Bähler et al. 1985; Dennis et al. 1984; Starr et al. 1985; Bennett et al. 1986. Figures in brackets indicate where the labelling is weak or where the data suggests more than one protein occurs

Muscle	fibre type	fast C-protein	slow C-protein	H-protein (86 kD protein)
Rabbit				
psoas	fast white	4–11	–	3
	fast intermediate	5–11	4	3
	fast red	–	3–11	–
	slow red	–	3–11	–
plantaris	fast white	(3) 5–11	3–4	–
	fast intermediate	(5–11)	3–4 (5–11)	–
	fast red	–	3–11	–
	slow red	–	3–11	–
soleus	slow red	–	3–11	–
Chicken				
pectoralis	fast	((3)) (5–11)	–	3–4 (5–11)
PLD	fast	(2) (3) (6–11)	(3) 4–5 (6–11)	?
ALD	slow	–	3–11	–

found except for those that contribute to the mass of stripes 1 and 2. The actual dispositions of the C-protein isoforms depend on the species, the muscle and the fiber type. They have only been investigated in any depth in the rabbit and the chicken skeletal muscle.

In rabbit, the pattern of labelling was observed in three muscles, the slow red soleus, the fast white psoas and the mixed plantaris (Starr et al. 1985; Bennett et al. 1986). At least 5 distinct patterns of labelling were observed (Table 2, Fig. 6). The simplest, in both the slow and fast red fibres of all three muscles was 9 stripes (3–11) of slow C-protein (X-protein) (Fig. 6c). In psoas, most of the fibers are fast white and have a pattern of 8 stripes (4–11) of fast C-protein and one (stripe 3) of H-protein (Fig. 6b and e). Other fibres in the psoas which were not so white, had a pattern in which slow C-protein on stripe 4 replaces fast C-protein (Fig. 6a and d). Similarly complicated patterns were seen in plantaris fibers, except here no H-protein was found in any of the fibers (Table 2). In some fibres, antibody stripes were of variable strength indicating that possibly more than one C-protein isoform could share stripes. Confirmation of this came from double labelling plantaris muscle with both fast and slow C-protein antibodies when all nine stripes had equal strength.

In the chicken similar observations have been made although the patterns seen are different (Dennis et al. 1984; Bähler et al. 1985b; Table 2). As in the rabbit, the slow fibres of the ADL contains only slow C-protein which is found on 9 stripes. On the other hand the fast fibres of chicken breast muscle have fast C-protein on seven stripes, but H-protein on all nine. The fibres in another fast muscle, the PLD, have a different arrangement. Most of their stripes are shared by both fast and slow C-protein.

The complexity of the patterns formed and the variation between muscles suggest that the patterns do not relate simply to fibre type but that there is considerable plasticity in C-protein isoform incorporation as is found for other sarcomeric proteins like troponin and myosin light chains (Dhoot et al. 1985). It is presently unclear what this degree of variability means. It certainly is suggestive of a fine tuning mechanism, such as, modulating thick-thin filament interaction.

6 Relationship to Thick Filament Structure

All of the data on the vertebrate striated muscle thick filament points to it being a very precisely contructed organelle of which C-protein is one of the necessary ingredients. Ever since the discovery of the association of C-protein with a specific location in the thick filament there have been attempts to ascribe a structural role to it. This role for C-protein is supported

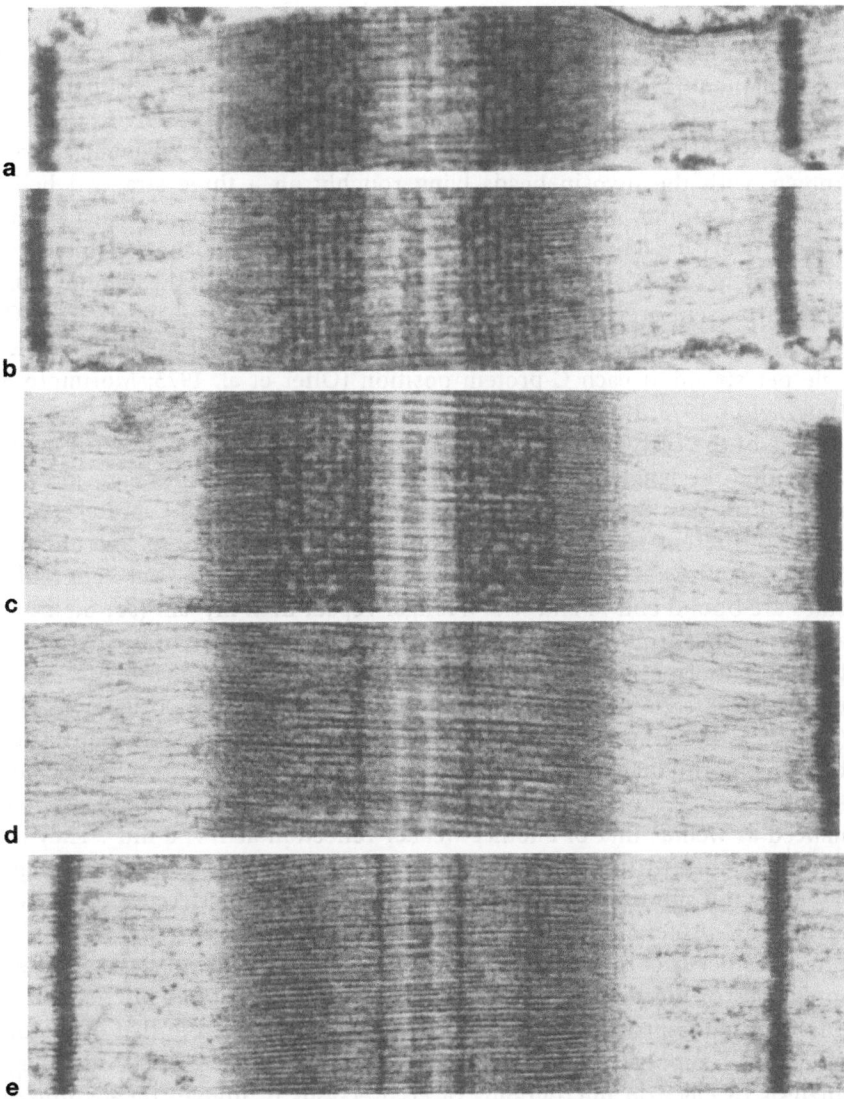

Fig. 6. Electron micrographs showing C-protein and H-protein labelling patterns in rabbit psoas fibres. a) and b) labelling of stripes with antibody to fast C-protein. **a)** stripes 5–11 and **b)** stripes 4–11. **c)** and d) labelling with antibody to slow C-protein (X-protein); c) stripes 3–11 and **d)** stripe 4 only. e) labelling of stripe 3 with antibody to H-protein. Magnification x 34 000. Micrograph from Bennett et al. (1986)

by the evidence described above that it is has a specific location in the filament, that it is expressed at the same time as myosin during development
and that at least the C-terminal third of the molecule is essential for correct
thick filament formation (Gautel et al. 1998; Bähler et al. 1985c; Gilbert et al.
1996).

The underlying arrangement of myosin in the thick filament is well
known, with the myosin heads lying roughly on a three stranded helix.
Along each strand, myosin molecules occur every 14.3 nm and each strand
repeats every 3 x 43 nm. In such a well defined organelle, the stoichiometry
of C-protein to myosin should be very precise. Estimates of the amount of C-
protein present are consistent with 3 molecules at each of the 18 axial C-
protein positions along each thick filament, that is, one C-protein-like molecule per strand at each C-protein position (Offer et al. 1973; Morimoto &
Harrington, 1973; 1974; Craig & Offer, 1976).

One of the early ideas about the function of C-protein was that it might
determine the length of the thick filaments by a vernier mechanism. This
mechanism says that two sets of proteins, e.g. myosin and C-protein, have
slightly different repeats which go in and out of phase with one another
(Huxley & Brown, 1967). In its simplest form, the rod-like C-protein molecules are bound end to end with a longer repeat than myosin (see Squire et
al. 1976). The strongest evidence for this was the presence in the X-ray diffraction pattern of muscle of a reflection on the meridion of $1/44$ nm^{-1} spacing which is significantly bigger than the repeat of the myosin molecules
(3 x 14.3 = 43 nm) (Huxley & Brown, 1967). Rome et al. (1973) demonstrated
that this meridional arose from the presence of C-protein by immunolabelling the muscle and showing that the intensity of the reflection was enhanced as well as that of another weaker reflection at $1/41.8$ nm^{-1}. They argued that the two reflections were a pair and that this could be brought
about by the splitting of a reflection at $1/43$ nm^{-1} by interference of the two
arrays of C-protein in each half of the A-band. It is accepted that such splitting does occur (Craig & Megerman, 1979), but since the observed relative
intensities of the two C-protein reflections are not quite as predicted it is
possible that C-protein has a spacing slightly different from that of the the
myosin. In support of this, Squire and his colleagues carried out image
analysis of electron micrographs of cryosections of human tibialis muscle
where interference effects can be avoided. They concluded that the C-
protein molecules do not lie exactly on the myosin repeat but that the difference between them (43.4 nm for C-protein compared to 42.9 nm for myosin)
is small (Sjöstrom & Squire, 1977; Squire et al. 1982). On the other hand,
Craig (1977) used a similar method of analysis to show that in the optical
diffraction patterns of electron micrographs of one half of A segments the

spacing of the meridional from the C-protein array was not significantly different than 3 times the myosin head repeat (no more than 0.25 nm). Similar results have been obtained in sections of freeze substituted frog sartorius muscle (Craig et al. 1992).

Several other factors mitigate against the simple vernier mechanism. Firstly, since C-protein has been shown to bind very strongly to myosin and to alter myosin filament structure *in vitro* (Moos et al. 1975; Koretz, 1979; Miyahara & Noda, 1980; Koretz et al. 1982; Davis, 1988), it seems unlikely that it can have a binding pattern that is entirely unrelated to myosin. Secondly, the arrangement proposed could not satisfy the electron microscopy data, since the small axial extent of the accessory protein stripes and the polyclonal antibody labelling suggest that the rod-like C-protein molecules are arranged perpendicular rather than parallel to the filament axis. Finally, a much better candidate for a template for thick filament length determination has been found in titin.

A number of possible arrangements of C-protein have been suggested which are consistent with the small axial extent of the stripe material at each level. A subset of these models accept that C-protein is associated with the filament backbone as has been argued from the appeerence of the stripes in sections of antibody labelled muscle (Offer, 1972). These range from an arrangement where three molecules wrap themselves around the filament to form a tight collar (Offer, 1972; Swan & Fischman, 1986), to a loose three membered ring in which the C-protein molecules are bound end to end (Bennett et al. 1985). Others have been influenced by the observation that C-protein can bind to actin, to suggest that some part of the molecule projects from the thick filament towards the thin (Moos et al. 1978). Recently, Luther (1998) has visualised lateral projections between thick and thin filaments at 43 nm intervals in relaxed muscle and suggested that they might be C-protein. Also, Magid et al. (1984) have suggested the presence of lateral struts contributing to lateral stiffness in the muscle and ascribed them to C-protein.

Recently, using the pan-C-protein antibody to the N-terminal domain of C-protein (Gautel et al. 1998) to immunolabel muscle and comparing these results to previous data has enabled us to obtain more information about the axial disposition of C-protein (Fig. 7). This antibody was found to label 9 stripes (3 to 11) in each half of the A-band in rat soleus muscle. The positions of the stripes are shown in Fig. 7c where they can be seen to be the same as the unlabelled stripes seen in the type 1 fibre of the human tibialis anterior (Fig. 7d; measurement taken from Fig. 2 of Sjöstrom & Squire, 1977). Thus the N-terminal half of C-protein is thought to lie here. However, the C-terminal is not at the same position. Firstly, antibody to H-protein, the molecule that is equivalent to the four C-terminal domains of C-protein, labels at stripe 3 in rabbit psoas at a position 15 to 20 nm closer to the M band than the stripes (Fig. 7a). In addition, a polyclonal antibody to slow C-protein labelled, in rabbit slow red muscle, a much broader stripe that can, in places, be distinguished as a doublet (Bennett et al. 1986; Fig. 7b). It was thought that this observation might imply two bands of C-protein within the 43 nm repeat. However, since the pan-C antibody only labels one stripe then this broad label can be thought to represent the axial extent of the C-protein along the filament. It is particularly compelling that one part of the doublet corresponds with the stripe position and the other is close to the H-protein position. To explain this observation, we propose that in the filament, C-protein is bent into an L shape similar to that adopted by the molecules in solution as revealed in electron micrographs of rotary shadowed C-protein (Hartzell & Sale, 1985; Bennett et al. 1985; Swan & Fischman, 1986; Fürst et

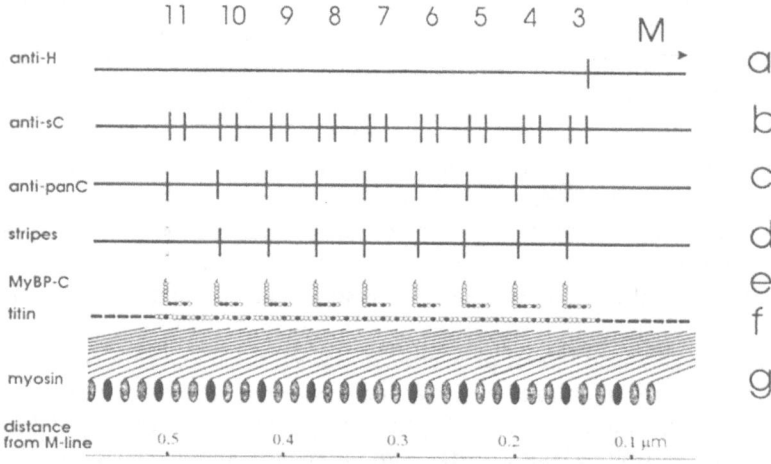

Fig. 7. Diagram illustrating the axial arrangement of C-protein and titin in the C-zone in one half of the thick filament. The stripe positions are labelled 3–11 at the top of the diagram and a scale bar is given at the bottom. a) position of the anti H-protein in rabbit psoas muscle (Bennett et al. 1986) b) positions of the doublets of antibody to slow C-protein in rabbit soleus muscle (Bennett et al. 1986) c) position of pan-C antibody in rat soleus muscle (our unpublished observations) d) Positions of stripes seen in crosections of human slow muscle.(measurements made on Fig. 2 of Sjöström & Squire, 1977) e) proposed arrangemnt of C-protein in a bent configuration with the carboxy-terminal closer to the M-line. f) seven superrepeats of titin arranged axially so as to allow the first Ig domain of the repeat to interact with the C10 domain of C-protein as suggested by the data of Freiburg & Gautel.(1996) g) linear arrangement of myosin molecules at 14.3 nm intervals

al. 1992; Fig. 5). The C-terminal myosin-binding domain is nearer to the M band and this part of the molecule lies approximately parallel to the filament axis and close to the backbone. The other half of the molecule is approximately perpendicular to the filament and contributes to the accessory protein stripes. This L-shaped conformation is similar to that proposed for the M band protein myomesin (Obermann et al. 1995; 1996).

The forces that restrain the N-terminal portion of C-protein have not been determined. However, there is evidence that in addition to LMM, C-protein can bind to myosin S2 (Starr & Offer, 1978) as well as to actin (Moos et al. 1978), so it is possible that the N-terminal portion of C-protein could be involved in holding the myosin heads back against the filament in relaxed muscle or in some interaction with the thin filament. In particular, the involvement of the MyBP-C motif between domains 1 and 2 is unknown. Since it is the site of phosphorylation in cardiac C-protein (Gautel et al. 1995) and this affects the acto-myosin interaction, it could be involved in this role.

So far the influence of titin has not been considered. The observation that myosin and C-protein are expressed shortly after titin implies that these three proteins are intimately associated in thick filament formation. Expression of a truncated C-protein indicates that the myosin binding domain is, itself, not sufficient (Gilbert et al. 1996). The three domain titin binding region is also essential, suggesting that myosin, C-protein and titin form a strong ternary complex every 43 nm along the filament.

For the part of its length associated with the C-protein region, titin has a repeating pattern of Ig-like and Fn-like domains. This pattern consists of 11 domains (Ig, 2 Fn, Ig, 3 Fn, Ig, 3 Fn) which at 4–4.5 nm / domain gives a length that corresponds closely to 43 nm, the myosin and C-protein repeat, as was early recognised (Labeit et al. 1992; Labeit and Kolmerer, 1995). Since there are 11 such superrepeats, this part of the titin molecule is supposed to dictate the pattern of accessory proteins in the A-band (Labeit & Kolmerer, 1995). In agreement with this is the observation that some titin antibodies recognise repetitive epitopes, and that the label coincides with the position of C-protein antibodies (Fürst et al. 1989). Nine of the 11 domain superrepeats are shown in Fig. 7. Binding data show that C-protein binds to titin (Fürst et al. 1992; Labeit et al. 1992; Soteriou et al. 1993). Freiburg & Gautel (1996) have investigated this interaction in more detail. They found that the C10 Ig domain of C-protein binds specifically to the first Ig domain of the 11 domain pattern of titin. Figure 7f shows nine of these superrepeats with their axial register with respect to C-protein as suggested by these findings.

There is little information about the binding site on myosin for C-protein or titin although Houmeida et al. (1995) have identified a position 20 nm from the end of the myosin tail where it binds to titin. However since this binding is found for all the A-band length of titin it is unlikely that it relates specifically to the C-protein interaction site.

One puzzle concerning the C-protein distribution is why it does not occur elsewhere in the A band since it binds so strongly to myosin. It is the presence of titin that offers an explanation. Because of the domain pattern, the interaction of titin with myosin will be different at each of the three levels of myosin within the 43 nm repeat. It is likely that only one of those levels is particularly favourable for C-protein binding. Elsewhere, titin may block the C-protein binding site on myosin. The presence of C-protein every 43 nm and the corresponding titin repeat also offers an explanation for the axial perturbation of the myosin heads from a strict 14.3 nm repeat in relaxed muscle (Huxley & Brown, 1967).

7 C-Protein: the Most Commonly Afflicted Gene Causing Familial Hypertrophic Cardiomyopathy

The complex and multiple protein-protein interactions in the sarcomere are highly interdependent. Mutations in a number of sarcomeric proteins have been identified which cause familial hypertrophic cardiomyopathy (Vikström and Leinwand, 1996; Towbin, 1998). C-protein was identified as one of the genes afflicted (Bonne et al. 1995; Gautel et al. 1995; Watkins et al. 1995), and a plethora of C-protein mutations have since been identified (Bonne et al. 1995; Watkins et al. 1995; Carrier et al. 1997; Rottbauer et al. 1997; Yu et al. 1998; Niimura et al. 1998). By now, it is emerging that mutations in the cardiac C-protein gene are the most common mutations leading to familial hypertrophic cardiomyopathy (Carrier et al. 1998).

The mutations in the C-protein gene are heterogeneous, and disease is associated with point mutations leading to amino acid exchanges, splice site mutations that result in mis-spliced and often truncated polypeptides, and deletions which selectively disrupt single domains while leaving the reading frame intact. Based on the functional characterisation of defined regions of C-protein described above, the mutations can be classified into two major groups: assembly-competent mutant proteins with single amino acid exchanges (containing at least the last three C-terminal domains) and mutant proteins with internal deletions but a predicted functional thick-filament binding site, or assembly-incompetent proteins which lack one or more functional C-terminal domains (compiled in Fig. 8).

The assembly competent mutations can been further subdivided into point mutants affecting domains far from the assembly region (Fig. 8, group A), and internal deletions predicted to result in a shortened protein with functional N- and C-terminal regions (group B). The largest group are C-terminally deleted proteins containing at least one truncated and mis-folded domain (group C in Fig. 8). Interestingly, most of the latter mutations are predicted to encode mutant proteins that are C-terminally deleted due to splice donor- or acceptor site mutations (Carrier, et al. 1997; Towbin, 1998) (Fig. 8). Mutant proteins of lengths between two and 9 domains have been predicted in many afflicted families. Since the N-terminal region, containing the MyBP-C motif with three phosphorylation sites, is probably involved in cardiac contraction regulation, it was suggested that the assembly-incompetent mutant proteins might act as poison polypeptides.

Since cardiac C-protein is crucially involved in myofibrillar assembly, reduced amounts of functional protein could equally lead to disturbed myofibril assembly resulting in sarcomeric disarray, which is the histological hallmark of FHC (Schwartz et al. 1995). Cardiac C-protein is the only isoform expressed in cardiac muscle throughout development (Fougerousse et al. 1998; Gautel et al. 1998). Whereas skeletal muscle coexpresses at least two C-protein isoforms, slow and fast (Dennis et al. 1984; Bennett et al. 1986), preceded by a possible embryonic isoform during development, a transcomplementation of isoforms can be envisaged for skeletal, but not for cardiac muscle (Gautel et al. 1998). This might explain why skeletal myopathies caused by mutations in the slow or fast C-protein genes have not been detected so far.

It is interesting that none of the short C-terminally truncated MyBP-C molecules in the transfection study by Gilbert et al. (1996) exerts a dominant-negative phenotype that would suggest a poison-polypeptide effect on myofibrillogenesis. However, very low levels of the delocalized N-terminal fragments of C-protein could act as poison peptides that might exert an adverse effect only over the long period of time that C-protein-associated FHC seems to take to manifest clinically (Watkins et al. 1995). It is noteworthy, however, that a french family develops FHC with a predicted mutant protein that even lacks the signalling MyBP-C motif (Mutant 19 in Fig. 8; Carrier et al. 1997). However, it is unknown whether, and at what levels, these mutants are actually present in the cardiomyocyte; this problem also applies to the point mutations which are even harder to detect. In the only study to date which investigated myocardium of an FHC patient, the deleted mutant protein could not be detected by Western blotting using C-protein specific antibodies (Rottbauer et al. 1997; similar results were observed in a second, independent family with a different mutation, M. Gautel and H.-P. Vosberg, unpublished observations). It was therefore proposed that the mutant protein was either translationally blocked or unstable and rapidly degraded. This could also be the case for some of the more discrete point mutations: for instance, the Asn755Lys mutation (Yu et al. 1997) is predicted to afflict a hydrogen bond crucial for Ig-domain stability. Rapid degradation of misfolded mutant C-protein could be similar to the rapid degradation of non-assembled, mutant cytosolic troponin T protein which was described in a 5' splice donor site mutation in intron 7 of Drosophila flight muscle troponin T, causing the upheld2 phenotype (Fyrberg et al. 1990). Mutations in the actin and myosin heavy chain genes of *D. melanogaster* have also been observed to result in changes in myofibrillar stoichiometry and altered myofibril turnover, ultimately leading to myofibril disarray (Beall et al.

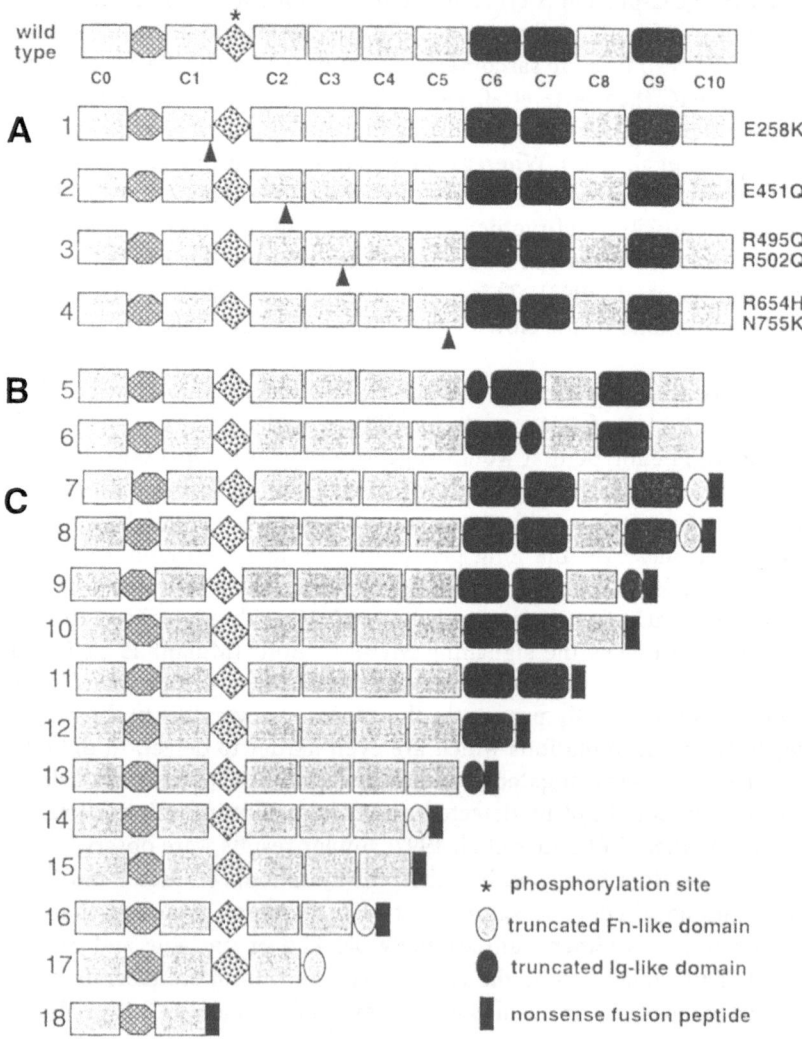

1989). Consequently, alterations in protein stoichiometry during myofibril-logenesis in mammalian muscle could well account for the chronical dete-rioration of myocardial function (Rottbauer et al. 1997) in a haplophenotype mechanism.

The molecular pathology of FHC associated mutations in MyBP-C is therefore plausibly quite different between the emerging subgroups of mu-tations, and both protein instability as well as direct competition effects may contribute to the development of disease.

8 Proposed Function of C-Protein

It is clear that C-protein plays an important role in all vertebrate striated muscles. It, or something like it is present in all A bands from Azolotl to human. That it is present at the very earliest time, with titin and myosin during myofibrillogenesis, and that if a curtailed molecule is present, sar-comere formation is disrupted points to its essential nature in this process. We have also discussed the genetic data that indicates that even when thick filaments are apparently formed correctly certain genetic defects in C-

---

Fig. 8. Summary of presently known MyBP-C mutations that lead to familial cardiac hypertrophy. Top: the domain pattern of cardiac MyBP-C. The biochemically de-fined regions are marked: phosphorylation by cAPK or a Ca^{2+}/Calmodulin depend-ent protein kinase is localised to the MyBP-C motif (Gautel et al. 1995), and myofi-brillar incorporation is directed by the four C-terminal domains (sorting signal) (Gilbert et al. 1996) which contain a myosin-binding site in C10 (Okagaki et al. 1993) and a titin binding region in C8–C10 (Freiburg and Gautel, 1996). C0 is a cardiac-specific Ig domain not found in the other known MyBP-C isoforms. Mutations in the MyBP-C gene encode mutant proteins which can be grouped in three distinct fami-lies. Representative mutations from several studies are shown. A: point mutations with unknown functional consequences but intact sorting signal; B: internal dele-tions with disrupted (and non-functional domains) and intact sorting signal; C: Mutations that, mostly via splice donor- or acceptor site mutations, lead to prema-ture termination of the polypeptide by stop-codons in a nonsense fusion peptide. These mutations are predicted to be assembly-incompetent due to the deleted sort-ing signal. Mutations 1–3 (Niimura et al. 1998); mutation 4 (Arg654His) (Moolman-Smook et al. 1998); mutations 5, 8, 12–14, 17, 18 (Carrier et al. 1997); mutations 7, 10 (Watkins et al. 1995); mutation 9 (Rottbauer et al. 1997); mutations 6 and 16 (Kimura et al. 1997); mutations 4 (Asn755Lys) and 11 (Yu et al. 1997), mutation 15 (Bonne et al. 1995)

protein can lead to a catastophic breakdown of the contractile behaviour of the muscle. This is clearest in cardiac muscle where it seems that phosphory-lation of C-protein plays a major role in this process. Phosphorylation is not, however, so important for skeletal muscle. Here it is possible to see the role of C-protein in contraction regulation independent of phosphorylation. There is evidence that during activation C-protein moves. X-ray diffraction shows that the 44 nm meridional which reveals the presence of C-protein with a repeating structure in relaxed muscle rapidly disappears almost en-tirely during activation both in skeletal (Bordas et al. 1991; Harford et al. 1991) and cardiac (J. J. Harford, personal communication) muscles. This implies that the order is lost, possibly due to the continual movement of the myosin heads during contraction. C-protein could become detached during this process or move with the heads. The order is restored when the muscle relaxes or goes into rigor. Further evidence for a role as a modulator of acti-vation is seen when C-protein is partially removed (Hoffmann et al. 1991a, b). Normally the onset of tension is a very sharp function of calcium concen-tration at a pCa of 7 to 6. However, when C-protein is extracted, the varia-tion with the calcium concentration is more gradual. This might be ex-plained if the C-protein plays a role – via its binding capabilty to myosin S2 – of keeping the myosin heads close to the backbone during rest. Phosphory-lation of C-protein could strengthen or weaken this function in cardiac muscle.

References

Alyonycheva, T. N., T. Mikawa, F. C. Reinach, and D. A. Fischman (1997) Isoform-specific interaction of the myosin-binding proteins (MyBPs) with skeletal and cardiac myosin is a property of the C-terminal immunoglobulin domain. J Biol Chem 272:20866–20872

Bähler, M., H. M. Eppenberger, and T. Wallimann (1985a) Novel thick filament protein of chicken pectoralis muscle: the 86K protein. I. Purification and charac-terization. J Mol Biol 186: 381–391

Bähler, M., H. M. Eppenberger, and T. Wallimann (1985b) Novel thick filament protein of chicken pectoralis muscle: the 86K protein. II. Distribution and local-ization. J Mol Biol 186:393–401

Bähler, M., H. Moser, H. M. Eppenberger, and T. Wallimann (1985c) Heart C-protein is transiently expressed during skeletal muscle development in the embryo, but persists in cultured myogenic cells. J Muscle Res Cell Mot 112:345–352

Beall, C. J., and E. Fyrberg (1991) Muscle abnormalities in Drosophila melanogaster heldup mutants are caused by missing or aberrant troponin-I isoforms. J Cell Biol 114:941–951

Bennett, P., R. Craig, R. Starr, and G. Offer (1986) The ultrastructural location of C-protein, X-protein and H-protein in rabbit muscle. J Muscle Res Cell Motil 7:550–567

Bennett, P., R. Starr, A. Elliott, and G. Offer (1985) The structure of C-protein and X-protein molecules and a polymer of X- protein. J Mol Biol 184:297–309

Bonne, G., L. Carrier, J. Bercovici, C. Cruaud, P. Richard, B. Hainque, M. Gautel, S. Labeit, M. James, J. Beckmann, J. Weissenbach, H.-P. Vosberg, M. Fiszman, M. Komajada, and K. Schwartz, K. (1995) Cardiac myosin binding protein-C gene splice acceptor site mutation is associated with familial hypertrophic cardiomyopathy. Nat Genet 11:438–440

Bordas, J., G. P. Diakun, J. E. Harries, R. A. Lewis, G. R. Mant, M. L. Martin-Fernandez, and E. Towns-Andrews (1991) Two-dimensional time resolved X-ray diffraction of muscle: recent results. Adv Biophys 27:15–33

Callaway, J. E., and P. J. Bechtel (1981) C-protein from rabbit soleus (red) muscle. Biochem J 195:463–469

Carrier, L., G. Bonne, E. Bahrend, B. Yu, P. Richard, F. Niel, B. Hainque, C. Cruaud, F. Gary, S. Labeit, J. B. Bouhour, O. Dubourg, M. Desnos, A. A. Hagege, R. J. Trent, M. Komajda, M. Fiszman, and K. Schwartz (1997) Organization and sequence of human cardiac myosin binding protein C gene (MYBPC3) and identification of mutations predicted to produce truncated proteins in familial hypertrophic cardiomyopathy. Circ Res 80:427–434

Carrier, L., G. Bonne, and K. Schwartz (1998) Cardiac Myosin-binding protein C and hypertrophic cardiomyopathy. Trends Cardiovasc Med 8:151

Craig, R. (1977) Structure of A-segments from frog and rabbit skeletal muscle. J Mol Biol 109:69–81

Craig, R., L. Alamo, and R. Padron (1992) Structure of the myosin filaments of relaxed and rigor vertebrate striated muscle studied by rapid freezing electron microscopy. J Mol Biol 228:474–487

Craig, R., and J. Megerman. (1979) Electron microscope studies of muscle thick filaments. In 'Motility in Cell Function'. Proc. J. M. Marshall Symp. Cell Biol, F. A. Pepe, J. W. Sanger and V. T. Nachmias, editors. Academic Press, NY, London pp 92–102

Craig, R., and G. Offer (1976) The localization of C-protein in rabbit skeletal muscle. Proc R Soc, London 192:451–461

Davis, J. S. (1988) Interaction of C-protein with pH 8.0 synthetic thick filaments prepared from the myosin of vertebrate skeletal muscle. J Musc Res Cell Motil 9:174–183

Dennis, J. E., T. Shimizu, F. C. Reinach, and D. A. Fischman (1984) Localization of C-protein isoforms in chicken skeletal muscle: ultrastructural detection using monoclonal antibodies. J Cell Biol 98:1514–1522

Dhoot, G. K., M. C. Hales, B. M. Grail, and S. V. Perry (1985) The isoforms of C protein and their distribution in mammalian skeletal muscle. J Musc Res Cell Motil 6:487–505

Einheber, S., and D. A. Fischman (1990) Isolation and characterization of a cDNA clone encoding avian skeletal muscle C-protein: an intracellular member of the immunoglobulin superfamily. Proc Natl Acad Sci USA 87:2157–2161

Fougerousse, F., A. L. Delezoide, M. Y. Fiszman, K. Schwartz, J. S. Beckmann, and L. Carrier (1998) Cardiac myosin binding protein C gene is specifically expressed in heart during murine and human development. Circ Res 82:130–133

Freiburg, A., and M. Gautel (1996) A molecular map of the interactions between titin and myosin-binding protein C. Implications for sarcomeric assembly in familial hypertrophic cardiomyopathy. Eur J Biochem 235:317–323

Fürst, D. O., R. Nave, M. Osborn, and K. Weber (1989) Repetitive titin epitopes with a 42 nm spacing coincide in relative position with known A band striations also identified by major myosin-associated proteins; an immunoelectron microscopical study on myofibrils. J Cell Biol 94:119–125

Fürst, D. O., U. Vinkemeier, and K. Weber (1992) Mammalian skeletal muscle C-protein: purification from bovine muscle, binding to titin and the characterization of a full length human cDNA. J Cell Sci 102:769–778

Fyrberg, E., C. C. Fyrberg, C. Beall, and D. L. Saville (1990) Drosophila melanogaster troponin-T mutations engender three distinct syndromes of myofibrillar abnormalities. J Mol Biol 216:657–675

Gautel, M., D. O. Fürst, A. Cocco, and S. Schiaffino (1998) Isoform transitions of the myosin binding protein C family in developing human and mouse muscles: lack of isoform transcomplementation in cardiac muscle. Circ Res 82:124–129

Gautel, M., O. Zuffardi, A. Freiburg, and S. Labeit (1995) Phosphorylation switches specific for the cardiac isoform of myosin binding protein-C: a modulator of cardiac contraction? EMBO J 14:1952–1960

Gilbert, R., M. G. Kelly, T. Mikawa, and D. A. Fischman (1996) The carboxyl terminus of myosin binding protein C (MyBP-C, C-protein) specifies incorporation into the A-band of striated muscle. J Cell Sci 101–111

Hanson, J., E. J. O'Brien, and P. M. Bennett (1971) Structure of the myosin-containing filament assembly (A-segment) separated from frog skeletal muscle. J Mol Biol 58:865–871

Harford, J. J., M. W. Chew, J. M. Squire, and E. Towns-Andrews (1991) Crossbridge states in isometrically contracting fish muscle: evidence for swinging of myosin heads on actin. Adv Biophys 27:45–61

Hartzell, H. C., and D. B. Glass (1984) Phosphorylation of purified cardiac muscle C-protein by purified cAMP-dependent and endogenous Ca^{2+}-calmodulin-dependent protein kinases. J Biol Chem 259:15587–15596

Hartzell, H. C., and W. S. Sale (1985) Structure of C-protein purified from cardiac muscle. J Cell Biol 100:208–215

Hartzell, H. C., and L. Titus (1982) Effects of cholinergic and adrenergic agonists on phosphorylation of 165,000-dalton myofibrillar protein in intact cardiac muscle. J Biol Chem 257:2111–2120

Hofmann, P. A., H. C. Hartzell, and R. L. Moss (1991a) Alterations in Ca2+ sensitive tension due to partial extraction of C-protein from rat skinned cardiac myocytes and rabbit skeletal muscle fibres. J Gen Physiol 97:1141–1163

Hofmann, P. A., M. L. Greaser, and R. L. Moss (1991b) C-protein limits shortening velocity of rabbit skeletal muscle fibres at low levels of Ca2+ activation. J Physiol 439:701–715

Houmeida, A., J. Holt, L. Tskhovrebova, and J. Trinick (1995) Studies of the interaction between titin and myosin. J Cell Biol 131:1471–1481

Huxley, H. E. (1967) Recent X-ray diffraction and electron microscope studies of striated muscle. J Gen Physiol 50:71–81

Huxley, H. E., and W. Brown (1967) The low-angle X-ray diagram of vertebrate striated muscle and its behaviour during contraction and rigor. J Mol Biol 30:383–434

Jeacocke, S. A., and P. J. England (1980) Phosphorylation of a myofibrillar protein of Mr 150 000 in perfused rat heart, and the tentative identifcation of this with C-protein. FEBS Lett 122:129–132

Kasahara, H., M. Itoh, T. Sugiyama, N. Kido, H. Hayashi, H. Saito, S. Tsukita, and N. Kato (1994) Autoimmune myocarditis induced in mice by cardiac C-protein. Cloning of complementary DNA encoding murine cardiac C-protein and partial characterization of the antigenic peptides. J Clin Invest 94:1026–1036

Kawashima, M., S. Kitani, T. Tanaka, and T. Obinata (1986) The earliest form of C-protein expressed during striated muscle development is immunologically the same as cardiac-type C-protein. J Biochem 99:1037–1047

Kimura, A., H. Harada, J. E. Park, H. Nishi, M. Satoh, M. Takahashi, S. Hiroi, T. Sasaoka, N. Ohbuchi, T. Nakamura, T. Koyanagi, T. H. Hwang, J. A. Choo, K. S. Chung, A. Hasegawa, R. Nagai, O. Okazaki, H. Nakamura, M. Matsuzaki, T. Sakamoto, H. Toshima, Y. Koga, T. Imaizumi, and T. Sasazuki (1997) Mutations in the cardiac troponin I gene associated with hypertrophic cardiomyopathy. Nat Genet 16:379–382

Kolmerer, B., N. Olivieri, C. C. Witt, B. G. Herrmann, and S. Labeit (1996) Genomic organization of M line titin and its tissue-specific expression in two distinct iso-forms. J Mol Biol 256:556–563

Koretz, J. F. (1979) Effect of C-protein on synthetic myosin filament structure. Biophys J 27:433–446

Koretz, J. F., L. M. Coluccio, and A. M. Bertasso (1982) The aggregation characteristics of column-purified rabbit skeletal myosin in the presence and absence of C-protein at pH 7.0. Biophys J 37:433–440

Labeit, S., and B. Kolmerer (1995) Titins: giant proteins in charge of muscle ultra-structure and elasticity. Science 270:293–296

Labeit, S., M. Gautel, A. Lakey and J. Trinick (1992) Towards a molecular under-standing of titin. EMBO J 11:1711–1716

Luther, P. K. (1998) Three dimensional structure of the A-band and localization of MyBP-C. Biophys J 74:A352

Magid, A., H. P. Ting-Beall, M. Carvell, T. Kontis, and C. Lucaveche (1984) Connect-ing filaments, core filaments, and side-struts: a proposal to add three new load-bearing structures to the sliding filament model. Adv Exp Med Biol 170:307–328

McCormick, K. M., K. M. Baldwin, and F. Schachat (1994) Coordinate changes in C protein and myosin expression during skeletal muscle hypertrophy. Am J Physiol C443–449

Miyahara, M. and H. Noda (1980) Interaction of C-protein with myosin. J Biochem 87:1413–1420

Moolman-Smook, J. C., B. Mayosi, P. Brink, and V. A. Corfield (1998) Identification of a new missense mutation in MyBP-C associated with hypertrophic cardiomy-opathy. J Med Genet 35:253–254

Moos, C., C. M. Mason, J. M. Besterman, I. M. Feng, and J. H. Dubin (1978) The binding of skeletal muscle C-protein to F-actin and its relation to the interaction of actin with myosin subfragment-1. J Mol Biol 124:571–586

Moos, C., G. Offer, R. Starr, and P. Bennett (1975) Interaction of C-protein with myosin, myosin rod and light meromyosin. J Mol Biol 97:1–9

Morimoto, K., and W. F. Harrington (1973) Isolation and composition of thick fila-ments from rabbit skeletal muscle. J Mol Biol 77:165–175

Morimoto, K., and W. F. Harrington (1974) Substructure of the thick filament of vertebrate striated muscle. J Mol Biol 82:83–97

Murakami, U., K. Uchida, and T. Hiratsuka (1976) Cardiac myosin from pig heart ventricle. Purification and enzymatic properties. J Biochem 80: 611–619

Niimura, H., L. L. Bachinski, S. Sangwatanaroj, H. Watkins, A. E. Chudley, W. McKenna, A. Kristinsson, R. Roberts, M. Sole, B. J. Maron, J. G. Seidman, and C. E. Seidman (1998) Mutations in the gene for cardiac myosin-binding protein C and late- onset familial hypertrophic cardiomyopathy. N Engl J Med 338:1248–1257

Obermann, W. M., M. Gautel, F. Steiner, P. F. M. van der Ven, K. Weber, and D. O. Fürst (1996) The structure of the sarcomeric M band: localization of defined domains of myomesin, M-protein, and the 250-kD carboxy-terminal region of titin by immunoelectron microscopy. J Cell Biol 134:1441–1453

Obermann, W. M., U. Plessmann, K. Weber, and D. O. Fürst (1995) Purification and biochemical characterization of myomesin, a myosin- binding and titin-binding protein, from bovine skeletal muscle. Eur J Biochem 233:110–115

Obinata, T., M. Kawashima, S. Kitani, O. Saitoh, T. Masaki, D. M. Bader, and D. A. Fishman (1985) Expression of C-protein isoforms during chicken striated muscle development and its dependence on innervation. In 'Molecular Biology of Muscle Development'. B. Nadal-Ginard and M. A. Q. Siddiqui, editors. Alan. R. Liss, New York

Obinata, T., F. C. Reinach, D. M. Bader, T. Masaki, S. Kitani, and D. A. Fischman (1984) Immunochemical analysis of C-protein isoform transitions during the development of chicken skeletal muscle. Dev Biol 101:116–124

Odermatt, E., J. W. Tamkun, and R. O. Hynes (1985) Repeating modular structure of the fibronectin gene: relationship to protein structure and subunit variation. Proc Natl Acad Sci USA 82:6571–6575

Offer, G. (1972) C-protein and the periodicity in the thick filaments of vertebrate skeletal muscle. CSH Symp. Quant Biol 37:87–95

Offer, G., C. Moos, and R. Starr (1973) A new protein of the thick filaments of vertebrate skeletal myofibrils. Extractions, purification and characterization. J Mol Biol 74:653–676

Okagaki, T., F. E. Weber, D. A. Fischman, K. T. Vaughan, T. Mikawa, and F. C. Reinach (1993) The major myosin-binding domain of skeletal muscle MyBP-C (C protein) resides in the COOH-terminal, immunoglobulin C2 motif. J Cell Biol 123:619–626

Pepe, F. A., P. K. Chowrashi, and P. R. Wachsberger (1975) Myosin filaments of skeletal and uterine muscle. In 'Comparative Physiology – Functional Aspects of Structural Materials'. L. Bolis, H. P. Maddrell and K. Schmidt-Nielsen, editors. North-Holland, Amsterdam. pp 105–120

Pepe, F. A. and B. Drucker (1975) The myosin filament. III. C-protein. J Mol Biol 99:609–617

Reinach, F. C., T. Masaki, and D. A. Fischman (1983) Characterization of the C-protein from posterior latissimus dorsi muscle of the adult chicken: heterogeneity within a single sarcomere. J Cell Biol 96:297–300

Reinach, F. C., T. Masaki, S. Shafiq, T. Obinata, and D. A. Fischman (1982) Isoforms of C-protein in adult chicken skeletal muscle: detection with monoclonal antibodies. J Cell Biol 95:78–84

Rome, E., G. Offer, and F. A. Pepe (1973) X-ray diffraction of muscle labelled with antibody to C-protein. Nature New Biol 244:152–154

Rottbauer, W., M. Gautel, J. Zehelein, S. Labeit, W. M. Franz, C. Fischer, B. Vollrath, G. Mall, R. Dietz, W. Kübler, and H. A. Katus (1997) Novel splice donor site mutation in the cardiac myosin-binding protein- C gene in familial hypertrophic cardiomyopathy. Characterization of cardiac transcript and protein. J Clin Invest 100:475–482

Schlender, K. K., and L. J. Bean (1991) Phosphorylation of chicken cardiac C-protein by calcium/calmodulin-dependent protein kinase II. J Biol Chem 266:2811–2817

Schultheiss, T., Z. X. Lin, M. H. Lu, J. Murray, D. A. Fischman, K. Weber, T. Masaki, M. Imamura, and H. Holtzer (1990) Differential distribution of subsets of myofibrillar proteins in cardiac nonstriated and striated myofibrils. J Cell Biol 110:1159–1172

Schwartz, K., L. Carrier, P. Guicheney, and M. Komajda (1995) Molecular basis of familial cardiomyopathies. Circulation 91:532–540

Schwarzbauer, J. E., R. S. Patel, D. Fonda, and R. O. Hynes (1987) Multiple sites of alternative splicing of the rat fibronectin gene transcript. EMBO J 6:2573–2580

Sjöström, M., and J. M. Squire (1977) Fine structure of the A-band in cryo-sections: the structure of the A-band of human skeletal muscle fibers from ultra-thin cryosections, negatively-stained. J Mol Biol 109:49–68

Soteriou, A., M. Gamage, and J. Trinick (1993) A survey of interactions made by the giant protein titin. J Cell Sci 104:119–123

Squire, J. M., M. Sjöström, and P. Luther (1976) Fine structure of the A-band in cryosections. II. Evidence for a length-determining protein in the thick filaments of vertebrate skeletal muscle. 6th Eur. Conf. Electron Microscopy, Jerusalem, pp 91–95

Squire, J., A. C. Edman, A. Freundlich, J. Harford, and M. Sjöström (1982) Muscle structure, cryo-methods and image analysis. J Microsc 125:215–225

Starr, R., R. Almond, and G. Offer (1985) Location of C-protein, H-protein and X-protein in rabbit skeletal muscle fibre types. J Muscle Res Cell Motil 6:227–256

Starr, R., and G. Offer (1971) Polypeptide chains of intermediate molecular weight in myosin preparations. FEBS Lett 15:40–44

Starr, R., and G. Offer (1978) The interaction of C-protein with heavy meromyosin and subfragment-2. Biochem J 171:813–816

Starr, R., and G. Offer (1982) Preparation of C-protein, H-protein, X-protein, and phosphofructokinase. Meth Enzymol 130–138

Starr, R., and G. Offer (1983) H-protein and X-protein. Two new components of the thick filaments of vertebrate skeletal muscle. J Mol Biol 170:675–698

Steiner, F., K. Weber, and D. O. Fürst (1998) Structure and expression of the gene encoding murine M-protein, a sarcomere-specific member of the immunoglobulin superfamily. Genomics 49:83–95

Swan, R. C., and D. A. Fischman (1986) Electron microscopy of C-protein molecules from chicken skeletal muscle. J Muscle Res Cell Motil 7:160–166

Sweeney, L. J., W. A. J. Clark, P. K. Umeda, R. Zak, and F. J. Manasek (1984) Immunofluorescence analysis of the primordial myosin detectable in embryonic striated muscle. Proc Natl Acad Sci USA 81:797–800

Takano-Ohmuro, H., S. M. Goldfine, T. Kojima, T. Obinata, and D. A. Fischman (1989) Size and charge heterogeneity of C-protein isoforms in avian skeletal muscle. Expression of six different isoforms in chicken muscle. J Muscle Res Cell Motil 10:369–378

Theiler, K. 1972. The house mouse. Springer Verlag Berlin Heidelberg New York

Towbin, J. A. (1998) The role of cytoskeletal proteins in cardiomyopathies. Curr Opin Cell Biol 10:131–139

Vaughan, K. T., F. E. Weber, S. Einheber, and D. A. Fischman (1993a) Molecular cloning of chicken myosin-binding protein (MyBP) H (86-kDa protein) reveals extensive homology with MyBP-C (C-protein) with conserved immunoglobulin C2 and fibronectin type III motifs. J Biol Chem 268:3670–3676

Vaughan, K. T., F. E. Weber, T. Ried, D. C. Ward, F. C. Reinach, and D. A. Fischman (1993b) Human myosin-binding protein H (MyBP-H): complete primary sequence, genomic organization, and chromosomal localization. Genomics 16:34–40

Vikström, K. L., and L. A. Leinwand (1996) Contractile protein mutations and heart disease. Curr Opin Cell Biol 8:97–105

Ward, S. M., D. K. Dube, M. E. Fransen, and L. F. Lemanski (1996) Differential expression of C-protein isoforms in the developing heart of normal and cardiac lethal mutant axolotls (Ambystoma mexicanum). Dev Dyn 205:93–103

Ward, S. M., L. F. Lemanski, U. N. Erginel, D. K. Dube, and N. Eiginel-Unaltuna (1995) Cloning, sequencing and expression of an isoform of cardiac C-protein from the Mexican axolotl (Ambystoma mexicanum) Biochem Biophys Res Commun 213:225–231

Watkins, H., D. Conner, L. Thierfelder, J. A. Jarcho, C. MacRae, W. J. McKenna, B. J. Maron, J. G. Seidman, and C. E. Seidman (1995) Mutations in the cardiac myosin-binding protein-C on chromosome 11 cause familial hypertrophic cardiomyopathy. Nature Genet 11:434–437

Weber, F. E., K. T. Vaughan, F. C. Reinach, and D. A. Fischman (1993) Complete sequence of human fast-type and slow-type muscle myosin- binding-protein C (MyBP-C) Differential expression, conserved domain structure and chromosome assignment. Eur J Biochem 216:661–669

Weisberg, A., and S. Winegrad (1996) Alteration of myosin cross bridges by phosphorylation of myosin-binding protein C in cardiac muscle. Proc Natl Acad Sci USA 93:8999–9003

Weisberg, A., and S. Winegrad (1998) Relation between crossbridge structure and actomyosin ATPase activity in rat heart. Circ Res 83:60–72

Williams, A. F., and A. N. Barclay (1988) The immunoglobulin superfamily – domains for cell surface recognition. Ann Rev Immunol 6:381–405

Wilson, F. J., and M. J. Irish (1980) The structure of segments of the anisotropic band of muscle. II. Preparation and properties of A segments from vertebrate skeletal muscle. Cell Tissue Res 212:213–223

Yamamoto, K. (1984) Characterisation of H-protein, a component of skeletal muscle myofibrils. J Biol Chem 259:7163–7168

Yamamoto, K., and C. Moos (1983) The C-proteins of rabbit red, white and cardiac muscles. J Biol Chem 258:8395–8401

Yasuda, M., S. Koshida, N. Sato, and T. Obinata (1995) Complete primary structure of chicken cardiac C-protein (MyBP-C) and its expression in developing striated muscles. J Mol Cell Cardiol 27:2275–2286

Young, O. A., and C. L. Davey (1981) Electrophoretic analysis of proteins from single bovine muscle fibres. Biochem J 195:315–327

Yu, B., J. A. French, L. Carrier, R. W. Jeremy, D. R. McTaggart, M. R. Nicholson, B. Hambly, C. Semsarian, D. R. Richmond, K. Schwartz, and R. J. Trent (1998) Molecular pathology of familial hypertrophic cardiomyopathy caused by mutations in the cardiac myosin binding protein C gene. J Med Genet 35:205–210

The Genetics and Molecular Biology of the Titin/Connectin-Like Proteins of Invertebrates

G. M. Benian[1], A. Ayme-Southgate[2], and T.L. Tinley[1]

[1]Departments of Pathology and Cell Biology, Emory University, Atlanta, GA 30322
[2]Department of Molecular Biology, Lehigh University, Bethleham, PA 18015

Contents

Introduction

The basic sarcomeric organization from both vertebrate and invertebrate muscles is very similar but shows a wide range of structural and physiological adaptations from the length of the individual filaments to the speed of contraction/relaxation. The mechanism of striated muscle contraction is universal and has been known, for quite some time now, to be the result of the sliding of actin and myosin filaments past each other (1). During the sliding process, the thick filament, made from the assembly of myosin heavy and light chain molecules "walks" on the thin filament composed of polymerized actin and regulatory proteins. The sliding process does not change the total length or sarcomeric position of these two filaments but rather their position relative to each other. How the complex myofibrillar structure is put together during myofibrillogenesis and maintained through cycles of contraction-relaxation is still largely a mystery, as is the molecular basis for physiological properties such as resting tension, elasticity and stretch activation.

To explain some of the properties of muscle structure and function, early models suggested the existence of a third filament system with elastic properties. In particular, electron microscopy studies of insect flight muscles revealed the presence of fine connections between the Z bands and the thick filaments (2–7). The search for component(s) of the third filament system lead to the identification and characterization of several very large proteins (size ranging from about 600 kDa to 3,000 kDa) found in both vertebrate and invertebrate muscles. These include "titin" or "connectin" (8, 9) in vertebrate muscles, "twitchin" in nematode (*Caenorhabditis*) and molluscan (*Aplysia*) muscles (10–12) and the insect protein, "projectin", also called "mini-titin" (13–22).

Based on antibody cross-reactivity and physical properties, it was suggested that twitchin, projectin and titin are related (18-24). The first member of this group from which amino acid sequence was determined was twitchin, encoded by the mutationally defined gene *unc-22* (11). Surprisingly, twitchin was discovered to consist almost entirely of multiple copies of immunoglobulin (Ig) and fibronectin type III (Fn) domains (11, 25). This made twitchin the first intracellular protein to join the Ig superfamily. Shortly thereafter, the sequences of both titin (26–28) and projectin (29–31) were shown to be very similar to twitchin, being composed of Ig and Fn domains, as well. In fact, in addition to these three proteins, there are 11 other proteins which fall into this intracellular, mostly muscle branch of the Ig superfamily. Nearly all of these proteins are discussed in this volume. They include the four proteins represented schematically in Fig. 1. These are twitchin of *C. elegans* (and *Aplysia*), projectin of *Drosophila*, *C. elegans* UNC-89 and *C. elegans* intestinal brush border twitchin. In addition to titin/connectin, the other known family members include smooth muscle (32) and non-muscle (33) myosin light chain kinase (MLCK); telokin (34, 35); myosin-binding proteins (MyBP)-C and H (36, 37); skelemin (38), M-protein (39) and myomesin (40); and kettin (41). The Ig-like sequences in this family are now known to form Ig folds. This was first demonstrated for telokin (42), but also shown for several domains from titin/connectin (43, 44) and one from twitchin (45). Ig and FnIII domains, when present in extracellular and cell surface proteins had been known to be involved in recognition or adhesion. Thus, it was postulated that similar motifs in the twitchin family would also function in protein-protein interaction. This hypothesis has been confirmed by numerous experiments. (i) Telokin, which is essentially one Ig domain, was shown to interact with myosin (46). (ii) MyBP-C was shown to have most of its myosin-binding activity localized to the C-terminal Ig domain (47). (iii) A kettin Ig domain, with flanking linker sequences, has been shown to bind to actin and α-actinin, but not to myosin (41). (iv) Two to seven Ig and FnIII domains of titin/connectin were expressed and purified from bacteria and shown to interact with the myosin rod and MyBP-C (27). (v) More recently, MyBP-C was shown to interact with only the 11-domain super-repeat of the A-band titin/connectin, and not with expressed portions from other regions of titin/connectin (48).

Arthropod Projectin

Projectin or mini-titin proteins were identified and characterized from several arthropod species, by principally two approaches: either through a biochemical isolation similar to ones used for the purification of vertebrate titin (19–22) or through the total analysis of myofibrillar and/or Z-line associated proteins (14, 17, 18). Four different insect species were used for these studies (*Lethocerus, Apis, Drosophila* and *Locusta*) and in each case a high molecular weight polypeptide between 700 and 1200 kDa estimated size was characterized and called either projectin or mini-titin. Crayfish muscles were also used for purification of an equivalent protein. Antibodies raised against the crayfish protein were shown to cross-react with *Drosophila* muscles (19, 21, 22). Some antibodies, raised against projectin from one insect species show cross-reactivity to the equivalent protein in other insects (17, 18, 20). Some antibodies directed against titin were also able to recognize some of these purified mini-titin proteins (19, 23). Cross reactivity between *Drosophila* projectin and nematode twitchin was also demonstrated (24). From the antibodies cross-reactivity and the biochemical profile of these arthropod

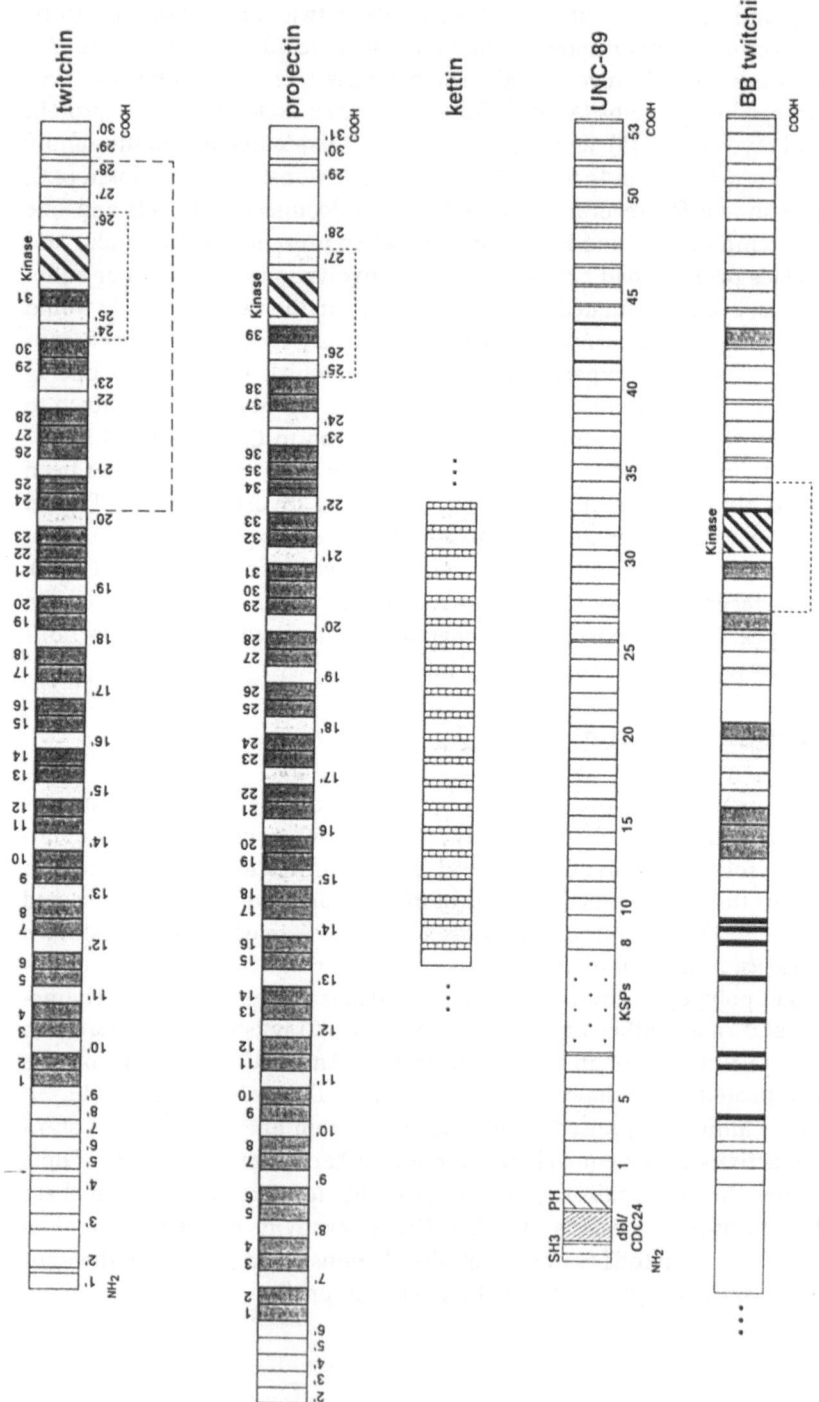

proteins, it seems clear they are equivalent proteins in the different species considered and are related to titin and twitchin.

The localization of these large proteins was determined by immunofluorescence microscopy on insect muscles and in most reports, the location of projectin is the I-Z-I region of insect flight muscles (17, 18, 20, 21). Electron microscopy of total stretched myofibrils or purified Z disks of insect flight muscles have shown the presence of a filament extending or "projecting" from the Z band towards the myosin filament, just overlapping the tip of the A band (14, 17). In honeybee flight muscle, connecting filaments can be extended to well over ten times their rest length. When muscles are stretched in rigor and then released, the recoil forces of the connecting filaments cause the sarcomere to shorten, leading to the crumpling of thin filaments held in rigor (5). Electron microscopy of purified "mini-titin" fractions have demonstrated the presence of an elongated protein with a length between 0.2 and 0.3 μ and a width of 4 nm (19, 20). The length of one molecule would be sufficient to cover in insect flight muscle the distance between Z and A bands, ranging between 0.1 and 0.2 μ (49). Saide demonstrated unequivocally by antibody staining and biochemical analysis the third connecting filament of honeybee flight muscles to be composed of projectin (14). These data establish clearly the existence, in the flight muscles of several insect species, of a large elongated protein, forming a filament-like structure between the Z band and the tip of the myosin thick filament. This protein, "projectin" or "mini-titin" was proposed to be foremost if not sole

----

Fig. 1. Five titin/connectin-like proteins found in invertebrates. Twitchin, UNC-89 and intestinal brush border twitchin (BB twitchin) sequences are from *C. elegans*, and the projectin and kettin sequences are from *Drosophila*. The longest known adult isoform of projectin is depicted. Ig domains are represented as lightly-shaded boxes, FnIII domains as darkly-shaded boxes, and the Ser/Thr protein kinase domains are designated "kinase". The finely dashed lines under twitchin, projectin and BB twitchin indicate the pattern of Ig, Fn and kinase domains conserved with vertebrate smooth muscle myosin light chain kinase. The wider dashed line under twitchin indicates the wider region of twitchin which is conserved in terms of domain organization with vertebrate titin/connectin. The arrow near the N-terminus of twitchin shows the position of a segment of 11 amino acids, 9 of which are glycine, which is hypothesized to form a flexible "hinge" in the protein. In kettin, the thin boxes with horizontal lines represent a 35 amino acid repeating sequence motif that has no similarities to other proteins in the databases. Although approx. 95% of BB twitchin is likely to have been determined, its N-terminus is unknown. In the leftmost region of the diagram of BB twitchin are thin black boxes representing the 8 copies of an ~30 residue motif which has no significant homology to other proteins in the databases. (Data on projectin are from (119) data on BB twitchin are from X. Tang and G.M. Benian, unpubl.)

component of an elastic filament involved in the stretch activation mechanism of insect flight muscles (14, 17).

In adult insects, the highly specialized indirect flight muscles (abbreviated as IFM) are powerful muscles adapted for the rapid repeated contractions necessary for flight. These muscles are stretch-activated, undergoing multiple rounds of contraction per nerve impulse (50) and because they lack a 1:1 relationship between nerve stimulation and contraction, the IFM are also referred to as asynchronous muscles. The other insect muscles, like most striated muscles in other organisms, undergo one contraction per nerve impulse and are classified as synchronous muscles (50). The stretch-activation mechanism is explained as a "delayed increase in tension due to stretch" which activates the muscle and results in contraction. The IFM are attached to the cuticle or exoskeleton and because they are organized as two sets of perpendicular muscles, their length will oscillate in response to stretch activation. The stretch activation mechanism has been shown to be an intrinsic property of the myofibrillar apparatus (51). An explanation, based on the matched axial repeats of the actin and myosin filaments, was put forward by Wray (52) but recently has been refuted (53). Another model suggests that the projectin filaments play an important role in the stretch activation mechanism (14). In this hypothesis, projectin, in asynchronous flight muscles, would be an elastic protein conferring high resting stiffness to the IFM and/or capable of transferring stress to the thick filament during stretching (14, 17, 18).

Since projectin was first proposed as the protein component of the third elastic filament and an essential participant in IFM stretch activation, it was explicitly proposed that projectin would be an IFM-specific protein. Further studies on the localization of projectin in different muscle types revealed, however, an interesting and unexpected pattern. In the IFM, antibody staining confirms the localization of projectin over the I band, whereas, in synchronous muscles, anti-projectin antibodies show a sarcomeric localization reciprocal of the distribution seen in IFM, where projectin is found only over the A band with the exclusion of the M line (24). Since stretch activation is not a property of synchronous muscles, projectin, in these muscles, might be involved in other functions requiring a different sarcomeric location.

Drosophila Projectin

Further studies suggest that *Drosophila* projectin is present as different isoforms, differing significantly in size within the various muscle types, as judged by mobility on SDS-polyacrylamide gels (see Fig. 2). The isoform from asynchronous IFM is smaller than either of the two forms detected from various synchronous muscles (24). Partial proteolytic digests of the different isoforms from various muscle types yield similar but not identical patterns and it seems likely the various isoforms differ by a few specific domains (24).

After antibody cross-reactivity between *C. elegans* twitchin and *Drosophila* projectin was established (24), the gene for projectin was cloned using PCR primers derived from twitchin conserved Fn motifs (11). *Drosophila* projectin is encoded by a single gene located on the fourth chromosome (position 102C/D) and transcripts of this gene are detected in both flight and synchronous muscles (29). Although there may be differences related to posttranslational modification, the different isoforms are, most probably, generated through alternative splicing of a primary transcript from a single gene.

Initial characterization of the *Drosophila* projectin sequence revealed the presence of Fn and Ig domains in a pattern similar to twitchin, Fn-Fn-Ig (29). To date, projectin is composed of 39 Fn domains and 31 Ig domains with a molecular weight of approximately 831 kDa. The central "core" region, consisting of the Fn-Fn-Ig pattern, contains 28 Fn and 14 Ig domains (see Fig. 1). Two additional patterns were identified, a Fn-Fn-Fn-Ig pattern and a Fn-Fn-Ig-Ig pattern, both found in the "intermediate region" (see Fig. 1; (119)). These two patterns are also present in twitchin at the same relative positions, COOH-terminal of the (Fn-Fn-Ig)n pattern but the potential significance of this shift for projectin assembly and interactions is unknown. Projectin Fn and Ig domains conform well with the consensus proposed for twitchin domains with an overall identity of 65% and 61.5%, respectively. At the COOH-terminus, arrays of Ig domains interspersed with unique sequences have been identified. The sequence of the unique regions found between the Ig domains is not conserved with twitchin. However, their lengths are comparable as if the distance, not the sequence is critical, for positioning and interactions (119). In the NH2-terminal region, we have also found Ig domains interspersed with unique domains. These NH2-terminal unique sequences, in contrast to the ones found at the COOH-terminus are well conserved with unique regions from both twitchin and titin. The Ig domains are more divergent from the twitchin Ig domains consensus (11), but interestingly, individual Igs in the projectin N-terminal region show a higher degree of similarity with the twitchin Ig domain in the equivalent position. One projectin Ig also shows a higher degree of identity to some

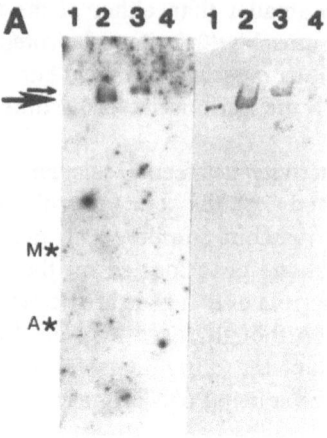

A

Auto-P Western

B

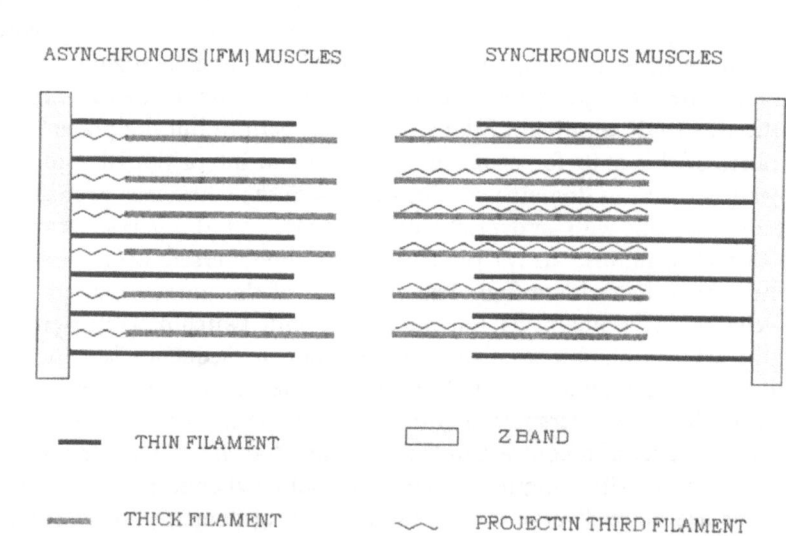

ASYNCHRONOUS (IFM) MUSCLES   SYNCHRONOUS MUSCLES

───── THIN FILAMENT   ☐ Z BAND

▬▬▬ THICK FILAMENT   ∿ PROJECTIN THIRD FILAMENT

titin Ig domains than to any twitchin Ig domain. These patterns of similarity might be significant, if these motifs serve specific functions in assembling and anchoring the projectin filament.

The projectin kinase domain is located at the same relative position as in twitchin and titin, towards the COOH-terminus (30). The catalytic core shares a high degree of similarity (61%) with twitchin kinase as well as smooth muscle myosin light chain kinase (30). The region immediately COOH-terminal of the kinase is well conserved with twitchin and, as shown for twitchin, could serve a regulatory function working as a kinase autoinhibitor (54, 55). Projectin has been shown to be phosphorylated *in vivo* on serine residues (21) and both the IFM and synchronous isoforms of projectin are capable of autophosphorylation *in vitro* (21, 30)—see Fig. 2A. The substrate(s) for projectin kinase are, however, still unknown.

The two projectin isoforms differ in their length and one of the proposed mechanisms is alternative splicing of the primary transcript. An interesting exon/intron pattern is present in the core region towards the NH2-terminus, where four exons, each containing the motif pattern (Ig-Fn-Fn)n, are present. Sequence and splicing pattern indicates each of these exons or any combinations of them could be alternatively used. This mechanism would generate a protein with varying length but probably not with a very different overall structure. RNA-PCR to test this idea is underway. It is interesting to note the central core domain of projectin is longer than twitchin's by four sets of Ig-Fn-Fn repeats. Recent evidence confirms the presence of alternative splicing at the COOH-terminus of the protein (119). This alternative pathway would generate two isoforms with different COOH-termini. In one pathway the protein would include two unique regions and three additional terminal Ig domains. In the other pathway, this entire region would not be used due to an early termination codon, generating a protein shorter by about 401 amino acids (or 40 kDa). The potential muscle-type specificity of these two alternative pathways is under investigation.

Fig. 2. Projectin isoforms: different sizes and locations A) Western detection and autophosphorylation analysis of projectin isoforms. Total muscle extracts from IFM (lanes 1 and 2) and larval synchronous muscles (lanes 3 and 4) were separated by SDS-PAGE and transferred to a membrane. The membrane was incubated with γ-P^{32}-ATP to detect autophosphorylation and then reacted with an anti-projectin antibody to identify projectin proteins. The two signals overlap perfectly, demonstrating that both isoforms of projectin are capable of autophosphorylation. The small arrow points to the larger synchronous isoform while the larger arrow points to the smaller IFM isoform. (Adapted, with permission, from Ayme-Southgate et al., 1995). B) Schematic representing the differential localization of projectin isoforms in the two muscle types

The projectin gene was mapped to polytene region 102C/D (29) and all of the available mutations and chromosomal rearrangements mapped to this region were analyzed for their effects on the projectin gene (56, 57). The bent^D mutation was originally described as a deletion, probably because the bent^D chromosome failed to complement two complementation groups on the fourth chromosome, l(4)2 and l(4)23 (56, 57). Genomic Southern analysis shows the bent^D chromosome does have a disruption in the projectin gene and one breakpoint of the bent^D rearrangement was localized either within or just N-terminal to the projectin kinase domain (30, 31). The pattern of hybridization and data from *in situ* hybridization to polytene chromosomes suggest the most probable rearrangement to be either a large insertion or an inversion. The position of the potential second breakpoint is not known. Presumably, it affects the locus of the lost l(4)23 complementation group.

In his mutational analysis of chromosome 4, Hochman described the lethal(4)2 complementation group as a "highly mutable locus" (56). Three of the 35 alleles Hochman isolated in this complementation group are still available and all three alleles are homozygous recessive lethal, with the heterozygous condition normal and capable of flight. The three l(4)2 alleles fail to complement each other in any combination and, consistent with the original data, the bent^D chromosome fails to complement any of these three l(4)2 alleles (56). The three l(4)2 alleles were tested by genomic Southern analyses to search for potential rearrangements in the projectin gene. Two alleles, bt$^{l\text{-}a}$ and bt$^{l\text{-}k}$, show alterations in the projectin region that are large enough to produce detectable changes in the restriction fragment pattern (30, 31). The bt$^{l\text{-}a}$ allele is a small insertion in the core region of the projectin gene which would cause premature termination (31). This region had previously been sequenced and shown to contain a regular arrangement of Fn and Ig domains (30). The second allele with a detectable rearrangement is bt$^{l\text{-}k}$. This rearrangement is either a large insertion or an inversion. The mutation would truncate the COOH-terminal region of projectin, reducing the number of Ig domains found at the end of the protein (30). More recently an EMS-induced F2 screen was carried out to isolate new mutations in the projectin gene. Three new mutants were recovered. All are homozygous recessive lethal and do not complement any of the previous projectin mutants including bent^D (M. Kulp and A. Ayme-Southgate, unpublished observations).

The time of death for all the alleles was determined and the analysis showed bent^D homozygotes die as late embryos, well after the formation of muscle, giving the impression they were unable to emerge from the chorion and vitelline membrane which protect the embryo. Muscle contractions can be seen in all embryos of this stock, even those that do not hatch and it is hypothesized that the contractions do not produce the amount of force necessary to open the chorion (30). Similar phenotypic assays show bt^{l-a} and bt^{l-f} homozygotes die as late embryos. In contrast to the $bent^D$ embryos, the bt^{l-a} homozygotes never show any of the spontaneous contractions. The other allele, bt^{l-k}, dies some time during the larval stage (30). Preliminary analysis to determine a more precise time of death points to the fact that the homozygous bt^{l-k} mutants do not all die at a specific time. Rather, the lethal period is staggered all through the three larval stages which is reminiscent of "wasting" degenerative defects, where the muscles are seemingly normal but degenerate as they are being used (A. Ayme-Southgate, unpublished observations). The lethality of the $bent^D$ embryos could have been due to the effect of the second breakpoint, presumably in the l(4)23 gene (see above), however recently, the $bent^D/ bt^{l-a}$ heterozygotes were shown to be embryonic lethal, arguing that the embryonic lethality of the $bent^D$ mutation is due, at least partially, to the mutation in the projectin gene. The newly recovered mutants are also all embryonic lethal, some showing muscle contractions while others do not (M. Kulp and A. Ayme-Southgate, unpublished observations).

Recent western analysis, using homozygous mutant embryos for all stocks, demonstrates the presence of the projectin protein in all mutants, implying that none of the available mutants is a true null. In some mutants a smaller truncated projectin protein is detected when compared to protein from wild type embryos (M. Kulp and A. Ayme-Southgate, unpublished observations). Immunofluorescence analysis of embryonic muscles in mutant stocks is still under progress but preliminary results suggest a strong correlation between lack of wild type projectin and a disorganized myofibrillar apparatus (M. Kulp and A. Ayme-Southgate, unpublished observation). This result is consistent with one of the proposed roles for the members of this protein family: to act as scaffold during embryogenesis (58–60). This phenotype has also been observed with some unc-22 (twitchin) alleles as adults (10, 61).

In the IFM, studies on the timing and pattern of assembly of various myofibrillar proteins reveal projectin and α-actinin assemble concomitantly and follow the same pathway of intermediate steps. To the contrary, myosin thick filaments appear later and seem to assemble into a regular sarcomeric pattern following a different route. Consistently, preliminary data suggest projectin to be organized normally in several myosin heavy chain flightless mutants (H Jacene and A Ayme-Southgate, unpublished observation).

C. elegans Twitchin

The identification of the thick-filament protein twitchin, encoded by the gene *unc-22*, marked the first time a new muscle component had been discovered through genetic analysis. Mutations in *unc-22* result in worms showing varying degrees of impaired movement and muscle structure disorganization (61). Weak alleles show nearly normal movement and nearly normal muscle structure. Strong alleles display slow movement and abnormal muscle structure in which thick and thin filaments, though present in normal numbers, are not organized into A and I bands. This information first suggested that *unc-22* is required for the final stages of sarcomere assembly. All *unc-22* alleles show a constant twitch of the body surface which originates in the underlying muscle. "Twitching" suggested that the *unc-22* product was somehow involved in regulating the contraction-relaxation cycle. *unc-22* displays conditional dominance. That is, heterozygous animals (*unc-22/+*) appear normal, but can be induced to twitch under certain conditions. These conditions include the presence of choline agonists such as levamisole or nicotine (10, 62), or the genetic background of being homozygous mutant for the gene *unc-89* (T. Tinley and G.M. Benian, unpublished observation). Having a phenotype with only one mutant copy of *unc-22* indicated that the *unc-22* product is required stoichiometrically, and as such, might serve more than an enzymatic function. After an extensive screen, all extragenic suppressors of *unc-22* were found to be certain missense mutations in the head region of myosin heavy chain (63). This suggested that the *unc-22* product might interact with myosin.

The *unc-22* gene was cloned by transposon tagging (64) and shown to encode a very large polypeptide (10), later called "twitchin" after the mutant phenotype. Polyclonal antibodies were used in immunofluorescence microscopy to localize twitchin to muscle A-bands, with the exclusion of the M-line region (10)–see Fig. 5. Analysis of the complete *unc-22* gene sequence (11, 25) revealed that twitchin is a 753,494 Da polypeptide that consists of a protein kinase domain, most similar to the protein kinase domain of myosin light chain kinase (MLCK), 31 fibronectin type III (Fn) and 30 immunoglobulin (Ig) domains (see Fig. 1). In the central portion of twitchin these domains are arranged as 10 consecutive copies of the triplet Fn-Fn-Ig. This is in contrast to the predominant repeat in the A-band portion of titin, namely Ig-Fn-Fn-Ig-Fn-Fn-Fn-Ig-Fn-Fn-Fn. In twitchin, at the N- and C-termini, there are tandem arrays of Ig domains. The similarity between twitchin and smooth muscle and non-muscle MLCK extends beyond the protein kinase domain (11). The domain pattern, Ig-Ig-Fn-Kinase-Ig, is the same in twitchin and the much smaller MLCKs (indicated by a small dashed line in Fig. 1). Indeed, this pattern is conserved also in projectin, nematode brush border twitchin and titin. This arrangement has been conserved probably because it is used for the positioning of the protein kinase domains on the thick filament. The crystal structure of a fragment of twitchin that includes Kinase-Ig provides support for this hypothesis (see below). Furthermore, as first pointed out by Labeit et al. (27), there is an even more extended conservation in domain organization between twitchin and titin: the pattern of 16 Fn and Ig domains surrounding the protein kinase domains is the same in the two molecules (large dashed line in Fig. 1). The functional significance is unknown, but again, is likely to be important for the positioning of the kinase domains.

Using modifications of procedures designed for the purification of titin/connectin, both *C. elegans* twitchin (65, 66) and scallop mini-titins (22) have been observed by EM after rotary shadowing. Twitchin (or mini-titin) appears as flexible rods with contour length of 0.23–0.25 μ and about 3–4 nm wide, reminiscent of the appearance of titin. In our study of twitchin (65) and in the study by Vibert et al. (22) of molluscan mini-titin, most objects appear asymmetric, with a forked, looped or knob-like structure at one end. We found that in many molecules, there is a kink, located 0.04 μ (+/–0.008, N = 38) from the knobbed end. Beginning at residue 777 in the complete polypeptide sequence of twitchin (25), is a segment of 11 amino acids, 9 of which are glycine (indicated as an arrow in Fig. 1). It has been suggested that such glycine-rich regions are lacking in secondary structure and perhaps function as flexible "hinges" within a protein. If the knobbed end of twitchin is the N-terminus, it is quite possible that the "kink" corresponds to this glycine-rich region. That this knobbed end corresponds to the N-terminus is even more plausible when the following data is considered. An antiserum has been raised against the protein kinase domain of nematode twtichin (J. Lei, M. Valenzuela, and G. Benian, unpublished). This antiserum cross reacts with scallop mini-titin on a western blot, and decorates the end opposite from the knobbed end on EM images (22). The functional significance of an N-terminal knob and downstream hinge is unknown.

Single 1 μ long titin/connectin molecules are anchored at the M-line through their C-termini and are anchored at the Z-line through their N-termini. Thus, the titin protein kinase domain is localized near the bare zone (28). In contrast, the much smaller twitchin molecule cannot span much of the 10 μ long nematode thick filament. As mentioned above, twitchin has been localized by fluorescence microscopy (10) to the A-band (except in the middle). In addition, anti-mini-titin antibodies decorate native scallop thick filaments throughout their length (67). In order to span most of the length of a nematode thick filament, twitchin either forms a polymer, or individual molecules are periodically distributed with gaps between them. Thus, given the length of individual twitchin molecules and the length of thick filaments, it is unlikely that twitchin serves as a molecular ruler to determine the length of thick filaments, as has been proposed (59) for titin in vertebrate muscle. Because of the labeling pattern and analogy to titin/connectin, it is likely that most twitchin or mini-titin lies on the surface of thick filaments. This interpretation is also consistent with the phenotype of *unc-22* mutants: normal numbers of normal length thick filaments that are not organized into A-bands.

We know that not all 61 Ig and FnIII domains of twitchin need to be present in order for twitchin to have almost normal function. By examining revertants of *unc-22* resulting from imprecise transposon excision from one site in the middle of *unc-22*, Kiff et al. (68) have shown that a nearly wild-type phenotype can be seen when up to seven domains have been deleted from the polypeptide. Whether deletion of larger numbers of domains, or deletion from other regions of the protein can be tolerated, is not known. Indeed, an analogous phenomenon was later shown for a similarly large polypeptide, human dystrophin. In-frame deletions of small numbers of spectrin-like repeats in dystrophin result in the mild phenotype of Becker's muscular dystrophy, but out of frame deletions result in the severe phenotype of Duchenne's muscular dystrophy (69). The human giant proteins, nebulin and titin/connectin, are encoded by large genes, and thus could be frequent targets for mutation, although no genetic diseases have yet been associated. Complete loss of function of either nebulin or titin/connectin might be lethal. But, perhaps partial loss of function, through loss of small numbers of the polypeptide repeats found in nebulin or titin/connectin, might be silent polymorphisms, or result in clinical myopathies.

Activity and Structure of the Protein Kinase Domain of Twitchin

Of all the protein kinases with known substrates, the twitchin kinase sequence is most similar to vertebrate smooth muscle and skeletal muscle myosin light chain kinases (11). Although the physiological substrate for nematode twitchin is still unknown, it has been demonstrated that bacterially expressed twitchin kinase domain has protein kinase activity *in vitro* (54). Peptide substrates derived from myosin light chains yielded the highest activities [chicken smooth muscle myosin light chain residues 11–23 (kMLC11–23) gave a Vmax = 0.64 µmol per min per mg and a Km = 98 µM]. By comparing the activities of three constructs expressing the catalytic core and varying amounts of sequence C-terminal to the core, it was found that

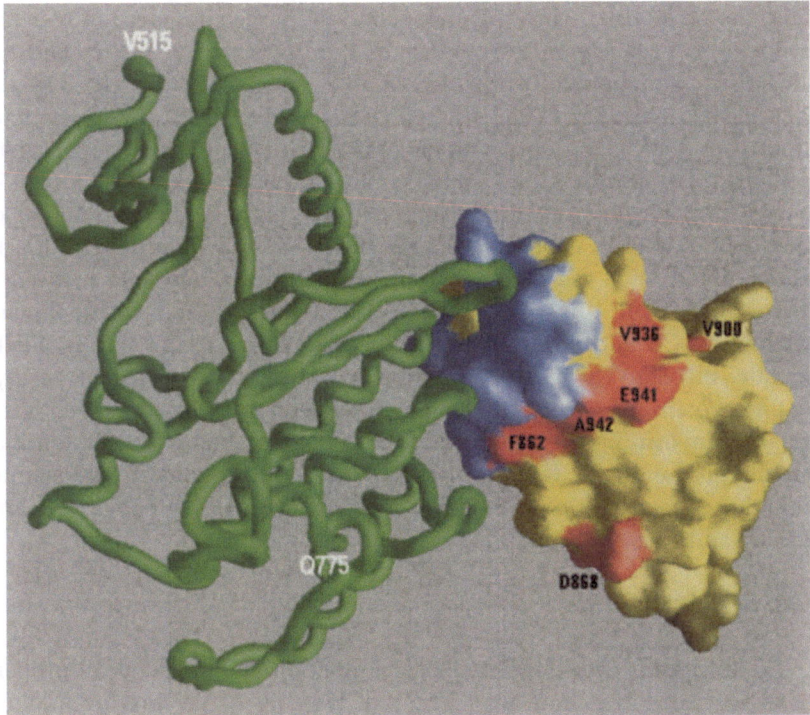

Fig. 3. Model of smooth muscle MLCK derived from the crystal structure of a twitchin fragment containing the protein kinase plus Ig domain 26'. The catalytic core (515–775) is shown in green in a "worm" representation and the Ig domain is shown in a surface representation. The surface formed by the residues of the Ig domain putatively interacting with the catalytic core is colored blue; the surface formed by the residues putatively involved in myosin binding is colored red. The autoregulatory sequence connecting the catalytic core to the Ig domain is omitted. Such a model can be extrapolated to twitchin, projectin, BB twitchin and possibly titin/connectin. (Reprinted, with permission, from Kobe et al. 1996)

the 60 residues C-terminal to the catalytic core act as an autoinhibitor (54). The crystal structure of the twitchin kinase domain (55) revealed the structural basis for this autoinhibition. In fact, this structure provided the first direct evidence for the intrasteric mechanism of protein kinase regulation, a common mechanism for regulating many different protein kinases. The twitchin kinase structure shows that the 60 residue autoinhibitory sequence extends up into the active site, wedged between the small and large lobes of the catalytic core. This allows for extensive contacts to be made with residues expected not only for substrate recognition, but also residues involved in the ATP-binding Gly loop, the activation loop and in catalysis. More recently, the crystal structure (70) of a somewhat larger fragment of twitchin that includes the protein kinase domain, autoinhibitory sequence and the following Ig domain (26'), provided another unexpected insight into the MLCK group of proteins. The Ig domain was found to make physical contact with the protein kinase domain on the opposite side from the catalytic cleft–see Fig. 3. In addition, of the 12 putative myosin binding residues in telokin inferred from a sequence comparison of several homologous Ig domains of MyBP-C (47), the six exposed residues cluster on the same face of the Ig domain and outside the kinase-Ig domain interface. Thus, this suggests a fixed, rather than diffusible, relationship between the protein kinase (twitchin, MLCK and possibly titin/connectin and projectin) and its likely substrate, myosin: Through the binding of the adjacent Ig domain to myosin, the protein kinase domain is held in position to make contact with the regulatory myosin light chain. This leads to the tantalizing speculation that there might be other functionally important inter-domain interactions within these molecules, even outside the relatively small regions of twitchin, projectin and titin/connectin that are analogous to MLCK.

Aplysia Twitchin

Important insights into the physiological function and biochemical properties of twitchin have come from study of twitchin in the marine mollusc *Aplysia*. This fairly large animal affords the opportunity to carry out both electrophysiology of nerve-muscle preparations and biochemistry on isolated muscle. It was found that a cAMP dependent protein kinase was involved in the mechanism by which two neuropeptide families modulate muscle contractions in the accessory radula closer muscle of this organism (12). An ~750 kDa protein was found to be the major substrate for this kinase. After purification of this large protein, partial amino acid sequence from 12 proteolytic peptides showed strong similarity to *C. elegans* twitchin including regions containing FnIII, Ig and kinase domains (12). Thus, this polypeptide is likely to be the *Aplysia* homolog of twitchin. Most importantly, Probst et al., have shown that an increase in cAMP-dependent phosphorylation of twitchin correlates with an increased rate of relaxation of the muscle (12). This suggests that the normal function of twitchin is to somehow inhibit the rate of relaxation, and cAMP dependent phosphorylation relieves this inhibition. The identity of the Ser/Thr residues in twitchin that are phosphorylated by the cAMP dependent kinase are unknown. How this phosphorylation alters the "activity" of twitchin is also unknown.

Purified *Aplysia* twitchin has been shown to have protein kinase activity, with autophosphorylation on threonine (71). Sequence analysis of a partial cDNA clone showed the protein kinase domain of the *Aplysia* protein is most similar to that of nematode twitchin (62% identical). Twitchin protein kinases belong to the Ca+2/calmodulin-regulated subfamily of protein kinases. Indeed, purified molluscan twitchin has been shown to bind to calmodulin with an EC50~70 nM (71). The calmodulin binding site has been limited to 16 amino acids within the N-terminal portion of the autoinhibitory region. This was shown by its ability as a synthetic peptide to inhibit calmodulin binding of whole twitchin. This sequence has the potential to form a basic amphiphilic α-helix, a structural common denominator for calmodulin binding sites. Indeed, the crystal structures of both *Aplysia* twitchin (70) and *C. elegans* twitchin (55) show that this sequence forms an amphiphilic α-helix.

The bacterially expressed *Aplysia* twitchin kinase domain, like its nematode counterpart, has protein kinase activity and is inhibited by the approx. 60 residues lying between the end of the kinase core and the following Ig domain (72). Addition of the recombinant non-inhibited *Aplysia* twitchin kinase to *Aplysia* actomyosin preparations resulted in phosphorylation primarily of the 19 kDa regulatory MLCs (72). This enzyme phosphorylated purified regulatory MLCs with a stoichiometry of 0.7 mol per mol protein, on threonine-15 in a region which shares a high level of homology with the vertebrate smooth muscle MLCs. Peptide analogs of the *Aplysia* light chain and a similar sequence from csmMLCs were phosphorylated with extremely high velocities (Vmax > 20 µmol per min per mg). Remarkably, under similar assay conditions, recombinant non-inhibited *C. elegans* twitchin kinase failed to phosphorylate *C. elegans* whole myosin or MLCs (73). A peptide that includes the expected phosphorylation site for nematode regulatory MLC is only poorly phosphorylated by the nematode twitchin kinase (54). Indeed, in the region of this putative phosphorylation site, nematode regulatory MLC is highly diverged from the *Aplysia* and csmMLC sequences. Recombinant constitutively active nematode and *Aplysia* twitchin kinase fragments differ in their catalytic activities, peptide-substrate specificities, and in their sensitivities to the naphthalene sulfonamide inhibitors ML-7 and ML-9 (73). To summarize, it is very likely that the physiological substrate for *Aplysia* twitchin kinase is regulatory MLC. The physiological substrate for *C. elegans* twitchin has not yet been demonstrated by either biochemical assay or by the analysis of mutants.

Thus, twitchin is a *bone fide* protein kinase, and at least for *Aplysia* the substrate is the regulatory MLC. But, the enzyme appears to be autoinhibited. So, what activates the enzyme? Strong homology to the MLCK group of enzymes suggested, that like MLCK, twitchin should be activated by Ca2+-calmodulin (CaM). Indeed, whole *Aplysia* twitchin (71), and bacterially expressed portions of the *Aplysia* (72, 74) and *C. elegans* twitchins (73) containing the kinase and autoinhibitory region, have been shown to bind to Ca2+-CaM. However, neither enzyme is fully activated by binding to Ca+2-CaM (54, 73). A strong candidate for physiological activator is Ca2+/S100A12, which Heierhorst et al. have demonstrated fully activates both *Aplysia* (75) and nematode twitchin kinase (73). S100A12 occurs in high concentrations in striated muscles (e.g. in human heart it comprises 0.2% of the total protein). S100A12 is a member of the S100 family of at least 16 members, each being ~10,000 Da and containing 2 EF-hand Ca2+-binding domains (76). Although S100 family members have been implicated in a variety of cellular processes and are expressed in many different tissues, twitchin was the first enzyme whose activity was shown to be dramatically enhanced by an S100 protein. The S100A12 binding site was localized to the same ~16 residue sequence within the autoinhibitory region as Ca2+-CaM had been mapped (75). Presumably, the interaction of Ca2+-S100 to a portion of the autoinhibitory sequence breaks a number of contacts between the autoinhibitory sequence and the active site, but details of this activation process are unknown. One note of caution should be made. That is, so far, S100 proteins have not yet been found in *Aplysia* or in a survey of predicted proteins encoded by the essentially complete genome sequence of *C. elegans*.

C. elegans UNC-89

The M-line is a structure in the center of A-bands that is likely to organize and maintain thick filaments in lateral registration. A model for the M-line, based upon high resolution EM of frog skeletal muscle, was proposed about 30 years ago (77, 78). Five, non-myosin proteins are now known to reside in the vertebrate M-line: (i) MM-creatine kinase (79), (ii) M-protein (39), (iii) myomesin (40), (iv) skelemin (38), and (v) the C-terminus of titin/connectin (80). All except MM-creatine kinase are members of the twitchin family.

Fig. 4. Most mutant alleles of *unc-89* lack M-lines, but several alleles do have M-lines. Transmission EM of body wall muscle from wild type and two *unc-89* mutants, *e2338* and *ad539* are shown. Closed arrows mark M-lines and open arrows mark dense bodies. *e2338* is an amber chain termination mutation resulting in very low levels of a slightly truncated UNC-89 polypeptide. As is typical of most *unc-89* alleles, *e2338* has a disorganized muscle structure with no M-lines. The nature of the sequence alteration in *ad539* has not been determined, but displays abnormal pharyngeal, but normal body wall muscle, and an abundant, wild type-sized UNC-89 polypeptide by immunoblot. As shown here by EM, *ad539* has fairly normal muscle structure and M-lines. Thus, it is likely that *ad539* bears a mutation in a pharyngeal-specific isoform of UNC-89. (Data from T.L. Tinley, R. Santoianni and G.M. Benian, unpubl.)

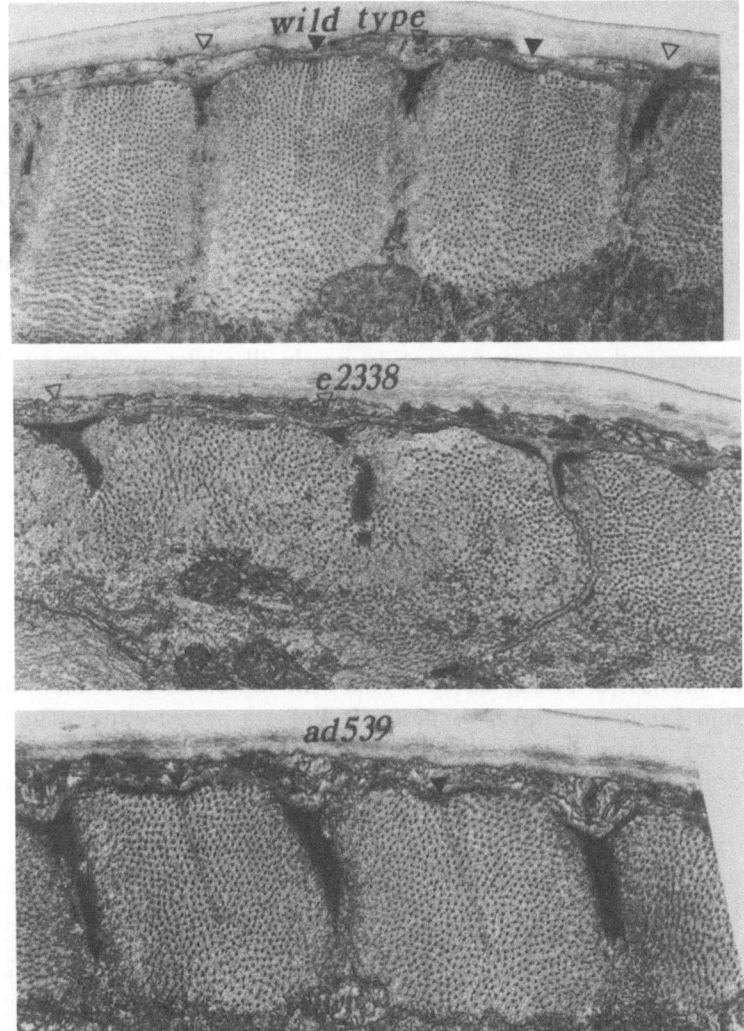

Recently, a group of 16 polyclonal and monoclonal antibodies generated to unique epitopes of known sequence from M-protein, myomesin and titin/connectin was employed in immunoelectron microscopy to locate the position of the epitopes at the sarcomere level (81). These data have been interpreted in a model for how these three molecules are associated and oriented. In contrast, very little is known about the composition of M-lines in invertebrate muscles. Other than UNC-89, only one other non-myosin component has been described. This is p400 of *Lethocerus* and *Drosophila* (82), located by immunoelectron microscopy at the M-line. P400 has a polypeptide molecular weight of 400,000 Da, but its sequence and loss-of-function phenotype is unknown (82).

Nematodes mutant in *unc-89* move almost as fast as wild type, but are thinner and more transparent. In nearly all alleles, polarized light microscopy reveals disorganized muscle structure in both body wall and pharyngeal muscle. For the most severe alleles, EM shows a normal number of thick filaments but they are not organized into A-bands, and there are no M-lines—see Fig. 4 (61; T. Tinley, R. Santoianni and G.M. Benian, unpublished observations). The *unc-89* gene has been cloned and sequenced and was shown to encode a 732 kDa polypeptide (85). A polyclonal antiserum, EU30, was raised to a portion of *unc-89* coding sequence. This antiserum localizes to the center of A-bands in body wall and pharyngeal muscles by immunofluorescence microscopy (85)—see Fig. 5. UNC-89 might also be present in the muscles of insects as EU30 labels the middle of the A-band in muscles from *Lethocerus* and *Drosophila* (B. Bullard, pers. comm.). The thick filaments of nematode body wall muscle contain two myosin heavy chain isoforms, myoB distributed along most of the length of the thick filament, and

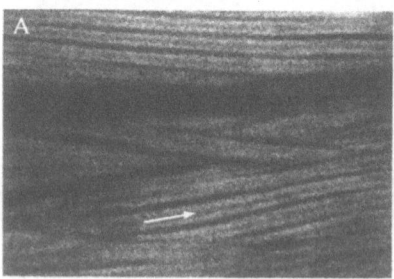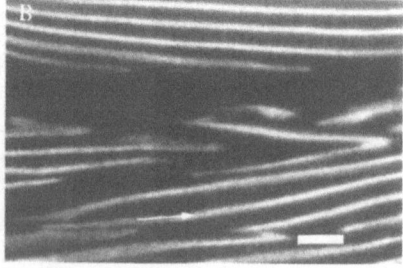

Fig. 5. UNC-89 and twitchin are located in reciprocal locations in A-bands. Immunofluorescent micrographs of the same body wall muscle cells stained with a polyclonal antibody to twitchin (shown in A) and a monoclonal antibody to UNC-89 (shown in B). Notice that twitchin is located throughout the A-band except for the middle, where UNC-89 is located. Bar represents 5 μ. (Reprinted, with permission, from Benian et al. 1996)

myoA located in the center (86). When viewed by immunofluorescence with monoclonals specific for each isoform, myoB localizes to the A-band with a gap in the middle, and this gap is filled with myoA. Anti-UNC-89 reacts with the center of A-bands, with the same width of distribution as myoA (85). The pattern of UNC-89 staining was examined in the strain *eDp23; unc-54(e190)* which produces no myoB (encoded by the *unc-54* gene) but forms thick filaments due to overexpression of myoA (87). Whereas anti-myoA localizes to the entire, broad A-band, anti-UNC-89 remains restricted to the center of A-bands (85). Thus, the most important determinants for UNC-89 localization appear to reside in proteins different from myoA.

Based on our sequencing of one genomic cosmid clone, RT-PCR products and several partial cDNA clones, we have deduced the complete sequence of the UNC-89 protein (85). It begins with 67 residues of unique sequence, SH3, dbl/CDC24, and PH domains, followed by 7 Ig domains, a KSP-containing putative multiphosphorylation domain, and another 46 Ig domains – see Fig. 1. The 44 KSPs (lysine, serine, proline) and related sequences are mostly arranged in a repeat of 10 residues (typically KSPTKKEKSP) and are present in 26 copies. Large numbers of KSPs have previously been found in two of three subunits of human neurofilament (NF) proteins. NF-M has 12 copies (88), and NF-H has 41 copies (89). Neurofilaments comprise the major cytoskeleton in axons and consist of parallel arrays of 10 nm filaments linked to each other by cross-bridges. These cross bridges are formed by the C-terminal tails of NF-H and NF-M (90, 91). The KSPs reside in the C-terminal tails of NF-M and NF-H and become phosphorylated at serines by a neuronal cdc2-like kinase (92). It is hypothesized that phosphorylation of the KSPs causes the C-terminal tails of NF-M and NF-H to project out perpendicular to the filament core, thus promoting cross-bridge formation (93). Interestingly, four KSPs have been found in the C-terminal, M-line portion of human cardiac titin/connectin, and are phosphorylated by a cdc2-like kinase in developing, but not differentiated muscle (80). Gautel et al. suggest that early in development, the C-terminal portion of titin/connectin, phosphorylated at the KSPs, might be inhibited from attaching to M-line proteins. Also, in the complete sequence of human cardiac titin/connectin (28), five MSPs (IRMSP(ARMSP)4) have been found in the Z-disc portion of titin/connectin. These MSPs are also phosphorylated by protein kinases in developing muscle tissues (94). The consensus phosphoryation site motif for p34cdc2-cyclin is $(S/T)P\_(K/R)$ (95). Thus, it is likely that the KSP_K sequences in UNC-89 are phosphorylated by cyclin-dependent kinases also. The phosphorylation of this domain possibly influences the interaction of UNC-89 with itself or with other, as yet unidentified, M-line proteins.

The most surprising domains encountered in the UNC-89 sequence are the SH3, dbl/CDC24 and PH domains, which are well known in signal transduction molecules (96). The human oncogene dbl, the yeast cell division cycle protein CDC24, and an expanding family of growth regulatory proteins share a homologous 238-amino acid sequence, generally termed a CDC24 domain. This domain from CDC24 (97) and dbl (98) has been shown to stimulate the exchange of GDP for GTP on Rho-like GTPases, thereby activating Rho-like GTPase activity. Activation of Rho has been shown to cause a reorganization of actin filaments via an unknown mechanism. This triggers a diverse set of cell processes including bud formation in yeast, cytokinesis in zygotes, maintenance of cell shape, formation of stress fibers and focal adhesions, cell aggregation, smooth muscle contraction, etc. (99). The presence of a possible Rho-like stimulator as a domain of the giant UNC-89 polypeptide suggests that Rho-like molecules might also trigger the assembly of M-lines in striated muscle. The CDC24 domain of UNC-89 is not merely a homologous sequence, this domain is likely to have the expected biochemical activity as well. A portion of UNC-89 containing the CDC24 and PH domains was produced and purified from E. coli. When this bacterially expressed protein was microinjected into mammalian tissue culture cells, a reorganization of the actin cytoskeleton, very similar to injecting activated Rac, was seen (E. Baraldi and M. Saraste, pers. comm.). The PH domain, first defined in the platelet protein called plekstrin, is a 100 residue sequence which has been found in many different proteins involved in intracellular signalling or the cytoskeleton (100). PH domains from several proteins have been shown to bind to inositol phosphates (101, 102). Hyvonen et al. (102) have proposed that this binding has two functions: (1) The anchoring of some proteins to membranes might occur through the interaction of a PH domain with membrane phosphoinositides. (2) In the case of CDC24 and its homologs, PH domains reside just C-terminal to the dbl/CDC24 domains, as is true for UNC-89. The binding of an inositol phosphate compound, such as Ins(1,4,5)P$_3$, to the PH domain, might regulate the nucleotide exchange activity of the neighboring CDC24 domain.

In the obliquely-striated muscle of C. elegans, M-lines and dense bodies are attached to the muscle cell membrane (see Fig. 4). At these positions, β-integrin (103) and α-integrin (B. Williams, pers. comm.) are located in the plasma membrane, and the UNC-52 protein, which is homologous to vertebrate perlecan (104), is located in the overlying basement membrane (84,105). Certain mutations in unc-52 and pat-3 (which encodes β-integrin) result in a "pat" embryonic lethal phenotype which is characteristic of mutations in at least 13 genes essential for muscle development in C. elegans (106). unc-52 and pat-3 mutants fail to assemble thin and thick filaments, although both actin and myosin are synthesized. This suggests that myofilament lattice assembly begins with positional cues laid down at the extracellular matrix and the cell membrane. Further support for this model comes from antibody staining of embryos at different stages of development (107). We hypothesize that the intracellular protein UNC-89 responds to these signals via the signal transduction domains, localizes, and then participates in assembling an M-line via the KSP region and Ig domains.

We have begun a complete molecular and phenotypic analysis of the 13 existing alleles of *unc*-89 (Tinley, Tang, Sanoianni and Benian, unpublished data). By polarized light microscopy, these alleles show varying degrees of body wall muscle structure disorganization. All of the alleles have been examined by immunoblot using the anti-UNC-89 antiserum. *e2338* is an amber chain termination mutant as demonstrated by the fact that its phenotype is suppressible by the amber suppressor mutation *sup-7*. This mutation results in a slightly truncated polypeptide of greatly reduced abundance on a western blot (85). *e2338* appears to be a null for UNC-89 as *e2338* / *deficiency* shows a phenotype of no greater severity than an *e2338* homozygote. The allele *st515* is a transposon-generated mutation which we have shown contains an 84 bp deletion and a 10 bp duplication, resulting in an in-phase TGA stop codon (85). On an immunoblot, anti-UNC-89 reacts with a single band of 120–135 kDa, of reduced abundance, from *st515*. This is the size of a polypeptide predicted from the position of the TGA stop codon created by the deletion/duplication in the *st515* sequence (85). By immunofluorescence microscopy, *st515* shows this truncated polypeptide is confined to the central region of the A-band (85), as it is in wild type. This suggests that at least some determinants for the localization of UNC-89 reside in the first 135 kDa of the UNC-89 polypeptide. Five of the alleles, however, show UNC-89 of normal size with reduced or nearly normal abundance. It is likely that most of these five alleles are missense mutations. Two of these five, *st79* and *ad539*, show the least severe structural disorganization by polarized light. Interestingly, by EM, both *ad539* and *st79* have M-lines, with *ad539* having a full length M-line in all sarcomeres and an almost normal muscle structure (see Fig. 4). This is in contrast to the originally described *e1460* allele and the null allele *e2338*, which have a more severely disorganized lattice and lack M-lines.

Certainly, UNC-89 is composed of many domains that have been implicated in protein-protein interactions. Indeed, the location of UNC-89 in the sarcomere and its domains suggest a number of possible interacting proteins. These include ligands for the SH3 domain, a rho-like GTPase activated by the CDC24 domain, a protein kinase that phosphorylates the KSPs, myosin heavy chain A, nematode homologs of known components of the vertebrate M-line, and finally, components of the muscle cell membrane to which the M-line is attached. We are taking advantage of the powerful genetics available for *C. elegans* and hope to identify these proteins by analysis of extragenic suppressor and enhancer mutations of *unc-89* mutants. To date, we have identified 8 extragenic suppressors of one allele and 4 extragenic suppressors of a second *unc-89* allele (TL Tinley, CA Alberico and GM Benian, unpublished observations).

C. elegans lintestinal Brush Border Twitchin

In PCR experiments designed to recover MAP-kinases from *C. elegans*, one clone was found to encode a protein kinase with highest BLAST score to twitchin (P. Winge and J. Fleming, personal comm.). This segment hybridized to YAC and cosmid clones on chromosome V. We determined the sequence of a 5.2 kb genomic fragment and waited briefly for the *C. elegans* Genome Project to determine the complete sequences of cosmids F12F3 and F17A9. The program GeneMark was used to predict exons within this genomic sequence. Exon-intron boundaries were confirmed by sequencing small RT-PCR products, and the 3' end of the gene was determined by sequencing an oligo-dT-derived cDNA clone (X Tang and GM Benian, unpublished data). We can account for ~95% of the coding sequence which specifies a 7,030 aa polypeptide, which we call "new twitchin". This coding sequence is distributed over ~37 kb and interrupted by 29 introns. GeneMark predicted several exons on the opposite strand of the 10th intron of this new twitchin. ACeDB (A *C. elegans* database) noted that this sequence is the mutationally-defined *exp-2* gene which encodes a potassium channel (W. Davis, I. Dent and L. Avery, pers. comm.).

The domain organization of new twitchin is shown in Fig. 1. It consists of 33 Ig domains (light grey), 7 FnIII domains (dark grey) and a protein kinase domain (X Tang and GM Benian, unpublished data). Thus, as compared to twitchin, new twitchin has far fewer FnIII domains (twitchin has 31) and the kinase domain is located more towards the middle of the molecule. New twitchin also has a 1426 residue region containing 8 copies of an ~30 residue motif (depicted as black bars). This motif has no significant similarity to any other sequences in the databases. The catalytic cores of the kinase domains of new twitchin, twitchin (nematode & *Aplysia*), *Drosophila* projectin and chicken smooth muscle myosin light chain kinase are ~53% identical to each other. In contrast, the kinase domain of titin/connectin is only 39% identical. The 60 residues lying just C-terminal to the end of the catalytic cores and shown to constitute an autoinhibitory region for sm. m. MLCK and the twitchins, are much less similar.

By RT-PCR, we have detected a splice variant of new twitchin in which an additional 15 aa are placed just N-terminal to the putative autoinhibitory sequence. The possible spatial and/or temporal specificity of this alternative splicing are unknown. The new twitchin kinase catalytic core was expressed as a GST-fusion protein (NTK1). NTK1 was tested for kinase activity against myosin light chain-related peptides (variants of chicken smooth muscle myosin light chain peptide, kMLC 11–23). The NTK1 Vmax was comparable to the Vmax of constitutively active twitchin kinase (pJK1), or ~600 nmol/min/mg. We then explored the effect on kinase activity of the sequences C-terminal to the catalytic core. Two more mini-new twitchins were expressed as GST-fusion proteins and their protein kinase activities towards a MLC-derived peptide were determined. Thus, adding 60 residues (in NTK3#B) only partially inhibits kinase activity. The comparable version of twitchin is completely inhibited. Adding 15 residues to the beginning of these 60 residues (in the alternatively spliced version NTK3#2), relieves much of this inhibition (X. Tang and G. M. Benian, unpublished observations). This seems to represent a novel mechanism of regulating protein kinase activity, but its functional significance *in vivo*, is unknown. In addition, all three versions of new twitchin were not stimulated by Ca^{+2}-S100. This is again in contrast to *C. elegans* and *Aplysia* twitchin.

A rabbit polyclonal antiserum (called EU48) was generated to a portion of new twitchin (X. Tang and G. M. Benian, unpublished). Unexpectedly, EU48 was seen to localize to the intestine, probably its inner surface. Previously, R. Francis generated about 40 monoclonal antibodies primarily to muscle cells, their underlying basement membrane and the hypodermis (84). One of these monoclonals, MH33, reacts with a series of 62-, 64- and 68-kDa polypeptides and localizes only to intestinal cells, specifically the terminal web (108; R. Francis, pers. comm.). Based on the size, pI and solubility, these polypeptides are probably intermediate filament proteins (R. Francis, pers. comm.). In double labeling experiments, EU48 and MH33 co-localize, suggesting that new twitchin also is located in the terminal web.

The intestinal brush border has been studied intensively for a number of years in vertebrates, especially the chicken (109–111). The brush border consists of two regions, the microvilli and the terminal web. Each microvillus contains a bundle of parallel actin filaments together with a host of actin binding proteins including fimbrin, villin and BB myosin I. The terminal web is a region where the microvillar actin bundles are anchored in a complex network of filaments containing myosin II, spectrins, tropomyosin, microtubules, and intermediate filaments. The terminal web also contains a circumferential ring consisting largely of actin and myosin II that originates from the zonula adherens and wraps around the entire cell. Contraction of this ring is regulated by myosin II light chain phosphorylation (112) and probably increases the permeability between epithelial cells (113).

Intriguingly, Eilertsen and Keller (114) have partially purified from chicken intestinal cell brush borders, a titin/connectin-sized polypeptide which has a similar morphology to muscle titin/connectin on EM of rotary-shadowed preparations. Antibodies to this "T-protein" react with muscle titin/connectin and localize by immunofluroescence to the terminal web. Both the sequence and the function of T-protein are unknown.

In *C. elegans*, the intestinal epithelium contains a dense layer of microvilli (115) and a very prominent terminal web (GM Benian, unpublished observation). We are currently seeking the null phenotype for new twitchin. Because new twitchin is expressed in intestinal epithelial cells and is located in the brush border (terminal web), we would expect that a null mutation would compromise the ability to absorb nutrients. This might yield animals which die soon after hatching, or animals which survive into adulthood, but are malnourished and small.

Kettin

Another invertebrate member of the titin/connectin-like family of proteins is represented by the *Drosophila* protein, kettin (41). Kettin is found in the Z line of both IFM and leg muscles and has been shown by in vitro binding assay to interact with actin and alpha-actinin but not myosin. Kettin is a high molecular weight protein, possibly present as two isoforms of distinct size, 500 kDa in the IFM and 700 kDa in other muscle types (41). As shown in Fig. 1, kettin also has a modular structure, composed, in the sequence so far available, of mostly Ig repeats with high similarity with Ig repeats found in the other family members described in this chapter (41, 116). Interestingly, each Ig domain is followed by a 35 amino acid region and each copy is highly conserved within the kettin molecule but does not show any homology so far with any other proteins in the database (41).

Kettin is an integral component of the Z band , crossing over its entire width and possibly protruding slightly on either side of adjacent sarcomeres (41, 117). Kettin is readily released from the Z disc following digestion by calpain as small peptides of molecular weight consistent with a single Ig domain (41).

Kettin is, very probably, an important structural protein of the Z band, potentially involved in anchoring the thin filament in concert with alpha-actinin (116, 117). The possible interaction between kettin and other Z band associated proteins remain to be investigated, in particular the association, in the IFM, between kettin and the IFM isoform of projectin.

Conclusions

A substantial amount of information has been gathered about the structure and function of twitchin/titin-related proteins in the invertebrates. This has been obtained through sequence analysis and the analysis of loss-of-function phenotypes in *C. elegans* and *Drosophila*. Nevertheless, a number of fascinating questions remain, including: (i) Why are these invertebrate proteins all of approx. 700–800 kDa? In terms of sarcomeric organization, what is the significance of this size? (ii) Why do three of these proteins consist of a mixture of Ig and Fn domains, whereas UNC-89 contains only Ig domains? This is even more interesting because the structures of Ig and Fn domains are very similar (118). What is the significance of the repeating pattern of groups of Ig and Fn domains (e.g. Fn-Fn-Ig)? (iii) How are twitchin and the synchronous muscle isoform of projectin situated on the surface of thick filaments? That is, do they form polymers or are they located at discrete locations with intervening gaps? (iv) What is the mechanism by which the fundamentally similar projectin isoforms get localized to different sarcomeric locations? (v) If the data on *Aplysia* twitchin can be extended to the muscles of other invertebrates, what is the mechanism by which twitchin inhibits the rate of relaxation? How does phosphorylation of twitchin relieve this inhibition? (vi) What are the substrates for the protein kinase domains of nematode twitchin and insect projectin? If rMLCs are indeed the substrates, how would and why does this phosphorylation take place for the IFM isoform of projectin, which resides primarily in the I band? If rMLCs are the substrates, given the stoichiometry of approx. 1:50 for twitchin:myosin, and the likely fixed position of twitchin along the thick filament, how does phosphorylation of just a few rMLCs result in a physiological effect (e.g. inhibition of relaxation)? What is the true activator for the twitchin and projectin kinases? (vii) How does UNC-89 participate in M-line assembly? (viii) What are the biochemical and physiological functions of intestinal brush border twitchin? A number of investigators will enjoy pursuing these and other questions for some time in the future.

Acknowledgments

GM Benian's lab is currently supported by the NIH and NSF, and was previously supported by the Muscular Dystrophy Association by the American Heart Association. A Ayme-Southgate's lab is supported by the National Science Foundation. We thank Robert Santoianni for performing the electron micrographs.

References

1. Squire J (1981) The structure and basis of muscular contraction. Plenum Press, New York
2. Auber J, Couteaux R (1963) Ultrasructure de la strie Z dans des muscles de Dipteres. J Microscopie 2:309–324
3. Saide JD, Ullrick WC (1974) Purification and properties of the isolated honey-bee Z-disc. J Mol Biol 87:671–683
4. Ashhurst DE (1977) The Z-line:its structure and evidence for the presence of connecting filaments. In:Tregear RT (ed) Insect Flight Muscle:Proceedings of the Oxford Symposium. Elsevier, Amsterdam, pp 57–73
5. Trombitas T, Tigyi-Sebe A (1977) Fine structure and mechanical properties of insect muscle gels to nitrocellulose sheets:procedure and some applications. In:Tregear RT (ed) Insect Flight Muscle:Proceedings of the Oxford Symposium. Elsevier, Amsterdam, pp 79–90
6. Candia Carnevali MD, De Eguileor M, Valvassori R (1980) Z line morphology of functionally diverse insect skeletal muscles. J Submicrosc Cytol 12:427–446
7. Deatherage JF, Cheng N, Bullard B (1989) Arrangement of filaments and cross-links in the bee flight muscle Z disk by image analysis of oblique sections. J Cell Biol 108:1775–1782
8. Wang K, McClure J, Tu A (1979) Titin:major myofibrillar components of striated muscle. Proc Natl Acad Sci USA 76:3698–3702
9. Maruyama K, Kimura S, Ohashi K, Kuwano Y (1981) Connectin, an elastic protein of muscle. Identification of "titin" with connectin. J Biochem 89:701–709
10. Moerman DG, Benian GM, Barstead RJ, Schreifer L, Waterston RH (1988) Identification and intracellular localization of the *unc-22* gene product of *C. elegans*. Genes and Devel 2:93–105
11. Benian GM, Kiff JE, Nickelmann N, Moerman DG, Waterston RH (1989) Sequence of an unusually large protein implicated in regulation of myosin activity in C. elegans. Nature 342:45–50
12. Probst WC, Cropper EC, Heierhorst J, Hooper SL, Jaffe H, Vilim F, Beushausen S, Kupfermann I, Weiss KR (1994) cAMP-dependent phosphorylation of Aplysia twitchin may mediate modulation of muscle contractions by neuropeptide cotransmitters. Proc Natl Acad Sci USA 91:8487–8491
13. Bullard B, Hammond KS, Luke BM (1977) The site of paramyosin in insect flight muscle and the presence of an unidentified protein between myosin filaments and Z line. J Mol Biol 115:417–440
14. Saide JD (1981) Identification of a connecting filament protein in insect fibrillar flight muscle. J Mol Biol 153:661–679
15. Hu DH, Kimura S, Maruyama K (1986) Sodium dodecyl sulfate-gel electrophoresis of connectin-like high molecular weight proteins of various types of vertebrate and invertebrate muscles. J Biochem 99:485–492
16. Locker RH, Wild DJC (1986) A comparative study of high molecular weight proteins in various types of muscle across the animal kingdom. J Biochem 99:1473–1484
17. Saide JD, Chin-Bow S, Hogan-Sheldon J, Busquets-Turner L, Vigoreaux JO, Valgeirsdottir K, Pardue ML (1989) Characterization of components of Z-

bands in the fibrillar flight muscle of Drosophila melanogaster. J Cell Biol 109:2157–2167

18. Lakey A, Ferguson C, Labeit S, Reedy M, Larkins A, Butcher G, Leonard K, Bullard B (1990) Identification and localization of high molecular weight proteins in insect flight and leg muscles. EMBO J 9:3459–3467

19. Hu DH, Matsuno A, Terakado K, Matsuura T, Kimura S, Maruyama K (1990) Projectin is an invertebrate connectin (titin):isolation from crayfish claw muscle and localization in crayfish claw muscle and insect flight muscle. J Muscle Res Cell Motil 11:497–511

20. Nave R, Weber K (1990) A myofibrillar protein of insect muscle related to vertebrate titin connects Z band and A band:purification and molecular characterization of invertebrate mini-titin. J Cell Sci 95:535–544

21. Maroto M, Vinos J, Marco R, Cervera M (1992) Autophosphorylating protein kinase activity of titin-like arthropod projectin. J Mol Biol 224:287–291

22. Vibert P, Edelstein SM, Castellani L, Elliot BW (1993) Mini-titins in striated and smooth molluscan muscles:structure, location and immunological cross-reactivity. J Muscle Res and Cell Motil 14:598–607

23. Matsuno A, Takano-Ohmuro H, Itoh Y, Matsuura T, Shibata M, Nakae A, Kaminuma T, Maruyama K (1989) Anti-connectin monoclonal antibodies that react with the unc-22 gene product bind dense bodies of Caenorhabditis nematode body wall muscle cells. Tissue and Cell 21:495–505

24. Vigoreaux JO, Saide JD, Pardue ML (1991) Structurally different Drosophila striated muscles utilize distinct variants of Z-band associated proteins. J Muscle Res Cell Motil 12:340–354

25. Benian GM, L'Hernault SW, Morris ME (1993) Additional sequence complexity in the muscle gene, unc-22, and its encoded protein , twitchin, of C. elegans. Genetics 134:1097–1104

26. Labeit S, Barlow DP, Gautel M, Gibson T, Holt J, Hsieh C-L, Francke U, Leonard K, Wardale J, Whiting A, Trinick J (1990) A regular pattern of two types of 100 residue motif in the sequence of titin. Nature 345:273–276

27. Labeit S, Gautel M, Lakey A, Trinick J (1992) Towards a molecular understanding of titin. EMBO J 11:1711–1716

28. Labeit S, Kolmerer B (1995) Titins:Giant proteins in charge of muscle ultrastructure and elasticity. Science 270:293–296

29. Ayme-Southgate A, Vigoreaux JO, Benian GM, Pardue ML (1991) Drosophila has a twitchin/titin-related gene that appears to encode projectin. Proc Natl Acad Sci USA 88:7973–7977

30. Ayme-Southgate A, Southgate R, Saide J, Benian GM, Pardue ML (1995) Both synchronous and asynchronous muscle isoforms of projectin (the Drosophila bent locus product) contain functional kinase domains. J Cell Biol 128:393–403

31. Fyrberg CC, Labeit S, Bullard B, Leonard K, Fyrberg EA (1992) Drosophila projectin:relatedness to titin and twitchin and correlation with lethal(4)102Cda and bent-dominant mutants. Proc R Soc Lond B 249:33–40

32. Olson NJ, Pearson RB, Needleman DS, Hurwitz MY, Kemp BE, Means AR (1990) Regulatory and structural motifs of chicken gizzard myosin light chain kinase. Proc Natl Acad Sci USA 87:2284–2288

33. Shoemaker MO, Lau W, Shattuck RL, Kwiatkowski AP, Martrisian PE, Guerra-Santos L, Wilson E, Lukas TJ, Van Eldik LJ, Watterson DM (1990) Use of DNA sequence and mutant analyses and antisense oligodeoxynucleotides to exam-

ine the molecular basis of nonmuscle myosin light chain kinase autoinhibition, calmodulin recognition and activity. J Cell Biol 111:1107–1125

34. Gallagher PJ, Herring BP (1991) The carboxyl terminus of the smooth muscle myosin light chain kinase is expressed as an independent protein, telokin. J Biol Chem 266:23945–23952

35. Collinge M, Martrisian PE, Zimmer WE, Shattuck RL, Lukas TJ, Van Eldik LJ, Watterson DM (1992) Structure and expression of a calcium-binding protein gene contained within a calmodulin-regulated protein kinase gene. Mol Cell Biol 12:2359–2371

36. Einheber S, Fischman DA (1990) Isolation and characterization of a cDNA clone encoding avian skeletal muscle C-protein:an intracellular member of the immunoglobulin superfamily. Proc Natl Acad Sci USA 87:2157–2161

37. Vaughan KT, Weber FE, Einheber S, Fischman DA (1993) Molecular cloning of chicken myosin-binding protein H (86 kDa protein) reveals extensive homology with MYBP-C (C-protein) with conserved immunoglobulin C2 and fibronectin type III motifs. J Biol Chem 268:3670–3676

38. Price MG, Gomer RH (1993) Skelemin, a cytoskeletal M-disc periphery protein, contains motifs of adhesion/recognition and intermediate filament proteins. J Biol Chem 268:21800–21810

39. Noguchi J, Yanagisawa M, Imamura M, Kasuya Y, Sakurai T, Tanaka T, Masaki T (1992) Complete primary structure and tissue expression of chicken pectoralis M-protein. J Biol Chem 267:20302–20310

40. Vinkemeier U, Obermann W, Weber K, Furst DO (1993) The globular head domain of titin extends into the center of the sarcomeric M band:cDNA cloning, epitope mapping and immunoelectron microscopy of two titin-associated proteins. J Cell Sci 106:319–330

41. Lakey A, Labeit S, Gautel M, Ferguson C., Barlow DP, Leonard K, Bullard B (1993) Kettin, a large modular protein in the Z-disc of insect muscles. EMBO J 12:2863–2871

42. Holden HM, Ito M, Hartshorne DJ, Rayment I (1992) X-ray structure determination of telokin, the C-terminal domain of myosin light chain kinase, at 2.8 A resolution. J Mol Biol 227:840–851

43. Pfuhl M, Pastore A (1995) Tertiary structure of an immunoglobulin-like domain from the giant muscle protein titin:a new member of the I set. Structure 3:391–401

44. Improta S, Politou A, Pastore A (1996) Immunoglobulin-like modules from titin I-band:extensible components of muscle elasticity. Structure 4:323–337

45. Fong S, Hammill SJ, Proctor M, Freund SMV, Benian GM, Chothia C, Bycroft M, Clarke J (1996) Structure and stability of an immunoglobulin superfamily domain from twitchin, a muscle protein of the nematode C. elegans. J Mol Biol 264:624–639

46. Shirinsky VP, Vorotnikov AV, Birukov KG, Nanaev AK, Collinge M, Lukas TJ, Sellers JR, Watterson DM (1993) A kinase-related protein stabilizes unphosphorylated smooth muscle myosin minifilaments in the presence of ATP. J Biol Chem 268:16578–16583

47. Okagaki T, Weber FE, Fischman DA, Vaughan KT, Mikawa T, Reinach FC (1993) The major myosin-binding domain of skeletal muscle MyBP-C (C-protein) resides in the COOH-terminal, immunoglobulin C2 motif. J Cell Biol 123:619–626

48. Freiburg A, Gautel M (1996) A molecular map of the interactions between titin and myosin-binding protein C:implications for sarcomeric assembly in familial hypertrophic cardiomyopathy. Eur J Biochem 235:317-323

49. Crossley AC (1978) The morphology and development of the Drosophila muscular system. In:Asburner M, Wright TRF (eds) The Genetics and Biology of Drosophila. Academic Press, London, pp 499-559

50. Pringle FRS (1978) Stretch activation of muscle:function and mechanism. Proc R Soc Lond B 201:107-130

51. Jewell BR, Ruegg C (1966) Oscillatory contraction of insect fibrillar muscle after glycerol extraction. Proc R Soc Lond B 164:428-459

52. Wray JS (1979) Filament geometry and the activation of insect flight muscle. Nature 280:325-326

53. Squire JM (1992) Muscle filament lattices and stretch activation:the match-mismatch model reassessed. J Muscle Res Cell Motil 13:183-189

54. Lei J, Tang X, Chambers T, Pohl J, Benian GM (1994) The protein kinase domain of twitchin has protein kinase activity and an autoinhibitory region. J Biol Chem 269:21078-21085

55. Hu S-H, Parker MW, Lei J, Wilce MCJ, Benian GM, Kemp BE (1994) Intrasteric regulation of protein kinases:insights from the crystal structure of twitchin kinase. Nature 369:581-584

56. Hochman B (1971) Analysis of chromosome 4 in Drosophila melanogaster. II. Ethylmethanesulfonate induced lethals. Genetics 67:235-252

57. Hochman B (1974) Analysis of a whole chromosome in Drosophila. Cold Spr Harb Symp Quant Biol 38:581-589

58. Trinick J (1994) Titin and nebulin:protein rulers in muscle? Trends in Bio Sci 19:405-409

59. Whiting A, Wardale J, Trinick J (1989) Does titin regulate the length of muscle thick filaments? J Mol Biol 205:263-268

60. Fulton AB, Isaacs WB (1991) Titin, a huge, elastic sarcomeric protein with a probable role in morphogenesis. BioEssays 13:157-161

61. Waterston RH, Thomson JN, Brenner S (1980) Mutants with altered muscle structure in *C. elegans*. Devel Biol 77:271-302

62. Moerman DG, Baillie DL (1979) Genetic organization in Caenorhabditis elegans:Fine structure analysis of the unc-22 gene. Genetics 91:95-104

63. Moerman DG, Plurad S, Waterston RH, Baillie DL (1982) Mutations in the unc-54 myosin heavy chain gene of C. elegans that alter contractility but not muscle structure. Cell 29:773-781

64. Moerman DG, Benian GM, Waterston RH (1986) Molecular cloning of the muscle gene unc-22 in C. elegans by Tc1 transposon tagging. Proc Natl Acad Sci USA 83:2579-2583

65. Benian GM, L'Hernault SW, Fox LA, Tucker CY, Sale WS (1991) Structure of twitchin: An unusually large sarcomeric protein of *C. elegans*. J Cell Biol 115 (3,2):29a

66. Nave R, Furst D, Vinkemeier U, Weber K (1991) Purification and physical properties of nematode mini-titins and their relation to twitchin. J Cell Sci 98:491-496

67. Vibert P, York ML, Castellani L, Edelstein S, Elliott B, Nyitray L (1996) Structure and distribution of mini-titins. Adv Biophys 33:199-209

68. Kiff JE, Moerman DG, Schriefer LA, Waterston RH (1988) Transposon induced deletions in unc-22 of C. elegans associated with almost normal gene activity. Nature 331:631-633

69. Koenig M, Beggs AH, Moyer M, Scherpf S, Heindrich K...(30)...Kunkel LM (1989) The molecular basis for Duchenne versus Becker muscular dystrophy:correlation of severity with type of deletion. Am J Hum Genet 45:498–506

70. Kobe B, Heierhorst J, Feil SC, Parker MW, Benian GM, Weiss KR, Kemp BE (1996) Giant protein kinases:domain interactions and structural basis of autoregulation. EMBO J 24:6810–6821

71. Heierhorst J, Probst WC, Vilim FS, Buku A, Weiss KR (1994) Autophosphorylation of molluscan twitchin and interaction of its kinase domain with calcium/calmodulin. J Biol Chem 269:21086–21093

72. Heierhorst J, Probst WC, Kohanski RA, Buku A, Weiss KR (1995) Phosphorylation of myosin regulatory light chains by the molluscan twitchin kinase. Eur J Biochem 233:426–431

73. Heierhorst J, Tang X, Lei J, Kemp B, Weiss K, Benian GM (1996) Substrate requirements, inhibition by naphthylene sulfonates, and distinct calmodulin affinities of twitchin kinases. Eur J Biochem 242:454–459

74. Heierhorst J, Mann RM, Kemp BE (1997) Interaction of the recombinant S100A1 protein with twitchin kinase, and comparison with other Ca2+-binding proteins. Eur J Biochem (in press)

75. Heierhorst J, Kobe B, Feil SC, Parker MW, Benian GM, Weiss KR, Kemp BE (1996) Ca+2 / S100 regulation of giant protein kinases. Nature 380:636–639

76. Zimmer DB, Cornwall EH, Landar A, Song W (1995) The S100 protein family:history, function and expression. Brain Res Bull 37:417–429

77. Knappeis GG, Carlsen F (1968) The ultrastructure of the M-line in skeletal muscle. J Cell Biol 38:202–211

78. Luther P, Squire J (1978) Three-dimensional structure of the vertebrate muscle M-region. J Mol Biol 125:313–324

79. Strehler EE, Carlsson E, Eppenberger HM, Thornell L-E (1983) Ultrastructural localization of M-band proteins in chicken breast muscle as revealed by combined immunocytochemistry and ultramicrotomy. J Mol Biol 166:141–158

80. Gautel M, Leonard K, Labeit S (1993) Phosphorylation of KSP motifs in the C-terminal region of titin in differentiating myoblasts. EMBO J 12:3827–3834

81. Obermann WMJ, Gautel M, Steiner F, van der Ven PFM, Weber K, Furst DO (1996) The structure of the sarcomeric M band:localization of defined domains of myomesin, M-protein and the 250 kD carboxy-terminal region of titin by immunoelectron microscopy. J Cell Biol 134:1441–1453

82. Bullard B, Leonard K (1996) Modular proteins of insect muscle. Adv Biophys 33:211–221

83. Waterston RH (1988) Muscle. In:Wood WB (ed) The Nematode Caenorhabditis elegans. Cold Spring Harbor Press, Cold Spring Harbor, NY, pp 281–335

84. Francis GR, Waterston RH (1985) Muscle organization in Caenorhabditis elegans:Localization of proteins implicated in thin filament attachment and I-band organization. J Cell Biol 101:1532–1549

85. Benian GM, Tinley TL, Tang X, Borodovsky M (1996) The C. elegans gene unc-89, required for muscle M-line assembly, encodes a giant modular protein composed of Ig and signal transduction domains. J Cell Biol 132:835–848

86. Miller DM, Ortiz I, Berliner GC, Epstein HF (1983) Differential localization of two myosins within nematode thick filaments. Cell 34:477–490

87. Maruyama IN, Miller DM, Brenner S (1989) Myosin heavy chain gene amplification as a suppressor mutation in C. elegans. Mol Gen Genet 219:113–118

88. Myers MW, Lazzarini RA, Lee VM-Y, Schlaepfer WW, Nelson DL (1987) The human mid-size neurofilament subunit:a repeated protein sequence and the relationship of its gene to the intermediate filament gene family. EMBO J 6:1617–1626

89. Lees JF, Shneidman PS, Skuntz SF, Carden MJ, Lazzarini RA (1988) The structure and organization of the human heavy neurofilament subunit (NF-H) and the gene encoding it. EMBO J 7:1947–1955

90. Hirokawa N, Glicksman MA, Willard MB (1984) Organization of mammalian neurofilament polypeptides within the neuronal cytoskeleton. J Cell Biol 98:1523–1536

91. Nakagawa T, Chen J, Zhang Z, Kanai Y, Hirokawa N (1995) Two distinct functions of the carboxyl-terminal tail domain of NF-M upon neurofilament assembly:cross-bridge formation and longitudinal elongation of filaments. J Cell Biol 129:411–429

92. Lew J, Wang JH (1995) Neuronal cdc2-like kinase. Trends Biochem Sci 20:33–37

93. Nixon RA, Sihag RK (1991) Neurofilament phosphorylation:a new look at regulation and function. Trends Neurosci 14:501–506

94. Gautel M, Goulding D, Bullard B, Weber K, Furst DO (1996) The central Z-disk region of titin is assembled from a novel repeat in variable copy numbers. J Cell Sci 109:2747–2754

95. Moreno S, Nurse P (1990) Substrates for p34cdc2:in vivo veritas? Cell 61:549–551

96. Cohen GB, Ren R, Baltimore D (1995) Modular binding domains in signal transduction proteins. Cell 80:237–248

97. Zheng Y, Cerione R, Bender A (1994) Control of the yeast bud site assembly GTPase CDC42. J Biol Chem 269:2369–2372

98. Hart MJ, Eva A, Evans T, Aaronson SA, Cerione RA (1991) Catalysis of guanine nucleotide exchange on the CDC42Hs protein by the dbl oncogene product. Nature 354:311–314

99. Takai Y, Sasaki T, Tanaka K, Nakanishi H (1995) Rho as a regulator of the cytoskeleton. Trends Biochem Sci 20:227–231

100. Gibson TJ, Hyvonen M, Musacchio A, Saraste M, Birney E (1994) PH domain:the first anniversary. Trends Biochem Sci 19:349–353

101. Harlan JE, Hajduk PJ, Yoon HS, Fesik SW (1994) Pleckstrin homology domains bind to phosphatidylinositol-4,5-bisphosphate. Nature 371:168–170

102. Hyvonen M, Macias MJ, Nilges M, Oschkinat H, Saraste M, Wilmanns M (1995) Structure of the binding site of inositol phosphates in a PH domain. EMBO J 14:4676–4685

103. Gettner SN, Kenyon C, Reichardt LF (1995) Characterizaiton of bpat-3 heterodimers, a family of essential integrin receptors in C. elegans. J Cell Biol 129:1127–1141

104. Rogalski TM, Williams BD, Mullen GP, Moerman DG (1993) Products of the unc-52 gene in C. elegans are homologous to the core protein of the mammalian basement membrane heparan sulfate proteoglycan. Genes Dev 7:1471–1484

105. Francis GR, Waterston RH (1991) Muscle cell attachment in C. elegans. J Cell Biol 114:465–479

106. Williams BD, Waterston RH (1994) Genes critical for muscle development and function in Caenorhabditis elegans identified through lethal mutations. J Cell Biol 124:475–490

107. Hresko MC, Williams BD, Waterston RH (1994) Assembly of body wall muscle and muscle cell attachment structures in Caenorhabditis elegans. J Cell Biol 124:491–506
108. Miller DM, Shakes DC (1995) Immunofluroescence microscopy. In:Epstein HF, Shakes DC (eds) C. elegans:Modern Biological Analysis of an Organism. Academic Press, San Diego, pp 365–394
109. Bretscher A (1991) Microfilament structure and function in the cortical cytoskeleton. Ann Rev Cell Biol 7:337–374
110. Heintzelman MB, Mooseker MS (1992) Assembly of the intestinal brush border cytoskeleton. Curr Top Dev Biol 26:93–122
111. Mamajiwalla SN, Fath KR, Burgess DR (1992) Development of the chicken intestinal epithelium. Curr Top Dev Biol 26:123–143
112. Keller T, Mooseker MS (1982) Ca+2-calmodulin-dependent phosphorylation of myosin, and its role in brush border contraction in vitro. J Cell Biol 95:943–959
113. Madara JL (1989) Loosening tight junctions. Lessons from the intestine. J Clin Invest 83:1089–1094
114. Eilertsen KJ, Keller TCS (1992) Identification and characterization of two huge protein components of the brush border cytoskeleton:evidence for a cellular isoform of titin. J Cell Biol 119:549–557
115. White J (1988) The Anatomy. In:Wood WB (ed) The Nematode Caenorhabitis elegans. Cold Spring Harbor Laboratory Press, Cold Spring Harbor, NY, pp 81–122
116. Bullard B, Leonard K (1996) Modular proteins of insect muscles. Adv Biophys 33:211–221
117. Cheng N, Deatherage JF (1989) Three-dimensional reconstruction of the Z disk of sectioned bee flight muscle. J Cell Biol 108:1761–1774
118. Main AL, Harvey TS, Baron M, Boyd J, Campbell ID (1992) The three-dimensional structure of the tenth type III module of fibronectin:an insight into RGD-mediated interactions. Cell 71:671–678
119. Daley J, Southgate R, Ayme-Southgate A (1998) Structure of the Drosophila projectin protein: isoforms and implication for projectin filament assembly. J Mol Biol 279(1):201–210